Developments in Petroleum Science

Volume 78

Fluid–Solid Interactions in Upstream Oil and Gas Applications

Edited by

Ibnelwaleed A. Hussein
Gas Processing Center, College of Engineering;
Department of Chemical Engineering, College of Engineering,
Qatar University, Doha, Qatar

Mohamed Mahmoud
Department of Petroleum Engineering,
College of Petroleum Engineering & Geosciences,
King Fahd University of Petroleum & Minerals,
Dhahran, Saudi Arabia

ELSEVIER

Elsevier
Radarweg 29, PO Box 211, 1000 AE Amsterdam, Netherlands
The Boulevard, Langford Lane, Kidlington, Oxford OX5 1GB, United Kingdom
50 Hampshire Street, 5th Floor, Cambridge, MA 02139, United States

ISBN: 978-0-323-99285-5
ISSN: 0376-7361

Publisher: Nikki P. Levy
Acquisitions Editor: Peter Llewellyn
Editorial Project Manager: Maria Elaine D. Desamero
Production Project Manager: Rashmi Manoharan
Cover Designer: Christian J. Bilbow

Typeset by STRAIVE, India

Developments in Petroleum Science
Volume 78

Fluid–Solid Interactions in Upstream Oil and Gas Applications

Series Editor
Baojun Bai
Zhangxing Chen

Contents

Contributors

Olalekan Alade (41), Center for Integrative Petroleum Research, King Fahd University of Petroleum and Minerals, Dhahran, Saudi Arabia

Mohammad Albeldawi (313), Gas Processing Center, College of Engineering, Qatar University, Doha, Qatar

Murtada Saleh Aljawad (75), Department of Petroleum Engineering, College of Petroleum Engineering & Geosciences, King Fahd University of Petroleum & Minerals, Dhahran, Saudi Arabia

Emad W. Al-Shalabi (181), Petroleum Engineering Department, Khalifa University of Science and Technology, Abu Dhabi, United Arab Emirates

Santiago Aparicio (277), Department of Chemistry, University of Burgos, Burgos, Spain

Muhammad Arif (247), Department of Petroleum Engineering, Khalifa University, Abu Dhabi, United Arab Emirates

Mohammed A. Ayoub (181), Petroleum Engineering Department, Universiti Teknologi Petronas (UTP), Seri Iskandar, Malaysia

Golibjon R. Berdiyorov (277), Qatar Environment and Energy Research Institute, Hamad Bin Khalifa University, Doha, Qatar

Giuliano Carchini (277), Gas Processing Center, College of Engineering, Qatar University, Doha, Qatar

Krishna Raghav Chaturvedi (21), Department of Petroleum Engineering and Geoengineering, Rajiv Gandhi Institute of Petroleum Technology, Jais, India

Satyajit Chowdhury (111), Gas Hydrate and Flow Assurance Laboratory, Petroleum Engineering Program, Department of Ocean Engineering, Indian Institute of Technology Madras, Chennai; Assam Energy Institute, A Centre of Rajiv Gandhi Institute of Petroleum Technology, Sivasagar, Assam, India

Elkhansa Elbashier (277), Gas Processing Center, College of Engineering, Qatar University, Doha, Qatar

Ahmed Hamza (1), Gas Processing Center, College of Engineering, Qatar University, Doha, Qatar

Amjed Hassan (41, 75), Department of Petroleum Engineering, College of Petroleum Engineering & Geosciences, King Fahd University of Petroleum & Minerals, Dhahran, Saudi Arabia

Anas M. Hassan (181), Petroleum Engineering Department, Khalifa University of Science and Technology, Abu Dhabi, United Arab Emirates

Ibnelwaleed A. Hussein (1, 21, 135, 277), Gas Processing Center; Department of Chemical Engineering, College of Engineering, Qatar University, Doha, Qatar

Muhammad Shahzad Kamal (41, 75), Center for Integrative Petroleum Research, King Fahd University of Petroleum and Minerals, Dhahran, Saudi Arabia

Syed Oubee Khadri (135), Gas Processing Center, College of Engineering, Qatar University, Doha, Qatar

Ravi Shankar Kumar (247), Department of Petroleum Engineering, Khalifa University, Abu Dhabi, United Arab Emirates

Yueliang Liu (135), School of Petroleum Engineering, China University of Petroleum (Beijing), Beijing, PR China

Mohamed Mahmoud (1, 41, 75), Department of Petroleum Engineering, College of Petroleum Engineering & Geosciences, King Fahd University of Petroleum & Minerals, Dhahran, Saudi Arabia

Mysara Eissa Mohyaldinn (181), Petroleum Engineering Department, Universiti Teknologi Petronas (UTP), Seri Iskandar, Malaysia

Mobeen Murtaza (41), Center for Integrative Petroleum Research, King Fahd University of Petroleum and Minerals, Dhahran, Saudi Arabia

Mustafa S. Nasser (21), Gas Processing Center; Department of Chemical Engineering, College of Engineering, Qatar University, Doha, Qatar

Mayank Rakesh (111), Department of Petroleum Engineering and Earth Sciences, University of Petroleum and Energy Studies, Dehradun, India

Klaus Regenauer-Lieb (135), WASM: Minerals, Energy and Chemical Engineering, Curtin University, Perth, WA, Australia

Hamid Roshan (135), School of Minerals and Energy Resources Engineering, UNSW Australia, Kensington, Sydney, NSW, Australia

Mohammed Saad (277), Gas Processing Center; Department of Chemical Engineering, College of Engineering, Qatar University, Doha, Qatar

Fadhil Sadooni (135), Environmental Science Center, College of Engineering, Qatar University, Doha, Qatar

Ahmad Sakhaee-Pour (135), Petroleum Engineering Department, Houston University, Houston, TX, United States

Jitendra S. Sangwai (111), Gas Hydrate and Flow Assurance Laboratory, Petroleum Engineering Program, Department of Ocean Engineering; Department of Chemical Engineering; Center of Excellence on Subsurface Mechanics and Geo-Energy, Indian Institute of Technology Madras, Chennai, India

Mohamed Shamlooh (21), Gas Processing Center, College of Engineering, Qatar University, Doha, Qatar

Mohammed Abdul Qadeer Siddiqui (135), School of Minerals and Energy Resources Engineering, UNSW Australia, Kensington, Sydney, NSW, Australia

Zeeshan Tariq (41), Ali I. Al-Naimi Petroleum Engineering Research Center, Physical Science and Engineering Division, King Abdullah University of Science and Technology (KAUST), Thuwal, Saudi Arabia

Japan Trivedi (21), School of Mining and Petroleum Engineering, University of Alberta, Edmonton, AB, Canada

Preface

Understanding fluid–rock interactions in oil and gas reservoirs has great significance in oilfield operations such as drilling, production, and hydrocarbon reservoir management. It is essential to maximize the efficiency and productivity and enhance the safety measures taken in the fields. This book presents a single source for readers interested in fluid–solid interactions in oil and gas reservoirs during different stages of well life and various locations. This book is mainly directed at academia and R&D in the oil industry, especially in the petroleum and chemical engineering departments.

In the first two chapters, the basics of reservoir fluids and rock properties and the rheological behavior of solid–fluid systems are introduced. The fluid–solid interactions are addressed in all stages of the reservoir, including the drilling and completion processes, acidizing, and fracturing. Then, interactions in tight and unconventional formations and enhanced oil recovery operations are highlighted. Also, the book describes the utilization of nanoparticles in different upstream applications. Molecular simulations have recently been introduced in simulating other processes, such as enhanced gas recovery, scale removal, and ultimate gas recovery. Therefore, a chapter is dedicated to molecular simulations in upstream applications. Finally, the last chapter highlights the environmental concerns and mitigations of offshore oil operations.

Authors from different universities (including Qatar University; King Fahd University of Petroleum Minerals, Saudi Arabia; Khalifa University, UAE; Universiti Teknologi PETRONAS, Malaysia; Indian Institute of Technology, India; China Petroleum University; University of Alberta, Canada; and Queen's University, Australia) have contributed to this book. Our thanks and appreciations are due to all of them. In addition, we are indebted to Elsevier for publishing this book. We extend our special thanks to Maria Elaine Desamero, the Editorial Project Manager at Elsevier. We hope that our readers find the book enjoyable and fruitful.

Prof. Ibnelwaleed A. Hussein

Chapter 1

Introduction to reservoir fluids and rock properties

Ahmed Hamza[a], **Ibnelwaleed A. Hussein**[a,b], **and Mohamed Mahmoud**[c]
[a]*Gas Processing Center, College of Engineering, Qatar University, Doha, Qatar*
[b]*Department of Chemical Engineering, College of Engineering, Qatar University, Doha, Qatar*
[c]*Department of Petroleum Engineering, College of Petroleum Engineering & Geosciences, King Fahd University of Petroleum & Minerals, Dhahran, Saudi Arabia*

1.1 Introduction

Hydrocarbons and nonhydrocarbon gases and liquids accumulate in a subsurface body of rocks to form a reservoir that can store and transmit these fluids. Understanding the properties of reservoir fluids and rock is essential for predicting the performance of the hydrocarbon reservoir. Hydrocarbons are preserved in sedimentary rocks because of the availability of high void space in these rocks compared to that in metamorphic and igneous rocks. Yet, the reservoir pressure and temperature conditions control the type of hydrocarbon formed in the pores. Moreover, the presence of different immiscible phases in contact with the reservoir rocks would trigger a force balance between the forces acting on the phases interface and between the fluids and the rocks. Hence, it is crucial to understand the properties of reservoir rocks and fluids and their interaction to forecast the reservoir performance and maximize the hydrocarbon recovery.

1.2 Mineralogy of reservoir rocks

1.2.1 Sandstone mineralogy

Clastic rocks are composed mainly of consolidated sediments resulting from the accumulation of rock fragments of preexisting rocks and particles separated and transported to the depositional area by mechanical agents such as wind, water, and gravity. The transportation mechanism affects the grains' mineralogy, shape, size, and surface texture. Thus, measurements of these variations may explain the transportation mechanism, which is vital for evaluating the sand quality and its distribution in the basin. The depositional environment

Developments in Petroleum Science, Vol. 78. https://doi.org/10.1016/B978-0-323-99285-5.00003-X

determines sand bodies' outer geometry and distribution in sandstone reservoirs. Mineralogy, texture, and diagenetic processes on the sediments affect the characteristics of sandstone reservoirs.

Sandstone reservoirs are clastic reservoirs that might contain quartz, feldspars, clay minerals, carbonate fragments, mica, and chert. Sand grains are sedimentary particles that have a diameter between 1/16 and 2 mm (Weimer and Tillman, 1982). Clays are defined as fine-grained sediments with less than 0.0039 mm in diameter (Gillott, 1984; Neuendorf et al., 2005). Sandstone rocks consist mainly of sand-sized grains and other sizes, transitioning to clay and silt-rich rocks. Sandstone rocks are classified based on the relationship between the composition and quantity of sand-sized grains and the silt and clay. The four-component diagram (Pettijohn classification) reveals the differences between clay and sand grains, which comprise rock fragments, feldspar, and quart. Pettijohn's grouping of sandstones (Fig. 1.1) is concerned with clays' existence, absence, and mineralogy. Sandstone is divided into arenites (contain grains only) and wackes (compose more than 15% matrix [clays]) (Bjørlykke and Jahren, 2015; Pettijohn et al., 1987).

Rock fragments involving microcrystalline quartz (counting chert) are categorized with the grains of quartz minerals. For example, sandstones that have a low percentage of rock fragments and more than 25% feldspar are called arkoses. Moreover, when the ratio of rock fragments is relatively high, the sandstone is called lithic sandstones, typically resulting from intrusive and basalt igneous rocks and very fine-grained sedimentary rocks. Hence, the sand grain might contain many minerals. A triangular plot that relates the clay percent and grain types (feldspar, rock fragments, and quartz) is used in the oil and gas industry to classify the sandstones.

1.2.2 Clays

Clays hold the grains in sandstone reservoirs as cementing material. Generally, clays are classified as authigenic or allogenic according to their existence in sandstones. Allogenic clays are available as individually dispersed particles such as laminae or matrices. Moreover, allogenic clays could exist in the form of pellets from clay flocculation or clay aggregates from preexisting shales outside the depositional area. On the other hand, authigenic clays have formed during or after deposition. Authigenic clays are formed by transforming allogenic clays into new types, diagenesis of nonclay materials, or precipitation of clay minerals from reservoir fluids. Clay minerals are categorized depending on the existence of octahedral (Al_2O_3) and tetrahedral (SiO_4) sheets (Kumari and Mohan, 2021). The main groups of clay minerals are listed below:

1.2.3 Kaolinite

Kaolinite has a structure of 1:1 layers stacked above each other. Kaolinite consists of Si^{4+} in the tetrahedral sites and Al^{3+} in the octahedral sites (Fig. 1.2).

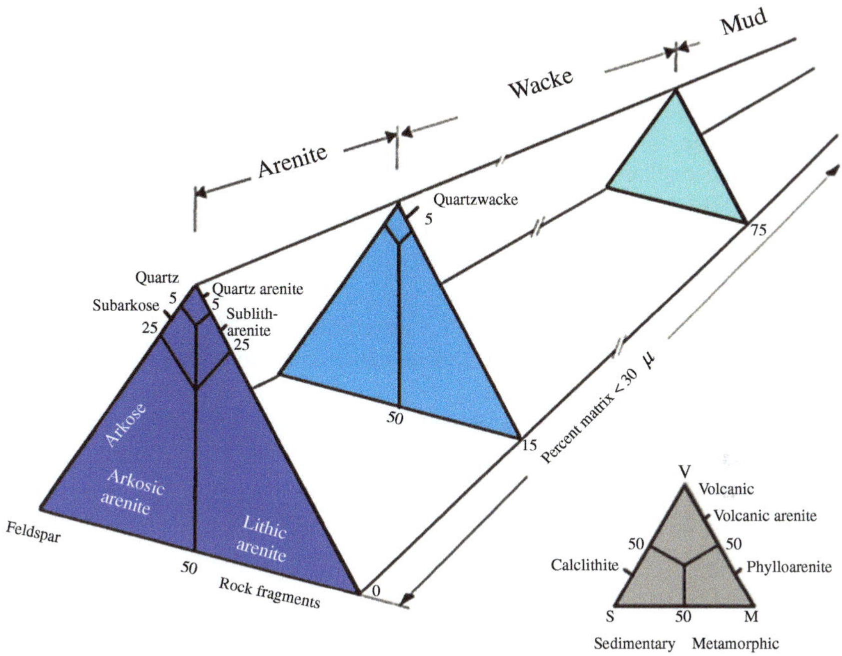

FIG. 1.1 Pettijohn's classification of sandstones.

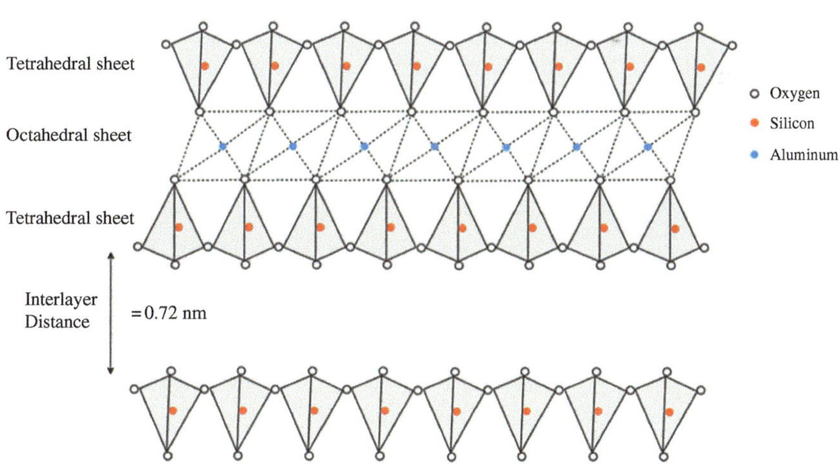

FIG. 1.2 Kaolinite structure.

The kaolinite layer is held by hydrogen bonding between the oxygen of the tetrahedral sheet and the hydroxyl of the octahedral sheet (Schulze, 2005).

The formula for kaolinite is $Al_2Si_2O_5(OH)_4$ because it has little isomorphous substitution in the octahedral or tetrahedral sheets. Moreover, kaolinite has a low surface area and exchange capacity because the 1:1 layer is neutral and has a low amount of substitution. Subsequently, surface areas and cation exchange capacities are typically low. Kaolinite can be formed in reservoirs due to releasing Si and Al by the weathering of other minerals such as feldspars. Kaolinite is a nonswelling clay with a small exchange capacity and the ability to exchange K^+ cation. Kaolinite can easily disperse and move (Civan, 2007).

1.2.4 Smectite

Smectite comprises a 2:1 structure with a general formula $M^+_{0.33}Al_2(Si_{3.67}Al_{0.33})O_{10}(OH)_2$, where M^+ points out to the exchangeable cations such as Ca^{2+} and Mg^{2+} (Fig. 1.3).

Smectite has a high surface area and swelling clay because K^+ is not fixed. The negative charge of chlorites is less than smectite (Kumari and Mohan, 2021). Smectite has a large base exchange capacity and readily adsorbs Na^+, resulting in a high degree of swelling and dispersion (Civan, 2007).

1.2.5 Chlorite

Chlorite minerals have a 2:1 layer structure with an interlayer of hydroxide and octahedral sheets (Fig. 1.4). The octahedral sheets can be either dioctahedral or trioctahedral and can contain Cr^{3+}, Fe^{3+}, Al^{3+}, Mg^{2+}, Ni^{2+}, Mn^{2+}, and Fe^{2+}, giving many different mineral species. The chlorite structure has a negative charge that is balanced by the positive charge of the interlayer hydroxide sheet instead of K^+. The hydroxide sheet has a positive charge caused by substituting higher-valence cations for lower-valence ones. Chlorite can easily break apart, causing a blockage of the pore throat.

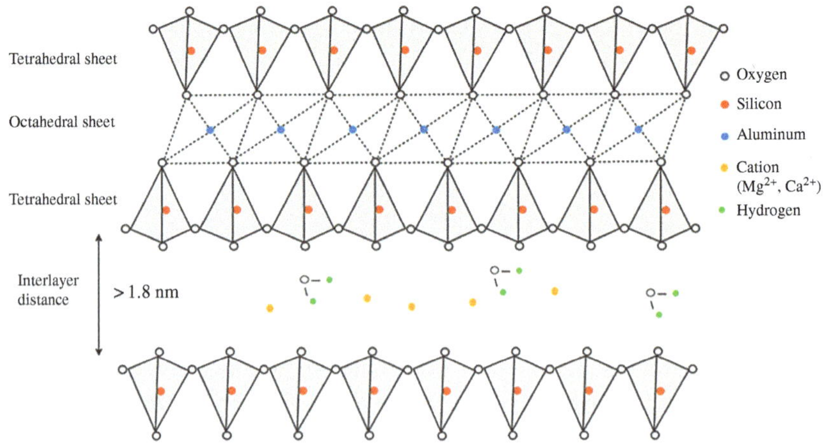

FIG. 1.3 Smectite structure.

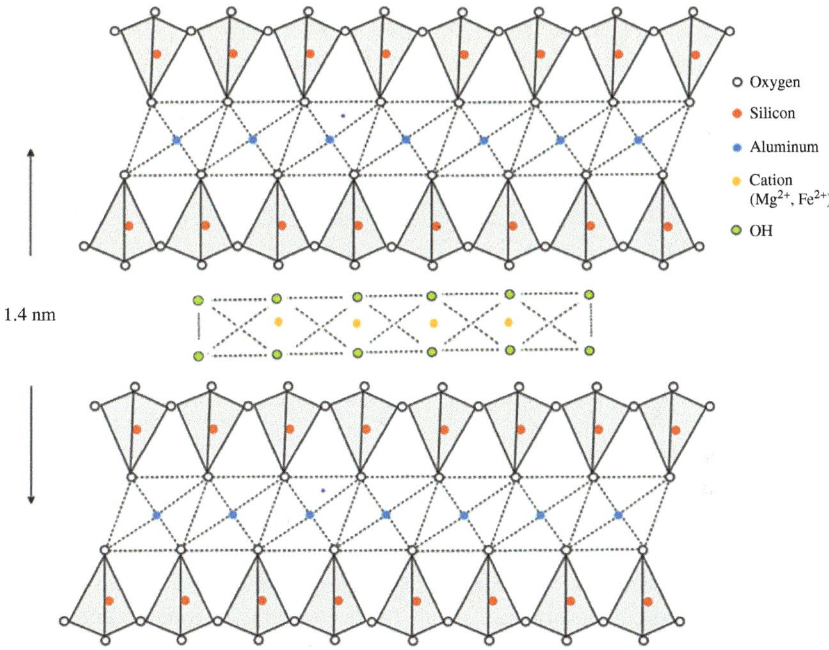

FIG. 1.4 Chlorite structure.

1.2.6 Illite

Illite minerals have a 2:1 layer structure with an interlayer structure and general formula $(K_{1-1.5}Al_4[Si_{7-6.5}Al_{1-1.5}O_{20}](OH_4)$. Al^{3+} occupies a quarter of the tetrahedral sites creating a negative charge in the layer structure, balanced by monovalent cations, commonly K^+, occupying interlayer sites (Fig. 1.5). Illites combine the properties of swelling and dispersible clays because of their interlayered structure. These unique characteristics make it challenging to stabilize illite clays (Civan, 2007). The morphology of illite is irregular with granules or elongated spines.

1.3 Carbonate mineralogy

The origin of carbonate rocks is the calcareous skeletons of organisms that lived in shallow marine water where algae were present. These organisms might fall and deposit deeper after dying. Carbonate precipitated from seawater cement bioclastic sediments fragments together to form the rock. All carbonates will be dissolved below a certain depth because of the high pressure. Carbonates originate within the basin of deposition, whereas clastic deposit needs the transport of grains to the sedimentary basin. Therefore, clastic

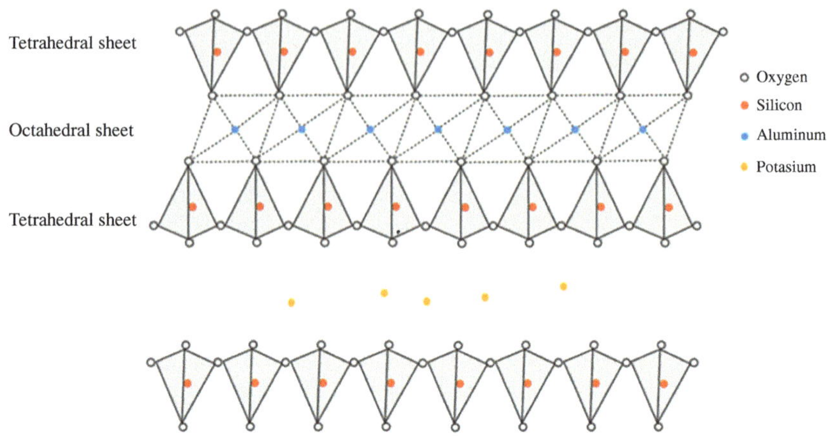

FIG. 1.5 Illite structure.

deposition disturbs the stability of the environment for organisms depending on photosynthesis, which explains that the coexistence of carbonate and clastic reservoirs is not possible. Nevertheless, variations in the depositional environment over geological time would result from overlying carbonate and clastic rocks (Mackenzie, 1978; Tiab and Donaldson, 2016a). Based on the depositional area, carbonate reservoirs comprise the following types (Darling, 2005):

a. *Shallow marine carbonates*: the skeletal production rate is high in this depositional environment. These skeletons break because of the action of turbulence, fish, and crustaceans. Later, the generated carbonate sediments might be transported to the final depositional area and altered by burrowing organisms. Fecal pellets may constitute grains, which create porosity.
b. *Reefs:* secreting calcium carbonate organisms developing on the remaining previously generated would form reefs.
c. *Deepwater carbonates:* these carbonates result from the deposition at a significant depth lower than where photosynthesis takes place. Usually, the sediments are made from oozes involving skeletons of pelagic organisms.

The original porosity of the matrices containing calcium carbonate grains (calcisands) is very high; however, diagenetic processes after deposition reduce the porosity. Therefore, porosity drops because of the grains' cementation, internal sedimentation, and compaction. Moreover, the dissolution of calcium carbonate in one place and redeposition in other positions might also reduce the rock porosity. On the other hand, some processes, such as leaching and dolomitization, may raise the porosity. The dolomitization, in which $CaCO_3$ is replaced by $CaMg(CO_3)_2$, creates intercrystalline or vuggy structures in nature. The dolomite mineral forms because Mg^{2+} ions replace Ca^{2+} ions in calcite. The source of magnesium is seawater evaporation in supratidal sabkha areas.

1.4 Unconventional reservoirs

Oil and natural gases are found not only in sandstone and carbonate but also in other mineralogy. Unconventional reservoirs need unique recovery processes and technologies to be economically viable. The unconventional reservoirs include coalbed, hydrate, shale, and tight reservoirs. Coal reservoirs are formed from the remaining vegetation deposited below layers of sediments that may contain sand, clay, and other minerals for long periods. Then, the reservoir pressure and temperature conditions squeeze the water out and convert the deposited vegetation into coal. Anaerobic bacteria produce methane in the early stage of the coalification process, which makes the composition of the gas different from the conventional gas reservoir as a higher methane content will be produced at the initial stages of the production. This phenomenon is explained by methane's lower affinity for coal surface adsorption than ethane and other gases (Zou, 2013, 2017). The mineral composition of tight gas reservoirs might be similar to that of sandstone and carbonate reservoirs; however, tight reservoirs have a permeability of less than 0.1 m-Darcy (mD). Shale consists mainly of silt and clay-sized particles and a considerable amount of organic material (Glorioso and Rattia, 2012). Moreover, the thermal maturity of the organic material determines the amount of heat and pressure applied to the rocks (Zou et al., 2013).

1.5 Types and composition of reservoir fluids

During the diagenesis of the reservoir rocks, some of the water initially found in the depositional area is trapped in the pore space between the rock grains. Then, hydrocarbons migrate from source rocks to the reservoir to invade some of the pores occupied by formation water. Therefore, the reservoir contains brines, hydrocarbons in liquid or gas states, and nonhydrocarbon gases. The initial reservoir conditions of temperature and pressure determine the composition of the hydrocarbons in the reservoir. Moreover, the fluid distribution in the reservoir depends on the density of each fluid. Gas is located on the top inside the pores, followed by oil and then water according to their density. Furthermore, gas is dissolved in oil depending on the pressure and temperature conditions.

Connate water is the seawater originally found where the sediments were deposited, with a high chloride concentration (on average 80,000 ppm). The existence of chloride indicates the seawater origin of the connate water in hydrocarbon reservoirs. However, the chloride concentration in connate water is higher than the average content in seawater as an additional concentration of chloride is believed to occur during the diagenesis of the sediments. Connate water surrounds rock grains and is hence called interstitial water, which is different from free water that is movable and can be produced from the reservoir.

Hydrocarbons in reservoirs are found as liquid crude oil, natural gas, or solid bitumen. Oil has a density less than seawater and hence floats on seawater. Natural gas comprises paraffin hydrocarbon, and the most abundant component in its composition is methane (CH_4). Dry gas refers to natural gas that consists of nearly completely methane. Moreover, natural gas is defined as a wet gas when the ethane content in natural gas in the reservoir exceeds 15%. On the other hand, heavy hydrocarbons and bitumens are in solid or semisolid forms because they contain albertite, pitch, tar, and asphalt.

Hydrocarbons are found in nature with impurities such as sulfur, oxygen, and nitrogen. Some sulfur may exist as hydrogen sulfide (H_2S) or in an elemental form in solution. Crude with a high concentration of sulfur (up to 3%) is known as sour crude, whereas sweet crude contains less than 0.1% sulfur. Similarly, natural gas that contains H_2S is known as sour gas. Moreover, nitrogen (N_2) as an impurity in crude oil is related to the asphalt content. The nitrogen content is considered high if the concentration exceeds 0.02%. Another major impurity in petroleum is oxygen compounds, forming acids in crude oil and carbon dioxide (CO_2) in natural gas. Furthermore, heavy metals such as nickel and vanadium exist as impurities in petroleum. Generally, nickel is related to crude with a low sulfur ratio, whereas vanadium is linked with high-sulfur crude oil.

The chemical composition of the hydrocarbons controls the physical properties of petroleum in the reservoir. Specific gravity (SG) is defined as the ratio of the weight of a given volume of a material to the weight of an equal volume of water at 4 °C, and its measuring unit is degrees. The specific gravity of oils lies between 0.73 and $\sim>1.0$, with paraffin-based oils being commonly light. In contrast, asphalt-based (naphthenic components) oils are almost always heavy. The petroleum industry uses an API (American Petroleum Institute) scale to grade crude oil. The high API gravity indicates the presence of light hydrocarbons, whereas low API gravity points out the presence of heavy hydrocarbons.

$$API^{o} = \frac{141.5}{SG} - 131.5 \tag{1.1}$$

Hydrocarbon reservoirs are categorized as oil or gas reservoirs based on the initial reservoir temperature and pressure. The initial reservoir temperature in gas reservoirs is higher than the critical temperature of the reservoir fluids. Hydrocarbon reservoirs are classified into black oil, volatile oil, dry gas, wet gas, and gas condensate.

Black oil reservoirs have a gas-oil ratio (GOR) of about 750 SCF/STB or less, whereas the GOR reaches 2500 SCF/STB in volatile oil reservoirs. Hydrocarbons in dry gas reservoirs are basically in the gaseous state, and no liquid drops out when the reservoir pressure decreases. Light and intermediate hydrocarbons are the main components of wet gas reservoirs, and hence, high API condensate is produced at surface conditions in low amounts

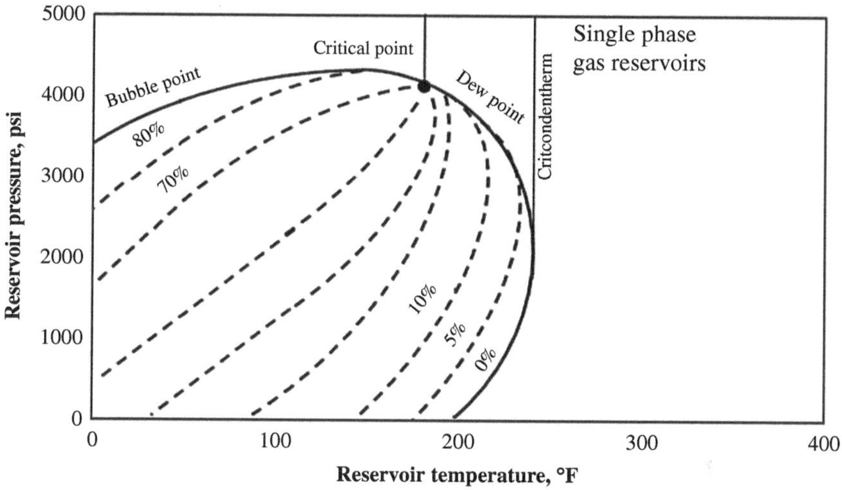

FIG. 1.6 Pressure–temperature diagram for a specific reservoir fluid composition.

(less than 5 STB/MMSCF). However, the GOR remains constant, and liquid condensate does not form inside the reservoir during pressure depletion. On the contrary, gas condensate reservoirs hold significant amounts of heptane plus fraction (C_{7+} components), and liquid condensate drops out in the reservoir (retrograde condensation). The amount of liquid condensate that drops out in the reservoir increases as the pressure decreases. The producing GOR in gas condensate reservoirs reaches about 2500–50,000 SCF/STB. Fig. 1.6 shows a pressure–temperature diagram for a specific reservoir fluid composition.

A reservoir containing both an oil zone and a gas cap implies that both fluids are in saturated conditions. Accordingly, the bubble point of the oil equals the dew point of the gas cap at initial pressure conditions.

1.6 Rock properties

Reservoir rocks might consist of unconsolidated, very loose or dense, and tight sand or carbonate minerals. Conventional reservoirs may contain grains of dolomite, limestone, or sand cemented by clays. Therefore, understanding the interaction between reservoir fluids and rocks is essential for predicting and evaluating the performance of a given reservoir.

1.6.1 Porosity

The deposition of the carbonate crystals and sand grains forms a microscopic space. The porosity (ϕ) of the rock measures the available volume to store fluids in the microscopic space between the intercrystalline carbonate crystals and the intergranular sand grains (Baker et al., 2015). The porosity can be estimated quantitatively by comparing the ratio of the pore volume to the total volume (bulk volume) of the rock.

$$\phi = \frac{\text{Void volume}}{\text{Bulk volume}} \tag{1.2}$$

Some void spaces might be interconnected, and others are isolated during the deposition of the sediments because of the cementing material between the grains. Therefore, absolute porosity is defined as the ratio of the total pore space in the rock to the bulk volume. Accordingly, a reservoir rock may have considerable absolute porosity and yet have no conductivity to fluid because of the lack of interconnection between pores. The ratio of the interconnected void volume to the bulk volume is known as effective porosity. Thus, the effective porosity represents the connected pore space that contains the recoverable hydrocarbon fluids. Based on the mode of the origin, porosity might be original during the sediment deposition or induced by a subsequent geological process such as fractures in shales and cavities in limestones. Some sources of diagenesis that might change the original porosity are fracture, dissolution, cementation, and compaction. For example, the dissolution of fossils might result in vugs in carbonate oolites. These types of pores created by diagenesis are known as secondary porosity.

1.6.2 Pore structure and permeability

The interconnected pores in the reservoir form the pore structure, which controls the flow of the fluids. Moreover, the pores and their connection, known as pore throat, are irregular in shape. The depositional environment and diagenesis control the shape of the pore structure. Permeability is defined as the ability of a rock to transmit fluids; the unit measure is in Darcies or milli-Darcies. Permeable rocks have well-interconnected pores, and hence, the pore throat size affects the quality of permeability. When a single fluid flows in the porous media, the measured permeability is the absolute permeability. A reservoir rock, on the other hand, contains immiscible fluids and thus preferentially allows the flow of one of them over the other fluids, known as effective permeability. The mineralogy and saturation of the fluids in the reservoir could affect the effective permeability. Relative permeability is defined as the ratio of the effective permeability of fluid at specific saturation to the absolute permeability of that fluid at total saturation. The existence of many fluids in the reservoir would generally restrict the flow of these fluids, and hence, relative permeability compares the ability of fluids to flow in multiphase conditions. Darcy's law, which relates the pressure gradient and flow rate, describes the flow of fluids in the reservoir (Fanchi, 2010).

$$q = -0.001127 \frac{KA}{\mu} \frac{\Delta P}{\Delta x} \tag{1.3}$$

where Δx = length, ft; μ = fluid viscosity, cP; P = pressure, psi; A = cross-sectional area, ft^2; K = permeability, md; and q = volumetric flow rate, bbl/day.

In radial coordinates, Darcy's law is expressed as follows:

$$Q = -\frac{0.00708Kh(P_w - P_e)}{\mu B \ln\left(\frac{r_e}{r_w}\right)} \tag{1.4}$$

where B = formation volume factor, RB/STB; μ = viscosity, cp; P_e = pressure at outer radius, psi; P_w = pressure at inner radius, psi; h = formation thickness, ft.; K = permeability, md; r_e = outer radius, ft.; r_w = wellbore or inner radius, ft.; and Q = liquid flow rate, STB/D.

$$q_s = -0.703\frac{Kh}{T_r\left(\ln\frac{r_e}{r_w}\right)}[m(P_e) - m(P_w)] \tag{1.5}$$

Darcy's law is valid for laminar flow.

However, many reservoir fluids such as gases might reveal a turbulent flow at high flow rates. Therefore, Forchheimer observed turbulent flow at high flow rates.

$$\frac{\Delta P}{\Delta x} = -\left(\frac{q}{0.001127A}\right)\frac{\mu}{K} + \beta\rho\left(\frac{q}{A}\right)^2 \tag{1.6}$$

Gas molecules exhibit a slippage effect on the wall of the pores. Klinkenberg showed that the slippage of gas molecules affects the permeability measurements depending on the pressure. The slippage effect increases with increasing the pressure.

$$k_g = k_{abs}\left(1 + \frac{b}{\overline{P}}\right) \tag{1.7}$$

where b = Klinkenberg's effect; $\overline{P}$ = mean flowing gas pressure in the flow system; k_g = apparent permeability calculated from gas flow tests; and k_{abs} = true absolute permeability of the rock.

1.6.3 Rock mechanical properties

The lithologies above the reservoir rocks create overburden pressure because of the weight of the grains. The overburden pressure increases during grain sedimentation because of compaction and cementation of the grain and loss of pore fluids. On the other hand, hydrostatic pressure, also known as normal pressure, is the predicted pressure of the freshwater column from sea level to a certain depth. Therefore, overburden pressure (lithostatic pressure) is always higher than hydrostatic pressure (Fig. 1.7). Meanwhile, the formation at a specific depth can withstand a certain pressure before breaking down.

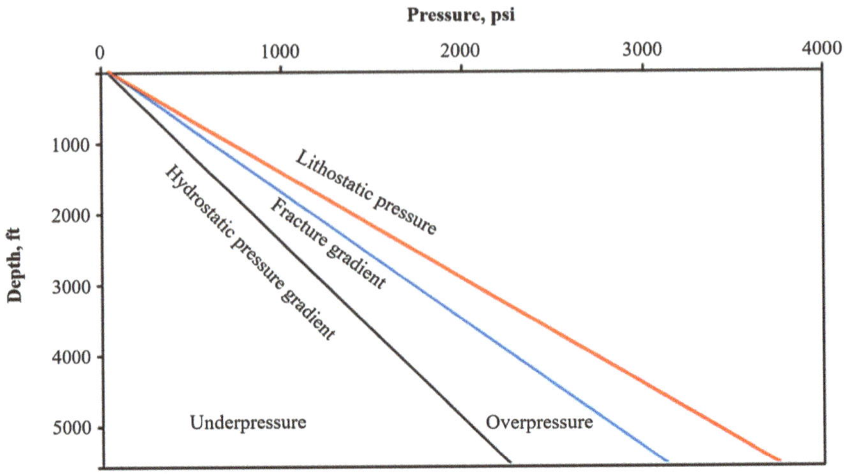

FIG. 1.7 Overburden pressure.

$$P(z) = P_o + g \int_0^z \rho(z) dz \tag{1.8}$$

where P_o is the initial pressure and $\rho(z)$ is the density of the gas at distance z.

As the reservoir fluids are being produced, the pore volume will change due to decreasing pore pressure. This change in the pore volume with respect to the pore pressure is known as rock or formation compressibility.

$$c_f = \frac{1}{V_p} \left(\frac{\partial V_p}{\partial p} \right)_T \tag{1.9}$$

When the force balance is disturbed, the reservoir rocks might be subjected to deformation. For example, when the reservoir fluids are produced, the differential pressure between the pore pressure and overburden pressure will increase, and hence, the porosity will slightly decrease (Lyons, 2010). The stress–strain relationship of the reservoir rocks is influenced by many factors, such as mineral composition, the type of cementing material, the type and amount of pore fluids, rock compressibility, porosity, permeability, temperature, and pressure (Tiab and Donaldson, 2016b).

The external load applied to a rock develops internal stress on the rock. If the stress is high enough, it will deform the reservoir rocks, resulting in a change in their volume and shape. Based on the direction of the forces applied to the rock, stress types can be divided into tensile, shear, and compressive stress (Fig. 1.8) (Tiab and Donaldson, 2016b).

The permanent deformation resulting from the applied external forces on rocks, which changes their dimensions compared to the original ones, is known as strain. Generally, the rock body passes through four stages when exposed to directed forces: elastic, elastic-viscous, plastic, and rupture.

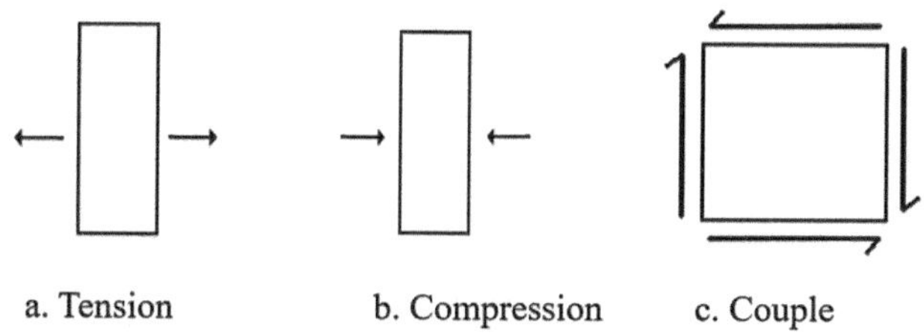

FIG. 1.8 Types of stresses on reservoir rocks.

These deformation stages depend on the pore pressure, the rigidity of the rock, stress history, and reservoir temperature (Tiab and Donaldson, 2016b). Inelastic deformation, in which the strain is a linear function of stress, is described by Hooke's law.

$$\sigma = E\varepsilon \tag{1.10}$$

where E is Young's modulus (elasticity modulus).

Young's modulus describes the ability of the rock to resist deformation. Moreover, Poisson's ratio relates the lateral strain to axial strain.

$$\upsilon = -\frac{\varepsilon_{\text{lat}}}{\varepsilon_{\text{ax}}} = \frac{\dfrac{\Delta d}{d_o}}{\dfrac{\Delta L}{L_o}} \tag{1.11}$$

where ε_{lat} is the lateral strain; ε_{ax} is the axial strain; d_o is the original diameter; L_o is the original core length; Δd is the change in diameter; and ΔL is the change in length.

Another elastic modulus is rigidity, which measures the ability of a rock to resist change in shape.

$$G = \frac{\text{shear stress}}{\text{shear strain}} = \frac{\tau}{\gamma} \tag{1.12}$$

1.6.4 Multiphase fluids and rock interaction

Pores of the reservoir rocks contain fluids, and hence, these multiphase fluids interact with each other and with the reservoir rocks. Different repulsive and attraction forces govern these interactions between the reservoir rock and fluids in the pores. Each phase (oil, water, and gas) occupies a fraction of the pore, and when compared to the total volume of the fluids in the pore, it is known as phase saturation.

1.6.5 Interfacial and surface tension

The reservoir contains multiphase fluids which are in contact with each other and with the rock surface. Therefore, the forces acting between the surface of two immiscible fluids and between reservoir fluids and rock considerably impact the fluids' distribution and production. The term "surface tension" describes the forces acting on the interface between liquid and gas. In contrast, the term interfacial tension is used in the case of two immiscible liquids. Pores in petroleum reservoirs contain fluids in contact with each other and are distributed according to their densities. The interference between these immiscible phases creates an interfacial tension and capillary pressure, which must be overcome to initiate flow. Capillary pressure is helpful in understanding the fluids' distribution in the reservoir.

1.6.6 Wettability

When two immiscible fluids exist within a rock surface, the surface tends to be in contact preferentially with one of them. This wetting phase will displace the nonwetting fluid from the surface and tends to spread over the rock surface. Measuring the contact angle between a droplet of a fluid with the rock surface would describe the wettability (Abdallah et al., 2007). The contact angle (θ) is measured through the liquid to the rock surface (Fig. 1.9). The contact angle decreases and might reach zero when the rock surface is completely wet by the liquid, whereas a contact angle of 180° would suggest the presence of a nonwetting liquid. However, intermediate wettability (contact angle between 60° to 90°) could be encountered on some rock surfaces, indicating the lack of a strong wettability preference. It is essential to differentiate between the intermediate wetting state and mixed wettability. Reservoir rocks contain many types of minerals, which have different tendencies toward reservoir fluids. This variation in wettability between minerals on the rock surface creates a mixed wetting state on the reservoir rocks.

The wetting state of reservoir rocks significantly impacts the distribution of the fluids in the porous media. The nonwetting phase tends to occupy the more open channels, whereas the wetting phase invades the smaller pores because of the attraction forces (Wu, 2016). Table 1.1 shows the range of contact angle for each wetting state.

1.6.7 Capillary pressure

The interaction between the immiscible fluids on the pore and the capillaries of the porous media results in pressure difference as the pressure exerted by each phase is different. Therefore, the interfacial and surface forces acting

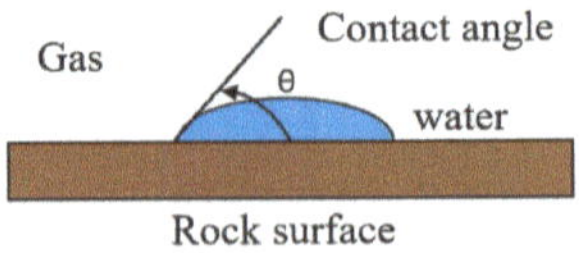

FIG. 1.9 Contact angle.

TABLE 1.1 Wettability and contact angel

Contact angle	Wettability state
0°–75°	Water wet
75°–105°	Intermediate wet
105°–180°	Oil wet

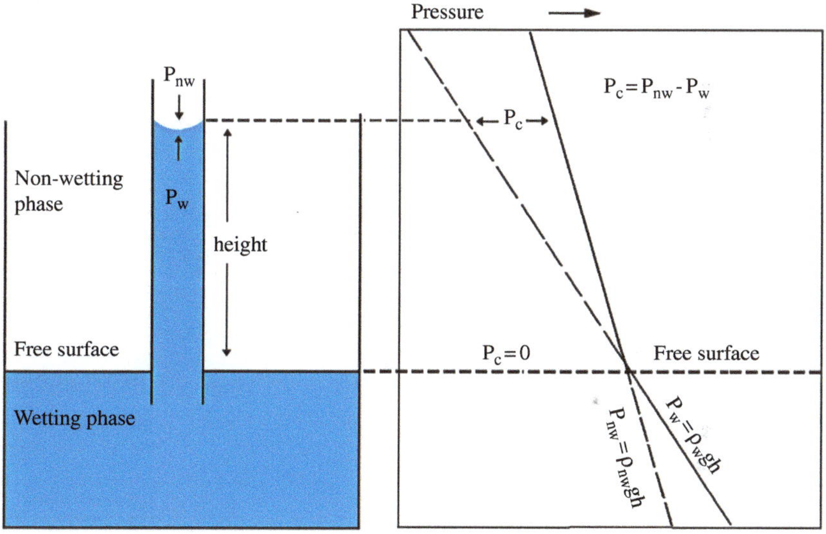

FIG. 1.10 Capillary pressure.

between the reservoir fluids and surrounding rocks require force balance which results in capillary pressure. Capillary pressure is the pressure that must be exceeded by the nonwetting phase to displace the wetting phase. Hence, it depends on the phase saturation of the phase saturations (Fig. 1.10).

$$p_{cow}(S_w) = p_o - p_w;\ \text{water wet} \tag{1.13}$$

$$p_{cwo}(S_w) = p_w - p_o;\ \text{oil wet} \tag{1.14}$$

$$P_c = \frac{2\sigma\cos\theta}{r_c} \tag{1.15}$$

$$P_C = (\rho_w - \rho_{nw})gh \tag{1.16}$$

where r_c is the capillary radius; θ is the contact angle between the fluids and the capillary tube; σ is the interfacial tension; h is the height above the free surface; g is the gravitational constant; ρ_{nw} is the density of the nonwetting fluid; and ρ_w is the density of the wetting fluid.

In a water-wet reservoir, the pressure exerted by the oil droplet must exceed a certain pressure required to displace the wetting phase, the water in this case. Therefore, the water saturation decreases sharply up to a value where no water will be expelled out of the pore in the presence of the oil phase, which is known as irreducible water saturation (Fig. 1.11). Hence, the interaction between the reservoir fluids (oil and water) and rocks creates irreducible water saturation. The decrease in wetting phase saturation represents the drainage process. Alternatively, the pores might be filled with the nonwetting phase, the oil in the case of water wet rocks, and the saturation of the wetting phase is allowed to increase, which is referred to as the imbibition process (Fig. 1.12). It is worth mentioning that both imbibition and drainage processes occur in the reservoirs. However, conducting both drainage and imbibition would not provide the same results, which indicates the impact of saturation history as hysteresis is noticed between the curves. The capillary pressure depends on the size of the pore size distribution (Li, 2010; Ling et al., 2014).

The Brooks-Corey model can estimate the capillary pressure:

$$p_{cow}(S_w) = p_o - p_w = p_e \left(\frac{S_w - S_{wir}}{1 - S_{wir}} \right)^{-\frac{1}{\lambda}} \tag{1.17}$$

where S_{wir} is the irreducible water saturation, fraction; S_w is the water (wetting) phase saturation, fraction; λ is the pore-size distribution, dimensionless;

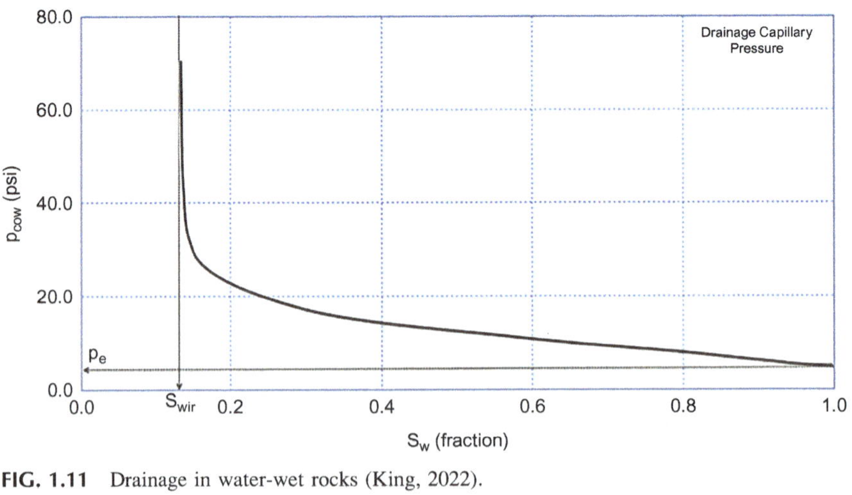

FIG. 1.11 Drainage in water-wet rocks (King, 2022).

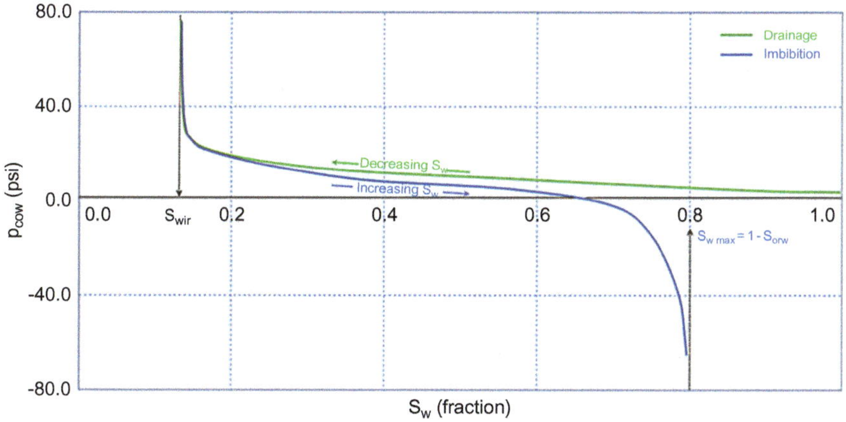

FIG. 1.12 Typical drainage and imbibition in water-wet rocks (King, 2022).

p_e is the entry pressure, psi; p_o and p_w are the pressures of the oil phase and water phase, respectively, psi; and p_{cow} is the oil–water capillary pressure, psi.

Most hydrocarbon reservoirs are initially water-wet, and then, a transition zone is created after the oil migration. The capillary pressure controls the distribution of the fluids in the transition zone based on the buoyancy differences. Different wetting states exist when a reservoir consists of different layers. Moreover, a tight zone might have a different wetting state from other surrounding layers. This heterogeneity in wettability would affect the recovery of hydrocarbons in pore and reservoir scales.

1.7 Conclusions

Oil and gas reservoirs contain multiphase fluids such as brine, water, oil, and gas within the pores of the rocks. This work reviewed the properties of reservoir fluids and rocks, which significantly impact the distribution of the different phases in the porous media and the recovery process. The mineralogy of the sandstone and carbonate rocks affects the wettability of reservoir rocks. Furthermore, the interfacial and surface tension are formed between the different phases in the reservoir. Moreover, the capillary forces are formed because of the difference in pressure between the wetting and nonwetting phase.

Acknowledgement

The authors would like to acknowledge the Qatar National Research Fund (a member of Qatar Foundation) for funding through Grant # NPRP13S-1231-190009. Al-Salam Petroleum Services Company, Qatar, is also acknowledged for cofunding this project. The findings achieved herein are solely the responsibility of the authors.

References

Abdallah, W., Buckley, J., Carnegie, A., John, E., Herold, B., Graue, A., Habashy, T., Seleznev, N., Signer, C., Hussain, H., Montaron, B., Ziauddin, M., 2007. Fundamentals of wettability. Oilf. Rev. 19 (2), 44–61.

Baker, R.O., Yarranton, H.W., Jensen, J.L., 2015. Rock and fluid properties. In: Practical Reservoir Engineering and Characterization. Elsevier, pp. 35–66. https://doi.org/10.1016/B978-0-12-801811-8.00002-X.

Bjørlykke, K., Jahren, J., 2015. Sandstones and sandstone reservoirs. In: Pet. Geosci. From Sediment. Environ. to Rock Physics, second ed, pp. 119–149, https://doi.org/10.1007/978-3-642-34132-8_4.

Civan, F., 2007. Mineralogy and mineral sensitivity of petroleum-bearing formations 11 Parts of this chapter have been reprinted with permission of the Society of Petroleum Engineers from Civan (1999a and 2001c). In: Reservoir Formation Damage. Elsevier, pp. 13–77, https://doi.org/10.1016/B978-075067738-7/50003-8.

Darling, T., 2005. Production geology issues. In: Well Logging and Formation Evaluation. Elsevier, pp. 137–154, https://doi.org/10.1016/B978-075067883-4/50010-3.

Fanchi, J.R., 2010. Porosity and permeability. In: Integrated Reservoir Asset Management, pp. 49–69, https://doi.org/10.1016/B978-0-12-382088-4.00004-9.

Gillott, J.E., 1984. Clay, engineering geologyclay, engineering geology. In: Applied Geology. Kluwer Academic Publishers, Dordrecht, pp. 45–59, https://doi.org/10.1007/0-387-30842-3_9.

Glorioso, J.C., Rattia, A., 2012. Unconventional reservoirs: basic petrophysical concepts for shale gas. In: Soc. Pet. Eng.—SPE/EAGE Eur. Unconv. Resour. Conf. Exhib. vol. 2012, pp. 748–785, https://doi.org/10.2118/153004-MS.

King, G., 2022. 3.4: Reservoir Rock—Fluid Interaction Properties | PNG 301: Introduction to Petroleum and Natural Gas Engineering [WWW Document]. URL https://www.e-education.psu.edu/png301/node/585 (accessed 1.5.22).

Kumari, N., Mohan, C., 2021. Basics of clay minerals and their characteristic properties. In: Clay and Clay Minerals. IntechOpen, https://doi.org/10.5772/intechopen.97672.

Li, K., 2010. Analytical derivation of Brooks–Corey type capillary pressure models using fractal geometry and evaluation of rock heterogeneity. J. Petrol. Sci. Eng. 73, 20–26. https://doi.org/10.1016/J.PETROL.2010.05.002.

Ling, K., Han, G., Shen, Z., He, J., Pei, P., 2014. Calculating pore size distribution by using capillary pressure. In: SPE—Eur. Form. Damage Conf. Proceedings, EFDC. vol. 2, pp. 782–794, https://doi.org/10.2118/168183-MS.

Lyons, W.C., 2010. Basic principles, definitions, and data. In: Working Guide to Reservoir Engineering. Elsevier, pp. 1–95.

Mackenzie, F.T., 1978. Carbonate mineralogy and geochemistry. In: Sedimentology. Springer Netherlands, Dordrecht, pp. 147–158, https://doi.org/10.1007/978-1-4020-3609-5_35.

Neuendorf, K.K.E., Mehl, J.P., Jackson, J.A., 2005. The Glossary of Geology. American Geological Institute, p. 800.

Pettijohn, F.J., Potter, P.E., Siever, R., 1987. Sand and Sandstone, Sand and Sandstone. Springer New York, New York, NY, https://doi.org/10.1007/978-1-4612-1066-5.

Schulze, D.G., 2005. Clay minerals. In: Encycl. Soils Environ. vol. 4, pp. 246–254, https://doi.org/10.1016/B0-12-348530-4/00189-2.

Tiab, D., Donaldson, E.C., 2016a. Introduction. In: Petrophysics. Elsevier, pp. 1–21, https://doi.org/10.1016/B978-0-12-803188-9.00001-2.

Tiab, D., Donaldson, E.C., 2016b. Effect of stress on reservoir rock properties. In: Petrophysics. Elsevier, pp. 483–582, https://doi.org/10.1016/B978-0-12-803188-9.00009-7.

Weimer, R.J., Tillman, R.W., 1982. Sandstone reservoirs. In: All Days. SPE, https://doi.org/10.2118/10009-MS.

Wu, Y.-S., 2016. Multiphase fluids in porous media. In: Multiphase Fluid Flow in Porous and Fractured Reservoirs. Elsevier.

Zou, C., 2013. Outlook on unconventional petroleum resources. In: Unconventional Petroleum Geology. Elsevier, pp. 355–362.

Zou, C., 2017. Meaning of unconventional petroleum geology. In: Unconv. Pet. Geol, pp. 49–95, https://doi.org/10.1016/B978-0-12-812234-1.00002-9.

Zou, C., Zhag, G., Yang, Z., Tao, S., Hou, L., Zhu, R., Yuan, X., Ran, Q., Li, D., Wang, Z., 2013. Concepts, characteristics, potential and technology of unconventional hydrocarbons: on unconventional petroleum geology. Pet. Explor. Dev. 40, 413–428. https://doi.org/10.1016/S1876-3804(13)60053-1.

Chapter 2

Rheology of upstream complex fluids

Mohamed Shamlooh[a], Ibnelwaleed A. Hussein[a,b], Mustafa S. Nasser[a,b], Krishna Raghav Chaturvedi[c], and Japan Trivedi[d]

[a]*Gas Processing Center, College of Engineering, Qatar University, Doha, Qatar*

[b]*Department of Chemical Engineering, College of Engineering, Qatar University, Doha, Qatar*

[c]*Department of Petroleum Engineering and Geoengineering, Rajiv Gandhi Institute of Petroleum Technology, Jais, India*

[d]*School of Mining and Petroleum Engineering, University of Alberta, Edmonton, AB, Canada*

2.1 Introduction

Rheology studies the deformation and flow of materials and their behavior in different environments. Studying rheology indicates important characteristics such as viscosity, plasticity, and elasticity. Rheological properties vary from one fluid to another depending on the surrounding conditions, such as temperature, pressure, and the rate and duration of the external forces on the material and fluid characteristics. Generally, the flow properties of a fluid are functions of the shear rate, time, and particular orientation. While it is essential in most engineering aspects nowadays, rheology is particularly important in petroleum engineering as the interaction between the different types of fluids (Fig. 2.1) and the formation governs the production operation of a particular reservoir.

Therefore, understanding the nature of formation and the fluids that exist in the reservoir helps the engineers choose and design suitable drilling fluids and injection fluids to optimize the operation. This chapter aims to introduce the applications of rheology and present some general guidelines for the design of complex fluidic systems in upstream applications. For example, in drilling, proper formulation of fluid rheology would prevent fluid loss, minimize fluid–rock interactions and consequently reduce formation damage. In fracturing, fluids with high viscoelasticity could reduce the plugging and retention in the formation. While in chemically enhanced oil recovery, the use of polymers to increase the viscosity of the sweeping fluid (water) enhances oil recovery and improves the sweep efficiency. Further, the use of surfactants to reduce interfacial tension is another approach to enhance oil recovery.

Developments in Petroleum Science, Vol. 78. https://doi.org/10.1016/B978-0-323-99285-5.00010-7

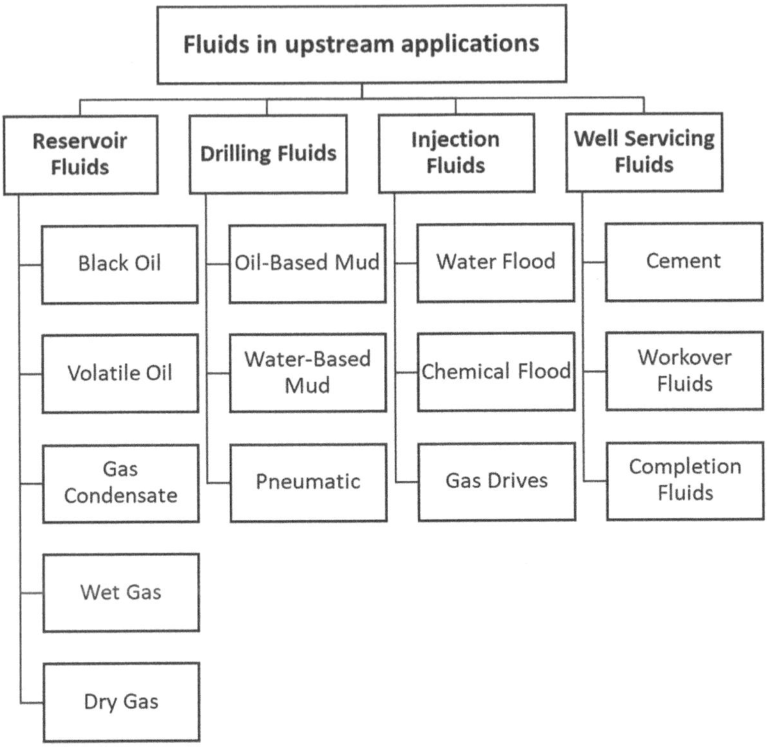

FIG. 2.1 Classification of the types of fluids in the upstream operations.

2.2 Basics of rheology

This section will lay the ground for some theoretical concepts that will be important to discuss the different upstream applications in the following sections. A detailed discussion about rheology can be found in Malvern's "A Basic Introduction to Rheology" report (Malvern, 2016). Fig. 2.2 illustrates the shear flow where the liquid is imagined as a system of multiple layers sliding on top of each other resulting from an external force. Shear stress is the force per unit area applied to the fluid system. The term "Shear Strain" refers to the displacement gradient, which can be calculated by dividing the lateral displacement over the overall height of the system. Shear rate, however, is the rate of change in the shear strain with time. Viscosity is the coefficient of proportionality between the shear stress and the shear rate, which measures resistance to flow resulting from the internal friction between the different layers of the fluid. Another important terminology in rheology is "Viscoelasticity," which describes the materials that exhibit both solid-like behavior (elastic) and liquid-like behavior (viscous), such as polymeric solutions or gels.

$$\sigma\ (shear\ stress) = \frac{F}{A}\ (Pa)$$

$$\gamma\ (shear\ strain) = \frac{x}{h}$$

$$\dot{\gamma}\ (shear\ rate) = \frac{d\gamma}{dt}\ (s^{-1})$$

$$\eta\ (viscosity) = \frac{\sigma}{\dot{\gamma}}\ (Pa.s)$$

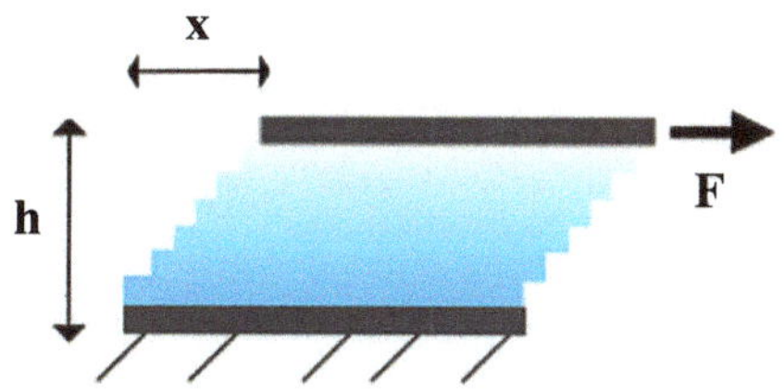

FIG. 2.2 Illustration of shear flow. *Reproduced from Del Giudice, F., 2022. A review of microfluidic devices for rheological characterisation. Micromachines 10.3390/mi13020167.*

Fluids can be categorized rheologically into two main categories; Newtonian and Non-Newtonian fluids. Newtonian Fluids are independent of the shear rate where the viscosity is constant since the shear rate is linearly correlated with the applied shear stresses at all points. Non-Newtonian fluids deviate from 'Newton's law of viscosity, where the viscosity is not constant. Non-Newtonian Fluids are complex fluids classified based on time-dependency, where the rheostable fluids are time-independent, and the rheounstable fluids are time-dependent. Non-Newtonian Fluids can exhibit either a shear-thinning behavior (Pseudoplastic and Thixotropic), shear thickening behavior (Dilatant and Antithixotropic), or none (Bingham), where the last deviates from the Newtonian behavior as it has a plastic behavior at low stresses (viscous behavior begins after the threshold stress is reached). The shear dependency of fluids is usually described mathematically using rheological models. Fig. 2.3 shows the rheological classification of fluids.

Numerous rheological models exist in the literature, and each has its window of application and limitations. Table 2.1 summarizes the most commonly used rheological models in upstream applications. Tao et al. (2020) illustrated graphically the responses of the most widely used rheological models (Fig. 2.4).

2.3 Rheology of reservoir fluids

Reservoir fluid systems gain their complexity from the wide range of existing fluids. The composition of the reservoir fluids covers a wide range from light hydrocarbons with few carbon molecules to heavy crude oil with thousands of carbon atoms and the existence of hydrocarbons. Other forms, such as gas hydrates, dissolved gas, paraffin, and asphaltenes, are also present. Analyzing the rheological behavior of the different combinations of hydrocarbon forms is essential in the production and transportation processes. Petroleum fluids, such as emulsions, suspensions, heavy oil, and asphaltenes, follow non-Newtonian behavior. In addition, multiphase systems, including wax in the oil, water in oil, and oil in water, are examples of non-Newtonian fluid systems. The same

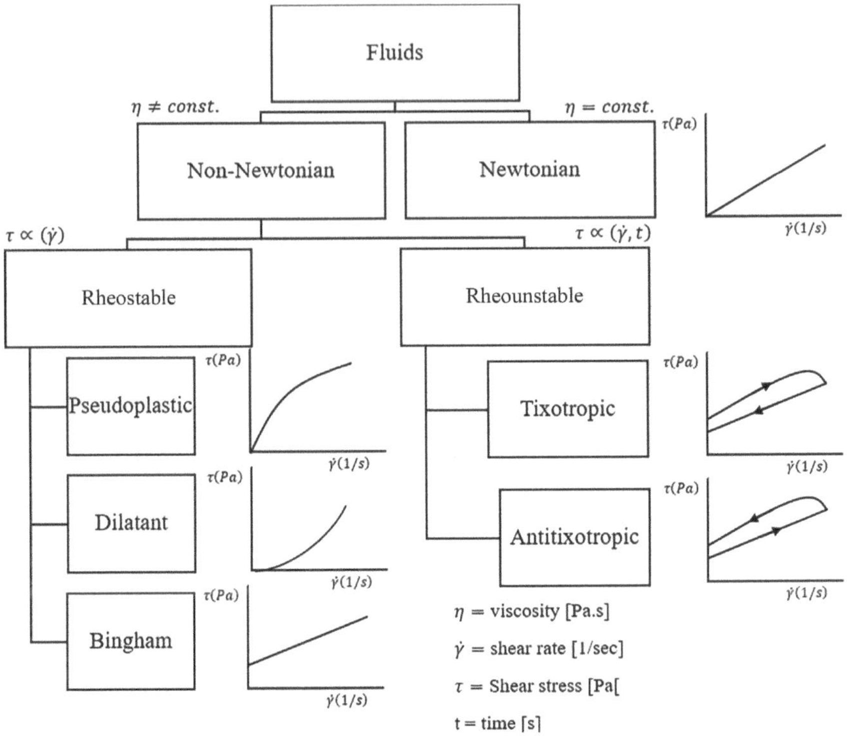

FIG. 2.3 Rheological classification of fluids (Leonowicz et al., 2014).

crude oil could behave as a Newtonian fluid at high temperature and as a non-Newtonian fluid at low temperature due to the formation of wax or asphaltenes depending on composition. One method of detecting the temperature at which wax starts to form in oil (cloud point) and the temperature below which the fluid will stop flowing (pour point) is by identifying the temperature when the oil rheology deviates from the Newtonian behavior (Rønningsen, 2012).

2.4 Rheological models

Moreover, waxy and paraffinic materials change the nature of the oil from viscous to viscoelastic. Ilyin et al. conducted a comparative study on the rheological behavior of light and heavy crude oil (Ilyin et al., 2016) (Fig. 2.5). The study demonstrated that the studied oil samples (from Russian reservoirs) had Newtonian behavior at elevated temperatures regardless of the type of oil. Nevertheless, heavy oil shows a viscoelastic behavior at subzero temperatures, indicating wax's presence. Therefore, light and heavy crudes are usually blended in specific ratios to adjust the rheology to avoid wax formation in the pipelines.

TABLE 2.1 Models of non-Newtonian fluids used in upstream applications

Model	Equation	Applications	Limitations	Reference
Bingham	$\tau = \tau_0 + \mu_P \dot{\gamma}$	The most commonly used model in drilling muds as it is a simple two-parameters model with an acceptable accuracy at moderate shear rates	Fails to predict the behavior of fluids at very low and very high shear rates. Moreover, it often exaggerates in predicting the yield stress	Agwu et al. (2021)
Power Law	$\tau = k\dot{\gamma}^n$	Has wide applications in the petroleum industry such as drilling fluids and nanoemulsions	Assumes a zero yield stress leading to some inaccuracies. Cannot capture the linear behavior between shear and shear rate at very low and very high shears	Kumar et al. (2021)
Casson	$\frac{1}{\tau^{\frac{1}{2}}} = \frac{1}{(\tau_0)^{\frac{1}{2}} + (\mu_0 \dot{\gamma})^{\frac{1}{2}}}$	Mostly used for cement slurries. Can predict with some accuracy the behavior at high shear rates when only low-shear rate data are available	Not able to predict the limit for the maximum shear rate and can only estimate the infinite shear stress limit	Agwu et al. (2021)
Herschel-Bulkley	$\tau = \tau_0 + k(\dot{\gamma})^n$	Used for predicting the behavior of drilling mud with a relatively good accuracy better than Bingham and Casson models. Also can be applied with good accuracy in nanoemulsions	The model is, often, either accurate in the low shear range or the high shear range	Adewale et al. (2017), Herschel and Bulkley (1926), Kumar et al. (2021), Saasen and Ytrehus (2020)

Continued

TABLE 2.1 Models of non-Newtonian fluids used in upstream applications—Cont'd

Model	Equation	Applications	Limitations	Reference
Robertson-Stiff	$\tau = k(\dot{\gamma} + \dot{\gamma}_0)^n$	Predicts the behavior of drilling fluids with high accuracy compared to Bingham and Herschel-Bulkley. Commonly used for cement slurries	A complex mathematical model with three parameters	Khataniar et al. (1994), Mohamed et al. (2021)
Cross	$\tau = \dot{\gamma}\left(\mu_\infty + \frac{\mu_0 - \mu_\infty}{1 + \alpha\dot{\gamma}^{\frac{2}{3}}}\right)$	Used for viscosity prediction of shear-thinning fluids. Predicts with good accuracy the viscoelastic fluids with suspended solids (i.e., cuttings in drilling fluids)	A complex mathematical model with four parameters	Awad et al. (2021), Cross (1979)
Carreau	$\tau = \frac{\mu_c \dot{\gamma}}{\left(1 + \frac{\dot{\gamma}}{\dot{\gamma}_c}\right)^A}$	Used in various petroleum applications such as dispersions and flow in porous media. Also used in other applications such as food and blood rheology	Difficulties in parameter identification, especially at high shear rates, due to the complexity	Awad et al. (2021), Eberhard et al. (2019), Gallagher et al. (2019), Tabakova et al. (2015)
Hyperbolic	$\tau = \tau_y + A\sqrt{\left(\frac{\dot{\gamma} - \dot{\gamma}_{cp}}{B}\right)^2 - 1}$	The maximum shear stress parameter is introduced. Accurately predicts the behavior at low and high shear rates when the other models fail	A complex mathematical model with four parameters	Vipulanandan and Mohammed (2014)
API Model	$\tau_{Annulus} = k_a \dot{\gamma}^{n_a}$ $n_a = 0.657 \log\left(\frac{\theta_{100}}{\theta_3}\right)$ $k_a = \frac{\theta_{100}}{170.3^{n_a}}$	Used for the drilling mud in the annulus to overcome the error resulting from the power-law model at low shear rates. It uses the regular power-law model for the drill pipe	Not reported	Anawe and Folayan (2018)

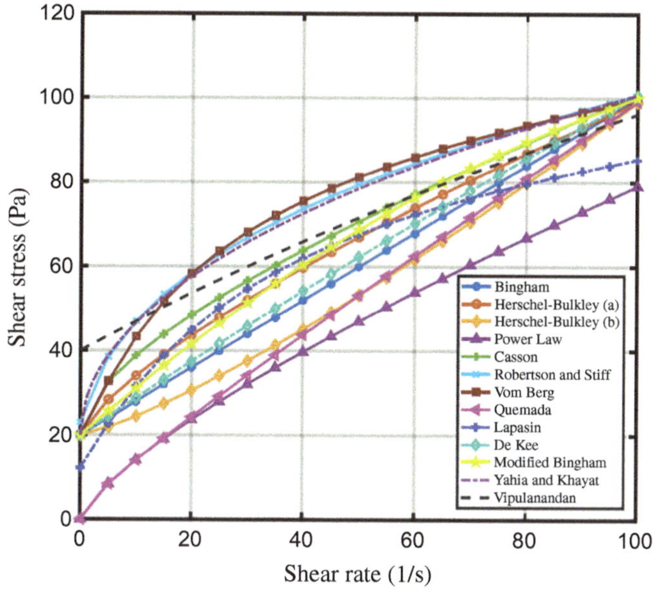

FIG. 2.4 Comparison of the rheological response between the most commonly used models (Tao et al., 2020).

Apart from the nature of hydrocarbons and the number of carbons in the chains, several factors affect the rheology of reservoir fluids. A common phenomenon that usually takes place in the transportation piping in the upstream facilities is the multiphase flow, where there is a mixture of oil and produced water. Due to the disturbances and the high turbulence during these operations, these mixtures usually form a dispersion of one form in another. Typically, oil becomes the continuous phase at low water cuts with suspended water droplets, which system is known as water in oil (w-i-o). At high water cuts, the system is inverted where water becomes the continuous phase in a system known as oil in water (o-i-w). Identifying the inversion range (the water cut range at which the inversion takes place from w-i-o to o-i-w) is crucial. The dominant phase highly influences the rheological properties essential in designing the transportation and storage network (Rønningsen, 2012). Generally, increasing the water cut increases both the viscosity and the yield stress of the system.

Dissolved gas is another factor that impacts the rheology of crude oil. Hu and Crawshaw conducted a rheological analysis on saturated crude oils with CO_2 at equilibrium conditions (Hu and Crawshaw, 2017). The study revealed that a significant impact is noticed upon the addition of CO_2 and a threshold pressure exists where the viscosity decreases initially when CO_2 pressure is below the threshold and then starts increasing above that. Furthermore, the

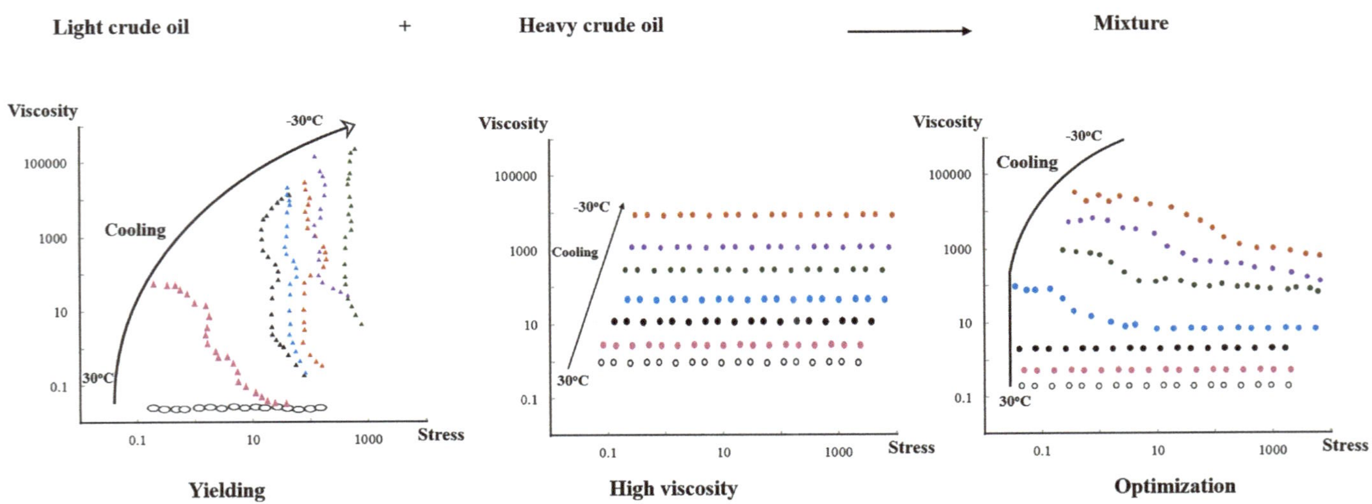

FIG. 2.5 Rheological behavior of light and heavy crude oils and their mixture (Ilyin et al., 2016).

non-Newtonian behavior of the crude oil is also affected where the shear-thinning behavior of the crude oil is weakened upon adding more CO_2. Overall, the rheological behavior of reservoir fluids is affected by several elements including, but not limited to:

a. Nature of hydrocarbons
b. Reservoir conditions (Temperature and Pressure)
c. Presence of water droplets (dispersions)
d. Presence of waxy and paraffinic materials
e. Dissolved gases (i.e., CO_2)

Besides the engineering importance of rheology in designing the piping and pumping systems for injection fluids, it is also vital in deciding the success of these fluids in achieving their objective. A slight change in the shear may result in a significant change in the viscosity and other rheological properties in some fluids; therefore, a good characterization of the used injection fluid plays a decisive role in the success of the operation.

2.5 Rheology of drilling fluids

While drilling oil and gas wells, drilling fluids (more commonly referred to as drilling mud/s) play a crucial role for a variety of reasons that range from transporting the drilled cuttings to the surface, cleaning, cooling, and lubricating the drilling bit, reducing friction, minimizing thermal heat stresses and most importantly, preserve borehole integrity (Ahmad et al., 2018; Maiti et al., 2019). Thus, the proper design and application of drilling mud is a critical component for the success of any drilling operation (William et al., 2014). Conventionally, the cost, environmental effect, and shale inhibitory properties of drilling mud are given the most important during the design phase, with water-based drilling muds and oil-based drilling muds being the two primary types of drilling muds most commonly used during oil and gas drilling operations with the difference being in the type of base fluid used for drilling fluid formulation (Agwu et al., 2021; Xu et al., 2018). Separately, using aerated/foam-based drilling fluid has also gained some applicability, especially in managed pressure drilling applications (Lyons et al., 2021). However, regardless of their origin, the drilling mud properties must be tracked continuously as the subsurface conditions (mainly pressure and temperature but also formation salinity, composition, and entrapped fluids) cause deviation from original values, which may compromise its performance and influence drilling success or failure (Agwu et al., 2021).

Furthermore, evaluating drilling mud rheological properties is essential for optimal wellbore hydraulics maintenance and management throughout any drilling operation. Rheology is a branch of physics that studies material flow, and deformation. Its proper understanding in the context of the drilling fluid can help explain its behavior under a variety of variables such as temperature, pressure, and external forces. Thus, the real-time tracking of drilling mud rheology has attracted considerable industrial and research attention in recent years (Chaturvedi et al., 2021; Kök and Bal, 2019). Thus, in recent years, the real-time tracking of drilling mud rheology has attracted considerable industrial and research attention. However, before delving deeper into the rheology of drilling fluids, it is essential to understand the key rheological parameters.

2.6 Essential rheological properties of drilling fluids

It is essential to note that drilling muds do not follow any particular fluid model exactly as they continuously undergo different types of flow during circulation. However, the most commonly used fluid model for drilling muds is the non-Newtonian Bingham Plastic model, for which standards have been published by the American Petroleum Institute (API) in the form of API BUL 13D (Clark, 1995). The most fundamental property of the drilling mud is its viscosity. While several advanced analytical types of equipment like viscometers and rheometers may also be used to obtain the viscosity of the drilling fluid, one of the widely used methods is the Marsh Funnel. The Marsh funnel is a simple device used to measure viscosity by obtaining the time a specific volume of liquid requires to flow from a cone via a short tube (Pitt, 2000). While simple, this method is widely used as it can quickly estimate the mud viscosity during ongoing drilling operations. A change in the values obtained by the Marsh Funnel can be used to identify mud loss, contamination, or worse, any reservoir fluid influx. However, a limitation of the Marsh Funnel is that the value of viscosity is measured at only one rate of shear, while the analysis also depends on ambient conditions during each investigation. Other ordinary drilling mud rheological properties include plastic and apparent viscosity, yield point, and gel strength which are conventionally obtained with the help of a viscometer, following standard API guidelines (Clark, 1995).

Conventionally, the resistance to the flow of a fluid is referred to as its plastic viscosity, which depends mainly on the viscosity of the base fluid, the size, and shape of any suspended particles in the base fluid, and the concentration of any additive (Bridges and Robinson, 2020). While the plastic viscosity represents mud viscosity when extrapolated to an infinite shear rate (considering the Bingham Plastic model), when viscosity is measured at a shear rate specified by API, it is referred to as apparent viscosity. Generally, the drilling muds with elevated plastic viscosity values are avoided as they exhibit increased equivalent circulation density (ECD) values. ECD refers to the effective density of the drilling mud after taking into account the pressure drop (in the annulus above the point of calculation) and is diligently monitored by the drilling crew to avoid mud losses and kicks, particularly in formations that have a low tolerance between fracture and pore-pressure gradient (Alsaihati et al., 2021). Furthermore, a fluid with high plastic viscosity tends to exhibit lower penetration rates (Agwu et al., 2021). Additionally, an increase in plastic viscosity also increases the energy required to run the mud circulation pumps.

The yield point of drilling mud denotes the amount of tension required to initiate deformation in the fluid to cause its flow (Gautam and Guria, 2020). Common wisdom holds that an increase in the yield point of mud denotes an increase in frictional pressure losses and, consequently, an increase in ECD (Amani, 2012; Murtaza et al., 2021).

Another essential characteristic to consider is the drilling fluid's gel strength, which denotes its ability to suspend drill cuttings when circulation has stopped (Agwu et al., 2021). To calculate the gel strength, the shear stress of the drilling mud is measured after being static for two independent periods (10 s and 10 min in the API standard). The cutting suspension property of the drilling mud is essential for hole cleaning as the suspended cuttings alter the rheological behavior of the mud (Wastu et al., 2019). The plastic viscosity and yield point values are also used to determine the cutting suspension ability of the mud via the yield point to plastic viscosity ratio (YP/PV) with a ratio of 0.75–1.5, indicating a high degree of suspension capacity.

2.7 Factors affecting mud rheology

Drilling fluids are non-Newtonian fluids whose behavior is strongly dependent on the base fluid used for synthesis, suspended solids, reservoir temperature, and pressure (Gautam and Guria, 2020; Hermoso et al., 2014). Table 2.2 illustrates the effect of the reservoir parameters on various types of drilling mud.

In previous studies, it has been observed that on change of temperature from ambient to seabed conditions (common in offshore operations), the

TABLE 2.2 Effect of reservoir parameters on the rheology of drilling muds

Reservoirs parameters	Oil-based muds	Water-based muds	Aerated muds/foams
Temperature	Less affected	Highly affected	Most affected
Pressure	Less affected	Less affected	Highly affected
Suspended solids	Less affected	Moderately affected	Less affected
Presence of shales	Less affected	Most affected	Highly affected
Gas influx	It depends on gas solubility, reservoir pressure, and temperature		

viscosity of mud tends to increase while an increase in temperature, reduces viscosity (Gautam and Guria, 2020; Murtaza et al., 2021). An increase in temperature tends to reduce the viscosity of the base fluid (especially true for water and foam-based drilling muds), which affects the drilling mud performance. This change in viscosity directly impacts cutting suspension ability, ECD, and well integrity (Murtaza et al., 2021). Aerated drilling muds, in particular, suffer from an increase in temperature as at higher temperatures, gas molecules display increased kinetic energy causing bubble destabilization and foam collapse (Raghav Chaturvedi et al., 2018). Oil-based drilling muds are comparatively less affected by increased temperature (Zhao et al., 2010). However, in previous work, increasing temperature from 100 to 400° F, the rheological characteristics of oil-based mud decreased but raising the temperature beyond 400–600°F, the properties started to increase again (Amani, 2012). On the other hand, the role of pressure is comparatively more muted as the base fluids exhibit little compression with an increase in pressure (Amani, 2012; Gokdemir et al., 2017; Rommetveit and Bjorkevoll, 1997). However, Foam-based drilling fluids exhibit inferior performance at elevated pressures as increasing pressure accelerates liquid drainage, leading, to bubble collapse and foam failure (Herzhaft, 2000).

Another factor influencing drilling mud rheology is the presence of suspended solids in the fluid phase. In a phenomenon most commonly observed with water-based muds, barytes and other additives tend to agglomerate and form large clusters within the fluid phase, which causes an increase in phase segregation and results in non-uniform mud viscosity (Basfar et al., 2019). This phenomenon known as baryte sag or solid sag can also occur in oil-based muds but is more commonly encountered in any type of mud used for high temperature/high deviation wells (Mohamed et al., 2019). Finally, one of the most important reservoir parameters influencing drilling mud rheology is the presence of shales in the subsurface. Shales tend to absorb water which reduces liquid content in the drilling mud (along with other wellbore integrity issues). This tends to increase the viscosity of water-based drilling muds, and thus, their use is not recommended for formations with shale presence (Lyu et al., 2015). One of the most important deductions from a continuous assessment of drilling mud rheology can be the amount of contamination. The influx of reservoir fluids, primarily gas, tends to reduce the viscosity of the drilling fluid by displacing the denser drilling fluid with comparatively much lighter gas, see Fig. 2.6 (Chaturvedi et al., 2021).

In water-based muds, while gas tends to reach the surface faster due to quicker slip velocity and upward migration, in oil-based muds, this depends majorly on reservoir pressure and temperature. Gas solubility in water is marginal but can be deceiving for well-operator in oil-based muds as gas may suddenly exfill from the encapsulating liquid and cause a run-away loss of well control, which may ultimately lead to a blow-out. To avoid such an occurrence, various rheological parameters of the drilling mud must be

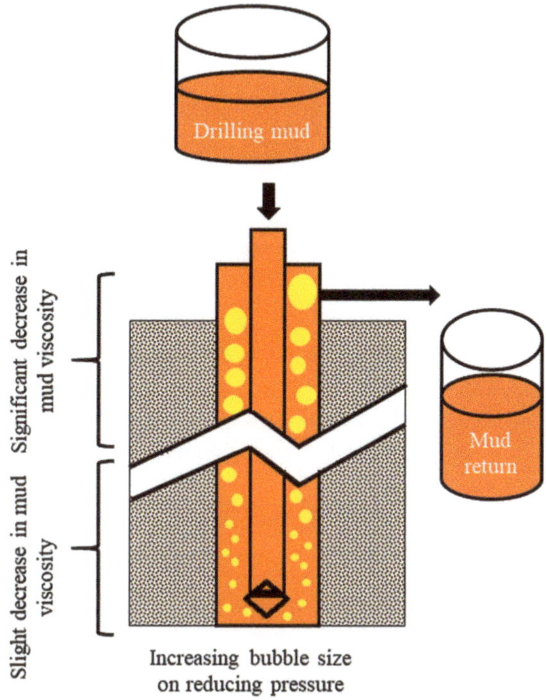

FIG. 2.6 Influence of gas influx on the viscosity of drilling mud. *Adapted from Chaturvedi, K.R., Fogat, M., Sharma, T., 2021. Low temperature rheological characterization of single-step silica nanofluids: an additive in refrigeration and gas hydrate drilling applications. J. Petrol. Sci. Eng. 204, 108742. doi:10.1016/j.petrol.2021.108742.*

monitored regularly as the importance of rheology cannot be more overstated for successful well drilling and control operations.

2.8 Rheology of emulsions

Emulsions are special fluids that exist naturally in oil reservoirs. Moreover, emulsions form due to the enhanced oil recovery treatments inside the reservoirs. Understanding the rheology of emulsions is essential to incorporate the best treatment that maximizes oil recovery. Table 2.3 below summarizes the most affecting factors on the rheology of emulsions.

2.9 Other applications in upstream

One of the most critical applications of rheology in upstream investigations is studying the behavior of cement slurries. Understanding the rheological behavior of such a system helps prevent and eliminate hydrocarbon migration-related problems (Yousuf et al., 2021). Cement slurries, in general, are reactive systems

TABLE 2.3 Factors affecting the rheology of emulsions and water/oil dispersions.

Factor	Remarks
Temperature	Emulsions behave similarly to other fluids with thermal thinning behavior. The temperature dependency of viscosity in emulsions can be described mathematically using the Arrhenius equation, which applies to most liquids, as given below: $$\mu = A \cdot e^{\frac{E_a}{RT}}$$ Furthermore, the emulsion stability is affected by the temperature
Water volume fraction	Water cut highly influences the rheology of the dispersion. Several correlations are proposed to estimate the viscosity as a function of the water fraction, with the equation below being the simplest for estimating the relative viscosity: $$\mu_r = e^{K\Phi}$$ where Φ represents the water fraction and K is system dependent constant. Another modification proposed by Broughton and Squires combines another system-dependent constant, A, as follows: $$\mu_r = Ae^{K\Phi}$$ More recently, new correlations have been proposed to incorporate the temperature with the water cut for more accurate estimations. Generally, the viscosity was found to increase exponentially with the increase in the water cut. Furthermore, The effect of dispersed water droplets is more pronounced at elevated temperatures
The viscosity of the continuous phase	Although the water dispersion affects the rheology of the system, the nature of the oil has the most dominant effect, especially at high temperatures. Generally, waxy oils tend to have more viscosities than a system of oil that has the same molecular weight with higher water cut and lower paraffinic content (Ronningsen, 1995). Hence, the average chain length in oil is a major factor in defining the rheology of the system. The following equation was found to approximate with reasonable estimations the viscosity as a function of the molecular weight (MW) of the oil: $$\ln(\mu) = r_1 + r_2 \cdot MW + \frac{r_3}{T} + \frac{r_4 \cdot MW}{T}$$ where r_{1-4} are constants that depend on the reservoir conditions

TABLE 2.3 Factors affecting the rheology of emulsions and water/oil dispersions.—Cont'd

Factor	Remarks
Type of flow	The effective viscosity is observed to drop in the turbulent flow compared to the laminar flow in emulsions due to the drag reduction. Hagen–Poiseuille equation and Blasius equation are used to estimate the effective viscosity for laminar flow and turbulent flow, respectively (Al-Yaari et al., 2014)
Other factors	Several other factors can affect the behavior of emulsions. 'Droplets' size of the dispersed phase, the presence of solids, and viscosifying agents are factors that can alter the non-Newtonian behavior of the system (Kumar et al., 2021)

Summarized from Ronningsen, H.P., 1995. Correlations for predicting viscosity of W/O-emulsions based on North Sea crude oils. In: All Days. SPE. doi:10.2118/28968-MS.

that undergo physical and chemical changes upon changes in the surrounding environment. Cement slurries show several complex rheological behaviors such as pseudo-plasticity, yield stress, and thixotropic effects. Hence, the design of a cement matrix based on these responses is vital in determining the success rate of the cementing job as it affects the stability and flowability of the uncured cement. For instance, a rule of thumb states that the yield stress and the plastic viscosity of the uncured cement slurries should be in the range of 10–100 N/m^2 and 0.01–1 Ns/m^2, respectively (Banfill, 2003). Several factors have been reported to influence the rheological response of cement, namely, time, water/cement ratio, nanoparticles, composition, and temperature.

Another important use of rheology in upstream applications is in polymeric gels. Polymeric gels have recently gained popularity in the petroleum industry. They have several applications such as water shut-off, conformance control, hydraulic fracturing, fracture sealing, and as a lost circulation material (Hamza et al., 2019; Shamlooh et al., 2020). Polymeric gels are a mixture of a crosslinkable base polymer and a crosslinker. The solution is designed to be injected as a liquid, then triggered by the 'reservoir's temperature, it forms a solid gel that seals the designated area. Rheology is a crucial element in the design of polymeric gelants for such applications as the different affecting factors should be considered in the design. The primary rheological responses are the gelation time and the final gel strength. The importance of gelation time is to ensure that the solution will flow until reaching the designated area to avoid the gelation in the transportation pipes, yet, the gelation time should not take very long after reaching its final destination to minimize the non-productive time of the operation (El-karsani et al., 2014). In addition, the final gel strength is as important as the gelation time to ensure that the gel can sustain the different stresses imposed inside the reservoir.

2.10 Conclusions

All in all, rheology plays a vital role in the oil and gas industry. Understanding the nature of fluid behavior helps tailoring the various operations to maximize efficiency and productivity. Moreover, it assists in enhancing the safety measures taken in the fields. The chapter presented the significance of rheology in different upstream operations. In reservoir fluids, understanding the flow behavior and the interaction with the other elements inside the reservoir helps in better designing the wells to maximize the profit. The rheology in drilling fluids is essential to ensure that it can efficiently achieve all the functions assigned to the drilling fluids, such as cutting transport and pressure maintenance. Emulsions are complex fluids that need to be understood to not disturb the operation by introducing fluids that can jeopardize productivity. Finally, rheology has other applications, such as cementing and gelation operations where curing is a vital factor that needs to be assessed well before the operation.

Acknowledgment

The authors would like to acknowledge the Qatar National Research Fund (a member of the Qatar Foundation) for funding through Grant # NPRP13S-1231-190009. Al-Salam Petroleum Services Company, Qatar, is also acknowledged for co-funding this project. The findings achieved herein are solely the responsibility of the authors.

References

Adewale, F.J., Lucky, A.P., Oluwabunmi, A.P., Boluwaji, E.F., 2017. Selecting the most appropriate model for rheological characterization of synthetic based drilling mud. Int. J. Appl. Eng. Res. 12, 7614–7629.

Agwu, O.E., Akpabio, J.U., Ekpenyong, M.E., Inyang, U.G., Asuquo, D.E., Eyoh, I.J., Adeoye, O.S., 2021. A critical review of drilling mud rheological models. J. Petrol. Sci. Eng. 203, 108659. https://doi.org/10.1016/j.petrol.2021.108659.

Ahmad, H.M., Kamal, M.S., Al-Harthi, M.A., 2018. Effect of thermal aging and electrolyte on bentonite dispersions: rheology and morphological properties. J. Mol. Liq. 269, 278–286. https://doi.org/10.1016/j.molliq.2018.08.058.

Alsaihati, A., Elkatatny, S., Abdulraheem, A., 2021. Real-time prediction of equivalent circulation density for horizontal wells using intelligent machines. ACS Omega 6, 934–942. https://doi.org/10.1021/acsomega.0c05570.

Al-Yaari, M., Al-Sarkhi, A., Hussein, I.A., Chang, F., Abbad, M., 2014. Flow characteristics of surfactant stabilized water-in-oil emulsions. Chem. Eng. Res. Des. 92, 405–412. https://doi.org/10.1016/j.cherd.2013.09.001.

Amani, M., 2012. The rheological properties of oil-based mud under high pressure and high temperature conditions. Adv. Pet. Explor. Dev. 3, 21–30. https://doi.org/10.3968/j.aped.1925543820120302.359.

Anawe, P.A.L., Folayan, J.A., 2018. Data analyses on temperature-dependent behaviour of water based drilling fluid rheological models. Data Br. 21, 289–298. https://doi.org/10.1016/j.dib.2018.09.100.

Awad, A.M., Hussein, I.A., Nasser, M.S., Karami, H., Ahmed, R., 2021. CFD modeling of particle settling in drilling fluids: impact of fluid rheology and particle characteristics. J. Petrol. Sci. Eng. 199, 108326. https://doi.org/10.1016/j.petrol.2020.108326.

Banfill, P., 2003. The rheology of fresh cement and concrete—a review. In: Proceedings of the 11th International Cement Chemistry Congress, pp. 50–62.

Basfar, S., Mohamed, A., Elkatatny, S., Al-Majed, A., 2019. A combined barite–ilmenite weighting material to prevent barite sag in water-based drilling fluid. Materials (Basel) 12, 1945. https://doi.org/10.3390/MA12121945.

Bridges, S., Robinson, L., 2020. A practical handbook for drilling fluids processing. In: A Practical Handbook for Drilling Fluids Processing. Elsevier, https://doi.org/10.1016/C2019-0-00458-X.

Chaturvedi, K.R., Fogat, M., Sharma, T., 2021. Low temperature rheological characterization of single-step silica nanofluids: an additive in refrigeration and gas hydrate drilling applications. J. Petrol. Sci. Eng. 204, 108742. https://doi.org/10.1016/j.petrol.2021.108742.

Clark, P.E., 1995. Drilling mud rheology and the API recommended measurements. In: Society of Petroleum Engineers—SPE Production Operations Symposium 1995, POS 1995. OnePetro, pp. 933–941, https://doi.org/10.2118/29543-MS.

Cross, M.M., 1979. Relation between viscoelasticity and shear-thinning behaviour in liquids. Rheol. Acta 18, 609–614. https://doi.org/10.1007/BF01520357.

Eberhard, U., Seybold, H.J., Floriancic, M., Bertsch, P., Jiménez-Martínez, J., Andrade, J.S., Holzner, M., 2019. Determination of the effective viscosity of non-newtonian fluids flowing through porous media. Front. Phys. 7, 71. https://doi.org/10.3389/fphy.2019.00071.

El-karsani, K.S.M., Al-Muntasheri, G.A., Hussein, I.A., 2014. Polymer systems for water shutoff and profile modification: a review over the last decade. SPE J. 19, 135–149. https://doi.org/10.2118/163100-PA.

Gallagher, M.T., Wain, R.A.J., Dari, S., Whitty, J.P., Smith, D.J., 2019. Non-identifiability of parameters for a class of shear-thinning rheological models, with implications for haematological fluid dynamics. J. Biomech. 85, 230–238. https://doi.org/10.1016/j.jbiomech.2019.01.036.

Gautam, S., Guria, C., 2020. An accurate determination of the shear rate for high-yield-stress drilling fluids using concentric cylinder Fann 35 viscometer data. SPE J. 25, 2984–3001. https://doi.org/10.2118/201238-PA.

Gokdemir, M.G., Erkekol, S., Dogan, H.A., 2017. Investigation of high pressure effect on drilling fluid rheology. In: Proceedings of the International Conference on Offshore Mechanics and Arctic Engineering—OMAE. American Society of Mechanical Engineers Digital Collection, https://doi.org/10.1115/OMAE2017-61449.

Hamza, A., Shamlooh, M., Hussein, I.A., Nasser, M., Salehi, S., 2019. Polymeric formulations used for loss circulation materials and wellbore strengthening applications in oil and gas wells: a review. J. Petrol. Sci. Eng. 180, 197–214. https://doi.org/10.1016/j.petrol.2019.05.022.

Hermoso, J., Martínez-Boza, F., Gallegos, C., 2014. Combined effect of pressure and temperature on the viscous behaviour of all-oil drilling fluids. Oil Gas Sci. Technol. 69, 1283–1296. https://doi.org/10.2516/ogst/2014003.

Herschel, W.H., Bulkley, R., 1926. Konsistenzmessungen von Gummi-Benzollösungen. Kolloid-Zeitschrift 39, 291–300. https://doi.org/10.1007/BF01432034.

Herzhaft, B., 2000. Rheology of aqueous foams: a literature review of some experimental works. Oil Gas Sci. Technol. Rev. IFP Energies Nouv. 54, 587–596.

Hu, R., Crawshaw, J., 2017. Measurement of the rheology of crude oil in equilibrium with CO2 at reservoir conditions. J. Vis. Exp. 2017 (124), 55749. https://doi.org/10.3791/55749.

Ilyin, S.O., Arinina, M.P., Polyakova, M.Y., Kulichikhin, V.G., Malkin, A.Y., 2016. Rheological comparison of light and heavy crude oils. Fuel 186, 157–167. https://doi.org/10.1016/j.fuel.2016.08.072.

Khataniar, S., Chukwu, G.A., Xu, H., 1994. Evaluation of rheological models and application to flow regime determination. J. Petrol. Sci. Eng. 11, 155–164. https://doi.org/10.1016/0920-4105(94)90036-1.

Kök, M.V., Bal, B., 2019. Effects of silica nanoparticles on the performance of water-based drilling fluids. J. Petrol. Sci. Eng. 180, 605–614. https://doi.org/10.1016/j.petrol.2019.05.069.

Kumar, N., Verma, A., Mandal, A., 2021. Formation, characteristics and oil industry applications of nanoemulsions: a review. J. Petrol. Sci. Eng. 206, 109042. https://doi.org/10.1016/j.petrol.2021.109042.

Leonowicz, M., Kozłowska, J., Wierzbicki, Ł., Olszewska, K., Zielińska, D., Kucińska, I., Wiśniewski, A., 2014. Rheological fluids as a potential component of textile products. Fibres Text. East. Eur. 103, 28–33.

Lyons, W.C., Stanley, J.H., Sinisterra, F.J., Weller, T., 2021. Stable foam drilling, in: air and gas drilling manual. Gulf Professional Publishing, pp. 315–333, https://doi.org/10.1016/B978-0-12-815792-3.00010-6.

Lyu, Q., Ranjith, P.G., Long, X., Kang, Y., Huang, M., 2015. A review of shale swelling by water adsorption. J. Nat. Gas Sci. Eng. 27, 1421–1431. https://doi.org/10.1016/j.jngse.2015.10.004.

Maiti, M., Ranjan, R., Chaturvedi, E., Bhaumik, A.K., Mandal, A., 2019. Formulation and characterization of water-based drilling fluids for gas hydrate reservoirs with efficient inhibition properties. J. Dispers. Sci. Technol. 42 (3), 338–351. https://doi.org/10.1080/01932691.2019.1680380.

Malvern, 2016. A Basic Introduction to Rheology Shear Flow. Malvern Instruments Limited.

Mohamed, A., Basfar, S., Elkatatny, S., Al-Majed, A., 2019. Prevention of barite sag in oil-based drilling fluids using a mixture of barite and ilmenite as weighting material. Sustainability 11, 5617. https://doi.org/10.3390/SU11205617.

Mohamed, A., Salehi, S., Ahmed, R., 2021. Significance and complications of drilling fluid rheology in geothermal drilling: a review. Geothermics 93, 102066. https://doi.org/10.1016/j.geothermics.2021.102066.

Murtaza, M., Alarifi, S.A., Kamal, M.S., Onaizi, S.A., Al-Ajmi, M., Mahmoud, M., 2021. Experimental investigation of the rheological behavior of an oil-based drilling fluid with rheology modifier and oil wetter additives. Molecules 26, 4877. https://doi.org/10.3390/molecules26164877.

Pitt, M.J., 2000. The marsh funnel and drilling fluid viscosity: a new equation for field use. SPE Drill. Complet. 15, 3–6. https://doi.org/10.2118/62020-PA.

Raghav Chaturvedi, K., Kumar, R., Trivedi, J., Sheng, J.J., Sharma, T., 2018. Stable silica Nanofluids of an oilfield polymer for enhanced CO2 absorption for oilfield applications. Energy Fuels 32, 12730–12741. https://doi.org/10.1021/acs.energyfuels.8b02969.

Rommetveit, R., Bjorkevoll, K.S., 1997. Temperature and pressure effects on drilling fluid rheology and ECD in very deep wells. In: Proceedings of the IADC/SPE Asia Pacific Drilling Technology Conference. APDT, pp. 297–305, https://doi.org/10.2118/39282-ms.

Ronningsen, H.P., 1995. Correlations for predicting viscosity of W/O-emulsions based on North Sea crude oils. In: All days. SPE., https://doi.org/10.2118/28968-MS.

Rønningsen, H.P., 2012. Rheology of petroleum fluids. Annu. Trans. Nord. Rheol. Soc. 20, 11–18.

Saasen, A., Ytrehus, J.D., 2020. Viscosity models for drilling fluids - Herschel-bulkley parameters and their use. Energies 13, 5271. https://doi.org/10.3390/en13205271.

Shamlooh, M., Hussein, I.A., Nasser, M.S., Magzoub, M., Salehi, S., 2020. Development of pH-controlled aluminum-based polymeric gel for conformance control in sour gas reservoirs. ACS Omega, 24504–24512. https://doi.org/10.1021/acsomega.0c02967.

Tabakova, S., Kutev, N., Radev, S., 2015. Application of the Carreau Viscosity Model to the Oscillatory Flow in Blood Vessels., 040019.

Tao, C., Kutchko, B.G., Rosenbaum, E., Massoudi, M., 2020. A review of rheological modeling of cement slurry in oil well applications. Energies 13, 570. https://doi.org/10.3390/en13030570.

Vipulanandan, C., Mohammed, A.S., 2014. Hyperbolic rheological model with shear stress limit for acrylamide polymer modified bentonite drilling muds. J. Petrol. Sci. Eng. 122, 38–47. https://doi.org/10.1016/j.petrol.2014.08.004.

Wastu, A.R.R., Hamid, A., Samsol, S., 2019. The effect of drilling mud on hole cleaning in oil and gas industry. J. Phys. Conf. Ser. 1402, 022054. https://doi.org/10.1088/1742-6596/1402/2/022054.

William, J.K.M., Ponmani, S., Samuel, R., Nagarajan, R., Sangwai, J.S., 2014. Effect of CuO and ZnO nanofluids in xanthan gum on thermal, electrical and high pressure rheology of water-based drilling fluids. J. Petrol. Sci. Eng. 117, 15–27. https://doi.org/10.1016/j.petrol.2014.03.005.

Xu, P., Xu, M., Tao, Z., Wang, Z., Huang, T., 2018. Rheological properties and damage-control mechanism of oil-based drilling fluid with different types of weighting agents. R. Soc. Open Sci. 5, 180358. https://doi.org/10.1098/RSOS.180358.

Yousuf, N., Olayiwola, O., Guo, B., Liu, N., 2021. A comprehensive review on the loss of well-bore integrity due to cement failure and available remedial methods. J. Petrol. Sci. Eng. 207, 109123. https://doi.org/10.1016/j.petrol.2021.109123.

Zhao, S.Y., Yan, J.N., Shu, Y., Zhang, H.X., 2010. Rheological properties of oil-based drilling fluids at high temperature and high pressure. J. Cent. South Univ. Technol. 15, 457–461. https://doi.org/10.1007/S11771-008-0399-7.

Chapter 3

Interactions of drilling and completion fluids during drilling and completion operations

Amjed Hassan[a], Mobeen Murtaza[b], Olalekan Alade[b], Zeeshan Tariq[c], Muhammad Shahzad Kamal[b], and Mohamed Mahmoud[a]

[a]*Department of Petroleum Engineering, College of Petroleum Engineering & Geosciences, King Fahd University of Petroleum & Minerals, Dhahran, Saudi Arabia*

[b]*Center for Integrative Petroleum Research, King Fahd University of Petroleum and Minerals, Dhahran, Saudi Arabia*

[c]*Ali I. Al-Naimi Petroleum Engineering Research Center, Physical Science and Engineering Division, King Abdullah University of Science and Technology (KAUST), Thuwal, Saudi Arabia*

3.1 Drilling and completion fluids components

3.1.1 Drilling fluids

Drilling fluid systems are designed and formulated to perform optimally in predicted wellbore conditions. The drilling fluid system is the only component of the wellbore construction process that keeps contact with the wellbore formations throughout the drilling operation. Drilling fluid technology improvements have enabled the implementation of a cost-effective, purpose-built system for each interval of the wellbore construction process, leading to a significant increase in drilling efficiency and well productivity. Drilling fluids are designed to regulate subsurface pressures, minimize formation damage, reduce the risk of lost circulation, prevent borehole erosion, and improve drilling parameters, including the penetration rate and hole cleaning (Rabia, 2001). The composition of drilling fluids varies depending on wellbore demands, rig capabilities, and environmental concerns. Furthermore, most modern wellbores have deviated, and directional profiles, in this case, drilling fluid systems, must assist in hole cleaning and stability issues unique to these wells.

Drilling fluids, often referred to as drilling mud, are a viscous, heavy fluid mixture used in oil and gas drilling operations to impart the following primary

Developments in Petroleum Science, Vol. 78. https://doi.org/10.1016/B978-0-323-99285-5.00009-0

functions that require regular observation and intervention (Fink, 2021; Hossain and Al-Majed, 2015):

a. Stop the influx of formation fluids oil, gas, or water by providing the required hydrostatic pressure.
b. Assist in removing the cuttings from the bottom of the bit to the surface. To ensure efficient transport inside the wellbore annulus and to aid in optimizing solids management equipment, the viscosity profile of the drilling fluid must be monitored and maintained.
c. Suspend solids, especially those with high specific gravity.
d. Create a thin and low-permeability mud cake able (Mahmoud and Elkatatny, 2017).
e. Ensure the borehole's uncased sections stay stable. To ensure the wellbore's integrity until the next casing setting point is reached, the mud weight and mud/formation chemical reactivity must be closely monitored.

In addition to the above-mentioned primary functions, certain functionalities do not require constant monitoring:

a. Reduce friction between the drilling string and the wellbore walls.
b. Allow sufficient time for the bit to cool.
c. Assist in collecting and analyzing drill cuttings, cores, and electrical logs.

In addition, it is crucial to ensure that drilling fluid is not dangerous or toxic to drilling personnel and the environment. It should not cause corrosion and wear and tear to the drilling equipment. All drilling and completion fluids are made up of three components: base fluids (brine, nonaqueous, and pneumatic) in which additives are mixed. Moreover, different drilling fluid additives are used to improve the drilling fluid properties. Active and inactive (inert) additives can be used to control one or more of the properties listed in Table 3.1. Also, the classification of different fluid systems based on the type of base fluids used to prepare the drilling fluids is shown in Fig. 3.1.

Drilling fluids can be classified into three main types: aqueous or water-based drilling fluids, nonaqueous or oil-based drilling fluid, and pneumatic fluids. In aqueous or water-based drilling fluids (WBDF), water is the primary medium in which additives are mixed. In nonaqueous drilling fluids (NADF), also called oil-based drilling fluid (OBDF), oil is a medium in which solid additives are mixed. The oil is the continuous phase in the emulsion of brine water or another low-activity liquid. Pneumatic fluids are used to remove drill cuttings from the wellbore using high-velocity streams of natural gas, air, carbon dioxide, nitrogen, or another gaseous fluid. Pneumatic systems are transformed into mist or foam systems in a slight water inflow. Overall, different subsurface conditions have led to many drilling fluid formulations over time. The optimal drilling fluid for the predicted conditions should be selected to keep costs down and avoid the danger of wellbore instability, loss of circulation, stuck drill pipe, and gas kicks. It is essential to keep in mind the need for adequate evaluation of the formation and optimum output.

TABLE 3.1 Drilling fluid properties and their functions

Properties	Functions
Mud weight (density)	Used to control subsurface pressures
Rheology	Essential for cuttings transport and drilling fluid hydraulics
Filtration	Used in wall building characteristics and fluid loss into the formations
Inhibition	Wellbore stability and strengthening
Chemical reactivity	pH, alkalinity, interfacial/surface activity, emulsification, contamination control, scavenging
Cation exchange capacity	Shale swelling and inhibition
Lubricity	Reduce friction between the drill string and walls of the wellbore

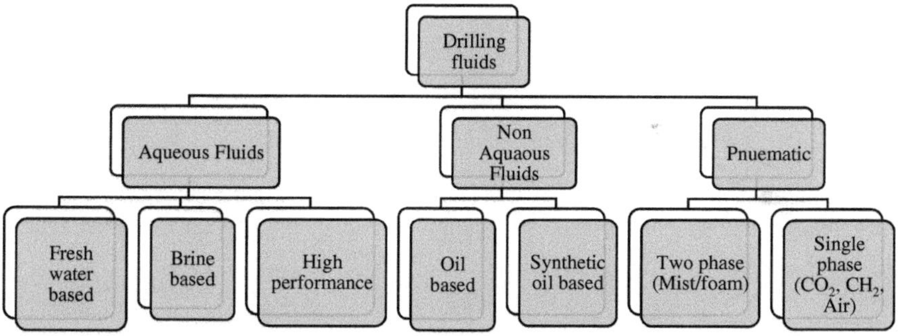

FIG. 3.1 Classifications of drilling fluids based on the type of base fluid.

3.1.1.1 Water-based drilling fluids (WBDFs)

WBDFs are the most used drilling fluids as they are environment friendly, cheap in cost, and easy to maintain. Nevertheless, there are some issues associated with WBDFs because of their poor shale swelling inhibition, lubricity, and thermal stability. Therefore, their composition is modified by using special additives that enhance their performance. WBDFs are categorized according to their compositions, functions, and application to pay zone or non-pay zone drilling. Conventional compositions can be used to prepare WBDF for non-pay zone drilling without fear of formation damage, leveraging the material's low cost. However, the most critical property should be the minimum amount of damage to the pay zone formation, if possible, for drilling fluid

TABLE 3.2 The commonly used additives in WBDFs

Functions	Common additives
Viscosifier	Bentonite, XC polymer
pH controller	KOH, NaOH
Alkalinity controller	Lime
Hardness removal	Na_2CO_3
Fluid loss controller	Starch, PAC, lignite, nanomaterials (nanosilica, graphene)
Deflocculant	Lignosulfonate
Shale swelling inhibitor	KCl, silicates, PHPA, ionic liquids, and surfactants
Scavengers	ZnO, $KMnO_4$
Bridging agent	$CaCO_3$
Weighting material	Barite, hematite

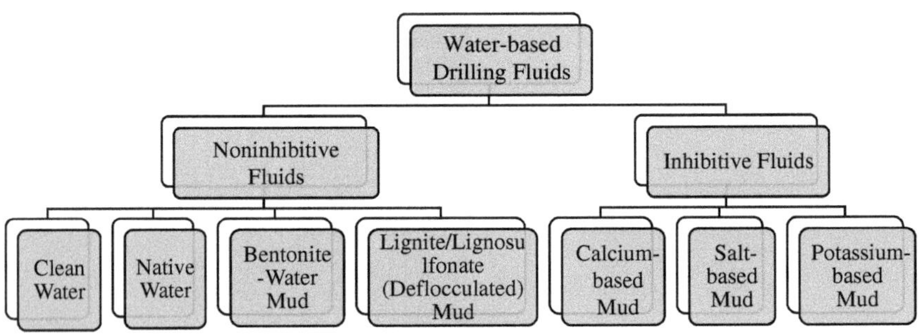

FIG. 3.2 Classification of water-based drilling fluids.

to be applied in pay zone drilling. WBDFs contain water, bentonite, and other additives such as weighting materials (barite), bridging materials ($CaCO_3$), viscosifiers (xanthan gum), fluid loss controllers (polyanionic cellulose [PAC] and starch), caustic soda, soda ash, H_2S scavengers (ZnO), and shale inhibitors as details mentioned in Table 3.2. Drilling fluid makes up 80% of water and 20% of solid additives. Fig. 3.2 provides the common classifications of water-based drilling fluids.

3.1.1.2 Oil-based drilling fluids (OBDFs)

An oil-based drilling fluid (OBDF) is prepared by using oil as the primary phase, while water will be dispersed within the oil phase. OBDFs contain viscosifier, fluid loss controllers, emulsifiers, and wetting agents. The water content in an oil base fluid indicates the type of fluid. These oil-based fluids can be made from crude oil, diesel, and mineral oils, or readily available non-petroleum organic fluids. These inert, nonaqueous, and synthetic fluids are now considered more eco-friendly fluids than diesel or mineral oils. These fluids are classed as invert emulsions or all-oil muds.

Contrary to the popular belief, traditional all-oil muds are formulated and water-free. Lacking water, asphaltic materials must regulate fluid loss and viscosity. Because no water is used during formulation and is avoided during drilling, only a small number of emulsifiers are required. Water can be tolerated in small amounts by all-oil muds but should be changed to an inverted emulsion if there is a chance of contamination. Invert emulsions are oil-based muds that contain a moderate to a high amount of water. Mud particles might become water-saturated if not emulsified quickly enough.

The invert emulsion requires water containing salts like calcium or sodium chloride. An inverted emulsion's liquid phase can include up to 60% water. The internal component of water is emulsified with special emulsifiers to prevent evaporation and aggregation. Water droplets can water wet already oil-wet particles, affecting the emulsion's stability. Special lignite derivatives are used to reduce fluid loss, while bentonite derivatives are used to increase viscosity and suspension.

Synthetic-based drilling fluids (SBDF) or oil-based drilling fluids (OBDFs) have shown excellent shale inhibition, easy cleaning, better lubricity, and high-temperature tolerance although they are more expensive than WBDF. On the contrary, OBDF application is limited due to their environmental footprints, high cost, and hindrances in the formation logging. WBDFs outweigh the performance of OBDFs in drilling operations as they are easy to maintain; environmentally friendly; have a better penetration rate, low cost, and toxicity; and have less impact on well logging. The main additives for oil-based drilling fluids and the functions for each additive are provided in Table 3.3.

In OBDF, specific properties are monitored and checked during the drilling operation, including mud weight, viscosity, emulsion stability, HPHT fluid loss, excess lime, oil–water ratio, and so on. Table 3.4 summarizes the optimum properties that are required for OBDF.

The design and control of drilling fluids are iterative processes influenced by both surface and subsurface conditions. The physical properties of drilling fluid change with respect to temperature and pressure conditions which change with depth. Continuous process engineering is required to fine-tune the properties of on-site drilling fluids in response to changing wellbore conditions. Various requirements, such as borehole stability density, thermal stability, and logistical and environmental considerations, should be obtained by the system. The drilling operations can be started using a simple drilling mud type. Simple bentonite water mixed drilling fluid is typically the starting fluid

TABLE 3.3 Oil-based drilling fluid additives and their functions

Additives	Functions
Primary emulsifier	Play a significant role in the water–oil emulsion. Emulsifiers provide stability to water droplets by surrounding them and preventing their coalescence
Secondary emulsifier	They are oil wetting chemicals used in support of primary emulsifiers and provide more emulsification to intruding water. These additives are polyamides
Organophilic lignite	They are used as fluid loss agents in HT conditions. They can produce stable emulsions. Amine is added to improve the lignite dispersion in oil
Organophilic gellants	Viscosity builders are made from bentonite, treated with an amine to make them oil dispersible
Wetting agents	Supplementary chemicals used to make oil-wet the drill solids and weighing agents
Polymeric viscosifiers	Compounds that improve the mud viscosity at high-temperature conditions
Rheological modifiers	Fatty acids with a low molecular weight are used to reduce or eliminate barite sag by giving a viscosity rise at low shear rates (3 and 6 rpm)
Weighting agents	They are utilized to raise the OBDFs density. Barite is the most common weighting material in drilling fluids

TABLE 3.4 The commonly used range for different fluid properties for OBM (Akpan et al., 2020; Gautam and Guria, 2020; Li et al., 2016; Murtaza et al., 2021)

Property	Value
Filter cake thickness, 1/32 in.	2/32
Electrical stability, V	≥400 V
Plastic viscosity (PV), cP	10–60 cP, preferably 15–40 cP
YP/PV ratio	0.75–1.50
Yield point (YP), lb./100 ft2	2.5–20, preferably 5–12.5
HPHT fluid loss°F, mL/30 min	<10
Gels 10 s, lb./100 ft2	4–10
Gels 10 min, lb./100 ft2	4–15

that is used to initiate the casing point. As the well depth increases, the pore pressure, the rising formation temperature, and different formations need high degrees of consideration to sustain the wellbore stability. As the drilling operations continue, the simple drilling fluid system can be changed into conventional mud types such as WBM or OBM.

Nanoparticles have wide-ranging applications in various industries, including biomedical technology, electronics, the coatings industry, and material composites. Nanotechnology can play a significant role in the drilling of oil and gas wells. The shortcomings of the drilling fluids can be handled by applying nanotechnology (Cheraghian, 2021; Ikram et al., 2021). Nanoparticles (NPs) can enhance the performance of drilling fluids due to their unique properties including electrical, physicochemical, and thermal characteristics (Amanullah et al., 2011). The required properties such as better rheology, low fluid loss, high inhibition, minimizing differential sticking, high thermal stability, and high shale inhibition can be achieved by incorporating nanomaterials in drilling fluids (Barry et al., 2015; Contreras et al., 2014; Hoelscher et al., 2012; Javeri et al., 2011; Li et al., 2012; William et al., 2014). Various nanomaterials have been added in drilling fluids to study their performance through rheology, fluid loss, and swelling inhibition (Cheraghian, 2021; Ikram et al., 2021). The nanomaterials include nanosilica, titanium oxide, zinc oxide, nanographene, carbon nanotubes, and copper oxide (Cheraghian, 2021). Recently, nanocomposites were introduced as a substitute for clays and polymers to enhance the mud rheological parameters. It has been determined that the application of NPs in the drilling and completion operations is novel, which can lead to improving the overall treatment efficiency.

The application of NPs in drilling fluids brings many economic and environmental benefits. The economic advantages of NPs can be seen in three different ways:

a. They reduce expensive materials in the drilling fluids by providing equal or higher benefits in low concentrations.
b. Their applications provide stability to drilling fluids and reduce the nonproductive time consumed handling wellbore problems.
c. They have the potential of solving critical problems faced during the drilling operation. Due to the unique property of having a high surface area to mass ratio, NPs are used in less concentration in drilling fluid, resulting in low environmental hazards than traditional drilling fluids.

3.1.2 Completion fluids

Completion fluids are the fluids used during completion or workover operations. The completion fluids are designed to prevent the formation damage. Completion fluids can be solids-free, clay-free, water-soluble, acid-soluble, or oil-soluble. Completion fluids are used for different treatments, such as wellbore cleaning, well control, drilling-in, and perforation. The initial cost of completion fluid is relatively high, but the obtained functions by completion fluids are very important and can be compensated by the overall cost. Typically, completion fluids are aqueous brines prepared by mixing bromides, chlorides, and formic acid. But they could potentially be different types of fluid with the appropriate rheological and flow properties. The completion fluid should be compatible with the rock and fluids of the reservoir system and is normally filtered extensively to avoid introducing particles into the reservoir section. The completion fluids can be prepared using water, oil, or brines.

Water-based completion fluid: Drilling fluids can be modified with fewer solids as completion fluids. The water-based fluid is economical, easy to maintain, and available. They have oil and acid-soluble dissolve additives. Drilling fluids as completion fluids can be harmful to the formation and cause severe formation damage as certain additives are challenging to remove chemically, such as barite. Due to the particles' concentration, pH, and ionic composition, conventional drilling fluids are rarely suited for completion operations. Drill-in fluids may be suitable for both purposes in some instances.

Oil-based completion fluids: The primary function of this type is to minimize clay damage. As a result of the minimal fluid loss, oil-based drilling fluids produce minor formation damage to the reservoir formation. Some chemicals can block and precipitate inside the formation pores. Although these fluids are less hazardous and more expensive than water-based fluids, they are typically more hazardous than solids-free fluids.

Brine-based completion fluids: The brines used to make solids-free completion fluids are listed in Table 3.5. These brines are used to maintain pressure control during the completion operations until the production tubing is installed. In addition, brine-based fluids are used to sustain the annual pressure at required levels during the different well operations. Both divalent

TABLE 3.5 Base fluid types in completion fluids

Brine-based completion fluids		
Inorganic brines		*Organic brines*
Monovalent	Divalent	Monovalent
NaCl	$CaCl_2$	$NaCHO_2$
KCl	$CaBr_2$	$KCHO_2$
NaBr	$ZnBr_2$	NaC_2HO_2
NH4Cl	—	CeC_2HO_2

and monovalent cations are present in inorganic brines. Monovalent ions are generally less harmful and have a better environmental impact than the divalent cationic brines. Chloride or bromide anions are not biodegradable and, in excess, impede the growth of vegetation.

3.2 Drilling and completion fluids stability

Drilling fluids are essentially complex fluids that comprise fluid and solid phases. The mixture is made up of clay particles suspended in a base of either water, diesel, or mineral oil and other materials such as lignite to stabilize the slurry and barite to increase its density. In general, complex fluids are sensitive to certain changes including mechanical and thermal forces and changes in composition (Larson, 1999). As illustrated in Fig. 3.3, in a typical shear thickening colloidal system (Pinto and Meo, 2017; Wagner and Brady, 2009; Wei et al., 2019), transitions from the equilibrium state of the particles in suspension to shear thinning and shear thickening occur due to change in the shear force and its effects on the repulsive interactions between particles (Pinto and Meo, 2017). This involves the change from an equilibrium ordered state to a disordered state when this fluid is subjected to a relatively low shear rate ($\dot{\gamma}$), which causes viscosity reduction (shear thinning effect). When the shear rate is increased above a critical point ($\dot{\gamma}_c$) at which the particles are squeezed together (forming the hydroclusters), shear thickening behavior is instigated with a sudden increase in the viscosity. Regarding the drilling fluids, it is well established that the sagging tendency can be significantly

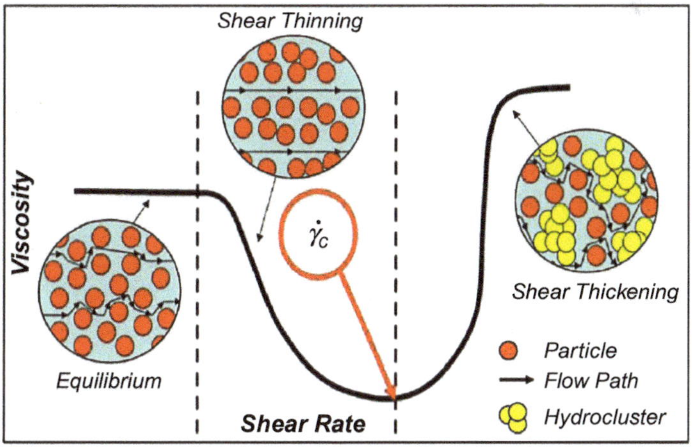

FIG. 3.3 Schematics of shear thinning and shear thickening effect. *Adapted from Pinto, F., Meo, M. (2017). Design and manufacturing of a novel shear thickening fluid composite (STFC) with enhanced out-of-plane properties and damage suppression. Appl. Compos. Mater. 24, 643–660. https://doi.org/10.1007/s10443-016-9532-1.*

affected by the change in the rheological properties (Dye et al., 2001, 2003; Elkatatny, 2018; Jose and Imqam, 2020; Nguyen et al., 2011).

The stability of drilling fluids has been a subject of discussion in the practices of drilling sequel to the increasing need for drilling under harsh conditions such as in deep wells under high temperature and pressure. Drilling under high-temperature and pressure conditions have been a challenging task due to changes that can be imposed on the rheological properties of the fluids and the tendency to promote particle segregation (Dahab, 1991; Elkatatny, 2018). At high-temperature and high-pressure conditions, the formation temperature and the pore pressure can exceed 300 °F and 10,000 psi, respectively.

Under such conditions, the mud may become thinner or thicker than it appears at the surface (Dahab, 1991), thus affecting the penetration and filtration rates as a result of the unstable viscosity. An indirect problem is the settling of the weighting material, usually, the barite particles. This phenomenon is commonly referred to as sagging and is essentially the occurrence of variation in drilling fluid density (Dahab, 1991), and change in viscosity, in different types of well, but the deviated wells show more changes (Elkatatny, 2018). Barite settling may cause many problems, including possibilities of mechanical and differential sticking, downhole drilling fluid losses, or wellbore instability (Khosravanian and Aadnøy, 2021; Jia et al., 2021).

The thermal stability of drilling fluids is totally dealing with their rheological properties at high temperatures along with active solid intrusion which is the major problem frequently encountered during geothermal drilling operations (Jose and Imqam, 2020). Thus, in order to mitigate the aforementioned situations, the drilling fluid is precisely designed to improve stability and rheological parameters at high-temperature and high-pressure conditions (Zhang et al., 2017). Several approaches have been investigated with the aim to mitigate this situation. These include using specially designed polymeric materials and nanoparticles as additives in the drilling fluids (Cheraghian, 2021). Different polymeric materials have been proposed as drilling fluid additives to enhance the rheological characteristics and prevent sagging under high-temperature/high-pressure (HT/HP) conditions. It has been suggested (Ahmad et al., 2018) that the intercalation of molecules of the polymers into the interspaces of the bentonite layer, through adsorption of polymer chains and interionic interactions, induces the reinforcement against thermal degradation. As illustrated in Fig. 3.4, this essentially involves the dissolution of clay minerals and dispersion of polymer into the spaces in the layers of clay.

Dahab (1991) investigated combinations of natural clay materials (Saudi palygorskite) and polymeric materials viz. polyanionic cellulose and bacterially produced biopolymer as additives which can assist to overcome gelation issues and fluid loss that might occur when drilling under high electrolytic environment and high-pressure and high-temperature wells. Their findings indicated that the combination of palygorskite in seawater yielded acceptable

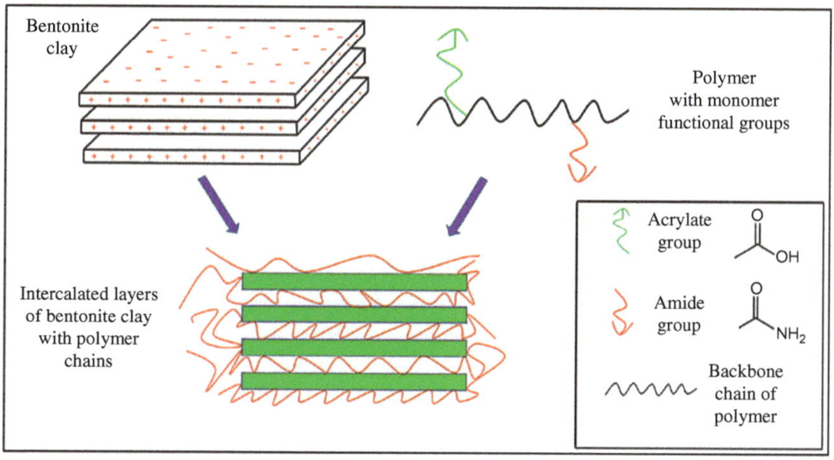

FIG. 3.4 Schematic illustration of bentonite clay intercalation with polymer chains in solution. *Adapted from Ahmad, H. M., Kamal, M. S., & Al-Harthi, M. A. (2018). High molecular weight copolymers as a rheology modifier and fluid loss additive for water-based drilling fluids. J. Mol. Liq., 252, 133–143.*

results. Similarly, Ahmad et al. (2018) investigated the influence of different parameters on the filtration and rheological behavior of bentonite dispersions.

Parameters such as the hydrolysis degree and copolymer types are studied. The output ultimately showed that the polymers enhanced the filtration and rheological properties of the drilling fluid. Also, they concluded that increasing the polymer molecular weight can improve the thermal stability and the rheological behaviors as well as improve the filtration characteristics.

In addition, Ofei et al. (2017) presented a combination of a water-based polymer and ionic liquid. Using 1-butyl-3-methylimidazolium chloride (BMIM-Cl) showed better rheological and stability behaviors. Also, it improves the filtration and solid suspension properties at temperatures up to 356°F. Similarly, Temple et al. (2004) utilized light polymers to minimize the barite sagging. Using polyalkyl methacrylate that has molecular weights of 40,000 to 90,000 led to reducing the barite sagging without much changes in the mud rheological parameters. To prevent barite sagging when drilling HP/HT wells, Elkatatny (2018) presented a mud formulation with a copolymer by using acrylic and styrene monomers to enhance the mud stability and rheological properties of invert emulsion type of drilling muds. Sag tests were conducted under static and dynamic situations to assess the performance of the copolymer in improving the solid suspension properties. Adding copolymer of 1 lb./bbl concentration showed considerable improvements in the sagging behavior, with slight impacts on the stability and rheological properties. The laboratory measurements were carried out at temperatures of more than 300°F. It was also reported that the addition of the copolymer was found to enhance the gel strength by around 50% and increase the storage modulus by more than 280%.

On the other hand, nanocomposites showed attractive performances in improving the rheological properties for drilling and completion fluids (Cheraghian, 2021; Ikram et al., 2021). Recently, graphene-based nanomaterials have been used in preparing the drilling fluids due to their thermal, electrical, chemical, and mechanical properties to improve mud cake stability and minimize the fluid loss that eventually reduces formation damage. Jose and Imqam (2020) evaluated the performance of silica nanoparticles (SiO_2-NPs) and graphene nanoplatelets (GNPs) to improve the WBM properties and minimize the clay swelling at high-temperature conditions. SiO2-NPs and GNPs showed significant improvements in the filtration properties, without much impact on the mud rheological parameters. NPs showed high stability and reduced the zeta-potential values to around −30 mV, indicating solution stability at different temperatures. The NPs were optimized to determine the NPs concentration, where 0.75 wt% showed the best performance. Also, using NPs improve the hole cleaning efficiency by around 44% compared to the conventional drilling muds. Moreover, NPs were able to stabilize the reactive clay minerals and avoid clay damage. Overall, NPs showed high performance in improving the drilling mud properties and lead to minimizing the clay swelling during drilling shale or unconventional formations.

Belayneh et al. (2021) reported the effects of molybdenum disulfide (MoS2) NPs on the thermal properties of different drilling fluids. At ambient conditions, 0.26 wt.% MoS2 led to an increase in the mud stability by around 25% as well as an increase in the mud electrical conductivity by 9% on average. Roldan et al. (2019) investigated SiO_2 nanoparticles as an additive and using the thermal rolling process tested their effects on thermal stability through the rheological behavior in bentonite-free, water-based mud. It was found that the nanoparticles interacted with the xanthan gum (XG) polymer inhibiting the polymer degradation caused by temperature. By employing two different types of SiO_2 nanoparticles, synthesized acidic surface modification silica (SiA) and fumed silica (SiC) were used. The laboratory investigations indicate that using SiC NPs can increase the mud viscosity by around 15% compared to the conventional polymer-based muds. Also, adding SiO_2 NPs showed slight changes in the mud properties including pH, solid content, and density. Moreover, SiC led to enhancing the yield stress, yield point, and apparent viscosity as well as reducing the mud filtration.

In addition, SiC NPs showed reductions of 49% and 68% in the mud filtration and spurt loss, respectively. Extensive research work has been presented by Agarwal et al. (2011) on using hydrophobic nanosilica particles and organically modified nanoclay to stabilize oil-based muds (OBM) through the thermal rolling process. The nanoparticles preserved emulsion stability; the water droplet size was conserved. In addition, the yield point (YP) was reduced by 23%, while the plastic viscosity (VP) was increased by 15%. Similarly, Mahmoud et al. (2016) aged a water-based mud (WBM) with 0.5 wt% Fe_2O_3 and SiO_2 nanoparticles at 200°C. It was reported that the drilling fluid with Fe_2O_3 resulted in minor changes in the yield stress (YS), increased by 42%, while SiO_2 nanoparticles presented the highest percentages by 130%, thereby increasing the PV by 166% and 211% for Fe_2O_3 and SiO_2 nanoparticles, respectively. Parizad et al. (2018) evaluated the effect of TiO_2 nanoparticles in a BWBM with 10 wt% of KCl after the aging process. The PV was reduced by 17% and 30% of YP. In addition, their results showed that the addition of 0.75 wt% of TiO_2 yielded a reduction in PV of 8% and YP decreased by 15%.

Moreover, Perween et al. (2018) synthesized $ZnTiO_3$ nanoparticles using the sol–gel method (SNP nanoparticles) and electrospinning technique (ENP nanoparticles) and evaluated the effect varying the concentration from 0.5% to 3%. They analyzed the addition of nanoparticles on the rheological and filtration properties of a BWBM with XG and PAC as the main polymers before and after the aging process. ENP nanoparticles presented a more pronounced impact on rheological and filtration properties. API filtration volume of the base mud, after the thermal rolling process, decreased by 32% and 36% by adding a 3.0% (w/v) concentration of SNP and ENP, respectively. Base drilling fluid reduced the apparent viscosity (AV) by 17.3% after rolling. However, this decrease in the AV was reduced by 6% by the addition of SNP at 3.0% (w/v) and 12% for ENP at the same concentration. Although nanoparticles could improve the thermal stability in BWBM, possible formation damage by bentonite invasion is an important concern. This shortcoming can be overcome by the use of BFWBM and nanoparticles. Li et al. (2015) also evaluated the improvements in filtration rheological properties by adding cellulose nanoparticles (CNPs) to drilling muds. The conventional bentonite water-based drilling fluid (BT-WBDF) was used. Adding CNCs to the drilling mud led to reducing the fluid loss and improving the mud rheological properties, at different temperatures. In the study conducted by Rasool et al. (2018), ZnO-interlinked chitosan (Ch) nanoparticles (CZNCs) were used to improve the mud stability. They reported that the developed green biocide additives can be used in drilling operations without much impact on the environment.

Overall, several instability issues can be induced in the drilling and completion fluids due to density variations and degradation in the rheological and filtration properties. The fluids' stability should be precisely examined to ensure successful drilling and completion operations. As shown in Table 3.6, various investigations have been conducted on the development of drilling fluid formulations using specially designed polymeric materials and nanoparticles as additives to enhance the rheological and filtration properties of the fluids. The outcomes from these research studies have shown promising results on the thermal stability of the drilling fluids. However, future research directions should focus on technological advancement which promotes the sustainability of the process. This includes improvement in the development of environmentally friendly green additives such as bio-based

TABLE 3.6 Summary of polymer and nanoparticle additives used in enhancing stability of drilling fluids at HT/HP

Drilling fluid additives	Stabilized fluid property at HT/HP	References
Polyanionic cellulose, biopolymer	Gelation issues and fluid loss (OBM/WBM)	Dahab (1991)
Acrylamide-based copolymers	Rheological and filtration properties (WBM)	Ahmad et al. (2018)
1-butyl-3-methylimidazolium chloride (BMIM-Cl)	Filtration and rheological behaviors (WBM)	Ofei et al. (2017)
Polyalkyl methacrylate	Rheological properties and barite sagging (WBM)	Temple et al. (2004)
Copolymers: styrene and acrylic monomers	Rheological properties and barite sagging (OBM)	Elkatatny (2018)
MoS_2 nanoparticle	Lubricity and electrical conductivity (OBM)	Belayneh et al. (2021)
Cellulose nanoparticle	Rheological and filtration properties (WBM)	Li et al. (2015)
SiO_2 nanoparticles and graphene nanoplatelets	Rheological properties and barite sagging (WBM)	Jose and Imqam (2020)
SiO_2 nanoparticles and xanthan gum polymer	Rheological and filtration properties (BFWBM)	Roldan et al. (2019)
Hydrophobic SiO_2 nanoparticles	Rheological properties and emulsion stability (OBM)	Agarwal et al. (2011)
TiO_2 nanoparticles	Rheological properties (BWBM)	Parizad et al. (2018)
$ZnTiO_3$ nanoparticles with XG and PAC polymers	Rheological and filtration properties (BWBM)	Perween et al. (2018)

polymers and nanoparticles. Moreover, economic sustainability should receive more attention from the research through process optimization.

3.3 Fluid/solid and solid/solid interactions in drilling and completion fluids

Understanding the fluid/solid and solid/solid interactions is the key factor to ensure the success of any drilling or completion operations (Caenn and Chillingar, 1996; Knight et al., 2010; Salehi et al., 2016). The presence of different solid and fluid types within the drilling mud and completion fluids can lead to severe interactions between the different components. These interactions can lead to severe impacts on the overall operation performance (Albdiry and Almensory, 2016; Khodja et al., 2010). This section of the chapter discusses the interactions between fluid and solid during the drilling and completion operations.

Nowadays, complex fluids are being used during the drilling and completion operations, due to the exploration in new challenging reservoirs (Barbati et al., 2016; Vryzas and Kelessidis, 2017). Therefore, more procedures are considered to ensure effective and safe drilling and completion fluids (Albdiry and Almensory, 2016). A large range of characteristics are examined before introducing the fluids into the wellbore (Knight et al., 2010; Salehi et al., 2016). The common properties include the rheological parameters such as the fluids' viscosity and density (Caenn and Chillingar, 1996). The interactions between solid and fluids can lead to a sagging effect; hence, a huge difference can be obtained in the drilling fluids' viscosity and density along the well. The dense particles will tend to go down leading to a higher fluid density at the bottomhole, whereas less fluid density will be induced at the top section of the drilled wells. Therefore, antisagging agents (such as polyethylene glycol [PEG], manganese tetra oxide [Mn_3O_4], and modified urea solution) are used to ensure homogenous distribution for the solid contents within the drilling fluids (Ahmed et al., 2021; Mohamed et al., 2020; Rafati et al., 2018). Moreover, less percentage of solid content can lead to poor filter cake due to the insufficient amount of solids. Also, it can result in severe fluid filtration into the near-wellbore area (Ikram et al., 2021). Furthermore, the poor design of completion fluids can result in severe problems at the downhole conditions, such as well instability or considerable invasion into the formations. The key role here is the proper selection of completion fluid components as well as the suitable ratio between the solid and fluids (Ali et al., 2020; Davis et al., 2004).

Drilling horizontal sections require more attention for designing the drilling and completion fluids (Davis et al., 2004; Mahmoud et al., 2020; Xindi and Baojun, 2017). The gravity forces play a significant role in the solid/fluid interactions, for example, considerable solid precipitations can be induced, especially for the large solid sizes (Pedrosa et al., 2021). Consequently, heavy solids will accumulate at the bottom portion of the horizontal section leading to considerable restrictions for the drilling operations. Reducing the wellbore size and borehole instability could be the main problems associated with the solid accumulation at the horizontal section. Also, the accumulated solids lead to an increase in the drilling torque which will significantly reduce the rate of penetration and result in more power requirements (Mahmoud et al., 2020; Pedrosa et al., 2021). Moreover, massive fluid filtration can be induced when the liquid–solid ratio is not properly selected, which can lead to borehole collapse in severe cases (Ikram et al., 2021; Xindi and Baojun, 2017).

In addition to the solid added to the drilling or completion fluids at the surface, more solids will be introduced into the drilling or completion fluids during the drilling and completion operations (Mahmoud et al., 2020; Salehi et al., 2016). These solids are coming basically from the formation due to the interactions between the used fluids and the formation rocks. Introducing more solids into the drilling fluids will lead to an increase in the fluid density (Jia et al., 2021). In some conditions, high drilling mud density results in reducing the rate of penetration and increasing the drilling torque, thereby decreasing the overall drilling performance (Siddig et al., 2022). Therefore, the interaction between drilling fluids and formation rocks should be extensively studied to ensure that changes in drilling fluids' properties will be within acceptable limits. The solid distribution within the liquid can affect the apparent viscosity, the Krieger–Dougherty law can be used to describe the changes in the apparent viscosity for different types of drilling fluids (Nazar et al., 2020). Increasing the solid ratio can lead to significant divergence in the mud viscosity; hence, more drilling and pumping energy will be required to compensate for the increase in fluid viscosity (Nazar et al., 2020).

Furthermore, drilling in high-pressure formations can result in a reverse invasion; the formation fluids will flow into the wellbore, leading to an increase in the liquid–solid ratio (Akpan et al., 2019; Elkatatny et al., 2020). As a result, the density and viscosity of the fluid will be reduced, which will affect the drilling efficiency. In most cases, the introduced liquid into the drilling fluid is saline water; therefore, the logging data can be used to estimate the location and percentage of the saline water. After that, the changes in drilling mud properties can be studied on the lab scale to come up with the most suitable drilling mud for different sections of the wells (Ahmed et al., 2021).

Finally, different additives can be added to the drilling and completion fluids to improve the solid/liquid and solid/solid interactions (Ali et al., 2020; Vryzas and Kelessidis, 2017). The rheological properties such as plastic viscosity or yield point can be improved by adding nanoparticles to the drilling muds. Also, clay swelling inhibitors (such as potassium chloride, polyethylene glycol [PEG], and hydroxyethylcellulose [HEC]) can be added to stabilize the clay minerals and avoid clay swelling, especially while drilling shale or clay-rich formations (Anderson et al., 2010). Overall, it becomes important to understand the interactions between the different components of drilling and completion fluids to ensure less damage and more efficient operations.

3.4 Compatibility of different additives used in drilling and completion fluids

Drilling and completion fluids are prepared by mixing different types of liquids and solids; thereby, it is very important to ensure that the mixed components are compatible (Siddig et al., 2020, 2022). Formation plugging or solid precipitation could be the main problems associated with using incompatible fluids during the drilling and completion operations (Al Jaberi et al., 2021). The compatibility analysis can be classified into two main parts; the first is the compatibility between all fluids components, and the second is the compatibility between the used fluids and reservoir or formation systems (Abbaszadeh et al., 2016). Failure to achieve compatible fluids will lead to poor drilling and completion operations. Different approaches such as laboratory measurements and numerical modeling are utilized to determine the optimum ranges for fluid components in terms of the proper solid and fluid types and the percentage of each component (Abbaszadeh et al., 2016; Mohammadi et al., 2021).

During the drilling operations, formation damage can be induced due to the fluid and fine solid invasion into the surrounding formations (Al-Otaibi et al., 2006). The filtrates, which are the liquids that pass into the formation from the wellbore, will interact with the formation fluids/rocks and could result in some permeability impairment. Therefore, different types of additives are used to control the amount of fluid invasions (Ahmad et al., 2018; Barry et al., 2015; Contreras et al., 2014; Ofei et al., 2017). In general, the formation damage caused during the drilling operations can be classified into two types: damage induced by the chemical incompatibility and damage induced by the pore blockage (Al Jaberi et al., 2021). The first type (fluids incompatibility) can be induced by several mechanisms, such as the chemical and physical reactions between the drilling filtrate and the formation fluids; hence, insoluble salt and scale issues will be triggered (Ali et al., 2021; Knight et al., 2010). Also, the rock wettability can be altered due to excessive fluid invasion, leading to different types of formation blockage such as liquid or emulsion blockages. The wettability alteration issues can be avoided by selecting the proper type of additives that will not induce nonfavorable rock wettability (Murtaza et al., 2021). Moreover, the contamination of the reservoir oil with the mud filtrate can lead to different types of precipitations such as heavy oil or asphaltic depositions within the reservoir pore spaces.

On the other hand, the second type of induced formation damage (pore blockages) can take place due to multiple mechanisms, such as fluid and solid invasions into the formation; thereby, the liquids and solids can plug the pore throat due to their incompatibility with the formation system (Al Jaberi et al., 2021). Also, the deposition of an ineffective mud cake on the formation face during the drilling operations can lead to plugging the pore throat and reduce the overall formation conductivity (Al Jaberi et al., 2021). Similar problems can be induced by using insufficient solid control equipment, where the volume of the solid will be more than the preferred limit (Jia et al., 2021; Knight et al., 2010). Therefore, solids will be accumulated within the borehole and induce more drilling problems. All these types of formation damages can be avoided or minimized by proper selection of the drilling additives to ensure that all mud components are compatible.

Moreover, the structure and the behavior of the drilling and completion additives are very crucial for optimizing the performance and cost of these operations (Davis et al., 2004; Gautam et al., 2020). The drilling and completion fluid additives should be able to improve the thermal stability, enhance the rheological properties, minimize and control the fluid loss, and optimize the lubrication behavior to reduce the overall friction (Belayneh et al., 2021). The common fluids that are used to enhance thermal stability are sulfonate, furan, methyl, lactam, phenyl, and pyrrolidine (Belayneh et al., 2021; Jose and Imqam, 2020). The rheological properties can be improved by using acrylates, amide, cellulose derivatives, organoclays, hydroxyl, and quaternary ammonium (Murtaza et al., 2021; Siddig et al., 2020). Also, the drilling filtration can be controlled by using several additives, such as cellulose derivatives and quaternary ammonium and pyridine (Li et al., 2015). More additives such as carboxylate, polyoxyethylene chains, and diallyl dimethyl ammonium chloride are utilized to minimize the friction forces by improving the lubrication behavior (Gautam et al., 2020). Ultimately, the type and percentage of drilling and completion fluid additives should be certainly examined and selected to ensure the used components are compatible and efficient.

3.5 Interaction of drilling and completion fluids solids with the formation rocks

Drilling and completion fluids can infiltrate the formation and react with formation minerals, resulting in formation damage (Fink, 2015). As defined by the oil and gas industry, formation damage occurs when particle retention, fluid invasion, or fluid/fluid interaction affects the oil and gas reservoir formation properties. The well deliverability and stability can be negatively affected by the formation damage (Hossain and Al-Majed, 2015). Drilling, completion, injection, and workover activities all have the potential to cause formation damage, which is both undesirable and challenging (Fink, 2015; Temple et al., 2004). Damage to a formation can be divided into primary damage and secondary damage (Al Jaberi et al., 2021). Filter cake generation and removal operations can cause primary and secondary formation damage. Fig. 3.5 depicts the development, removal, and formation damage of filter cake.

Because drilling and completion liquid systems contain fluids, solids, and chemicals, the primary mechanism of formation damage is solid and polymer clogging of pores. Damage caused by drilling and completion fluids happens in most oil and gas wells and typically occurs during the well's initial stages of production, leading to a significant reduction in the production rate. Additionally, because the fluid volume used is typically enormous and readily lost into the surrounding formations under overbalanced drilling operations, severe formation damage such as pore plugging and solid depositions can be induced. As a result, removing and mitigating the impact of formation damage caused by drilling and completion fluids is critical.

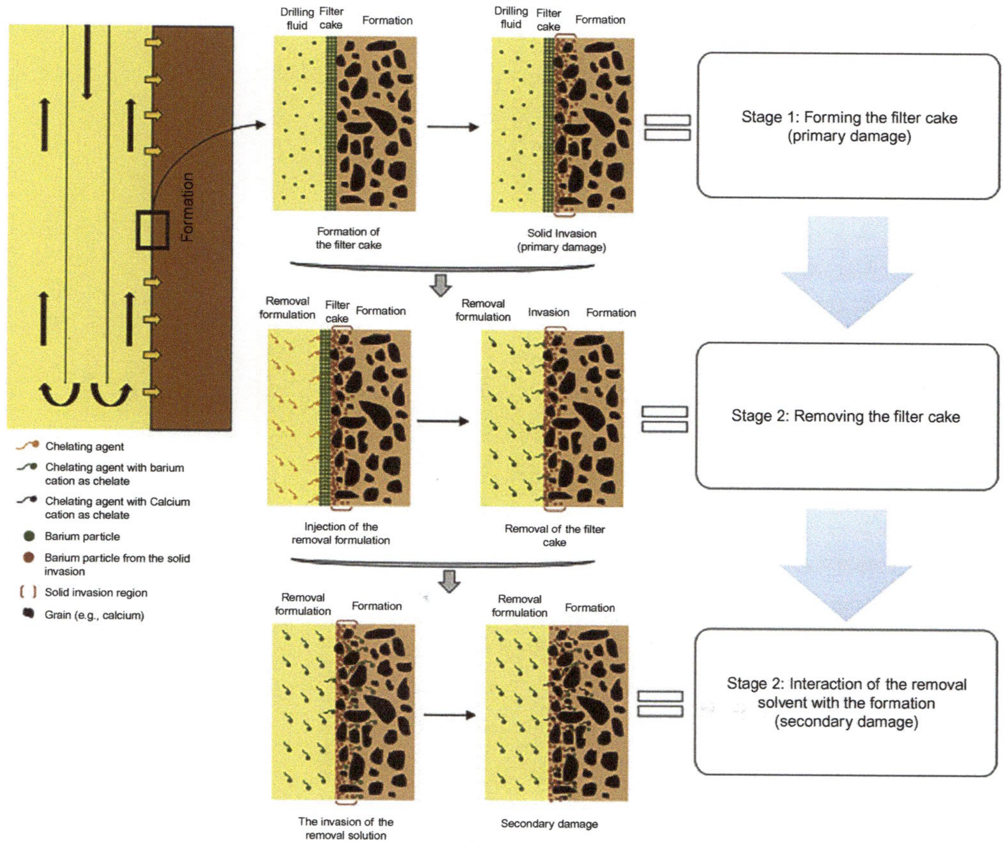

FIG. 3.5 Different stages of formation damage that are induced by the drilling and completion operations (Al Jaberi et al., 2021).

Because various fluids are utilized in the drilling and completion process, the methods by which fluids affect formations vary. Overall, two primary explanations can be applied to characterize the destructive mechanisms of drilling muds: (a) filtrate invasion (b) solids and polymer plug. When the wellbore pressure exceeds the pore pressure, filtrate invasion of drilling and completion fluids occurs. Drilling and completion fluids can cause clay swelling, and the disintegrating solids accumulate locally near pore throats, resulting in a decrease in permeability. Furthermore, filtrate infiltrated into the reservoir and reduced the flowability of oil and gas through water blockage.

The effectiveness of drilling and completion fluids is primarily determined by three factors: fluid viscosity, density, and solution pH (Shi et al., 2017). Precipitates from drilling and completion fluids can clog pore throats, which is more severe if no mud cake forms. As a result, fluids typically contain a variety of solid phases, such as drilled solids, weighing materials, and polymers, which can penetrate the formation and clog pore throats (Shi et al., 2017). Additionally, the water-based drilling fluid is often composed of bentonite and barium sulfate (barite) and calcium carbonate (calcite). Thus, drilling fluids combine subsurface fluids and rock, and if the two are incompatible, the precipitate can be formed (Shi et al., 2017). Drilling fluids containing a higher concentration of calcium chloride might cause formation damage when used in conjunction with connate water containing bicarbonate. Table 3.7 provides the types and mechanisms for different formation damage.

To regain the productivity of the formation following the well drilling, the formed filter cake on the borehole wall must be physically or chemically removed. Since the weighing agents account for 80%–90% of the filter cake, the removal solution must be capable of dissolving them (Ba Geri et al., 2017a). As a result, numerous chemical removal solutions for filter cake have been extensively explored and studied. Various chemicals, including acids,

TABLE 3.7 Types and mechanisms of formation damage (Shi et al., 2017)

Formation damages	Mechanism
Permeability reduction	Relative amount of fluids change
Wettability alteration	Surfactant invasion
Water blocking	Viscous fluid invasion
Pore throat clogging	Solids invasion
Scale	Inorganic precipitation
Mineral precipitation	Dissolution and minerals recrystallization
Particles migration	Dissolution of cement particles of formation
Rock dispersion	Loose formation

chelating agents, oxidizers, enzymes, converting agents, and emulsions, are used for dissolving the filter cake (Al Moajil et al., 2008; Ba Geri et al., 2017b). Chelating agent solutions such as EDTA and DTPA are special chemicals due to their high adsorption capacity for different types of minerals, their stability at elevated temperatures, and their low environmental toxicity (Barri et al., 2016; Mahmoud et al., 2018). Tariq et al. (2021) proposed the use of thermochemical fluid in drilling fluid formulations. The encapsulated thermochemical fluid can reduce fluid loss and generate a self-destructed barite filter cake. In addition to the destruction, the heat generated by the thermochemical fluid can create micro fractures to increase the flowability of the reservoir. Table 3.8 lists the common acid solvents that are usually used in removing different weighting materials for water-based drilling fluids.

The removal procedure may be a single-stage or a multistage process, depending on the compatibility of the various components of the removal solution (Al-Otaibi et al., 2010; Siddig et al., 2020). The term "single-stage" refers to a stage in which the polymer breaker and filter cake removal solutions are combined, whereas the term "multistage" refers to the process of

TABLE 3.8 Weighting materials and their filter cake removal solutions (Siddig et al., 2022)

Weighting material	Dissolver solution	References
Barite	Different chelating agents such as EDTA, DTPA, HEDTA, GLDA were used. Potassium carbonate was the common efficient catalyst. Thermochemical for barite self-destructive filter cake	Lakatos et al. (2002), Ba Geri et al. (2017a, 2017b), Mahmoud and Elkatatny (2017), and Tariq et al. (2021)
Calcium Carbonate	Organic acids and acid precursors	Al-Otaibi et al. (2006) and Nasr-El-Din and Al Moajil (2007)
Hematite	HCl, H_2SO_4, and nitric acid	Liu et al. (2003) and Matsumoto et al. (1994)
Ilmenite	HCl	Elkatatny et al. (2013)
Manganese tetraoxide	HCl, HCl/glycolic acid mixture, or HCl/lactic acid mixture	Al Moajil and Nasr-El-Din (2013)
Calcium sulfate	Glutamic diacetate acid (GLDA) chelating agent and potassium carbonate (K_2CO_3) as a convertor	Murtaza et al. (2021)

polymer breaking by using a chemical solution and then followed by filter cake removal using a dissolving agent (Akpan et al., 2020).

Oil-based drilling fluids showed several advantages compared to water-based drilling fluids, such as less fluid loss, stable rheology, improved hole cleaning ability, increased thermal stability, and efficient performance while drilling into a shale formation. These characteristics make OBDFs ideal for drilling through reservoir deposits. On other hand, the cleanup of OBDFs filter cakes is a difficult, costly, and time-consuming operation (Al-Otaibi et al., 2004; Davis et al., 2004). Treating WBDFs filter cakes is primarily determined by the availability of a polymer breaker and an acid solvent compatible with the weighing material. Given the presence of oil in the OBDFs filter cake, it is necessary to add material(s) that affect the wettability of the filter cake surface. To change the filter cake wettability, surfactant, mutual solvent, or both are added to the OBDFs filter cake removal solution. Siddig et al. (2022) compiled a detailed review on the treatment of various weighting materials used in oil-based drilling fluids, as mentioned in Table 3.9.

The characterization of the filter cake generated on the ceramic disc, filter paper, or core sample is essential before finding a suitable solution for its removal (Siddig et al., 2022). Usually, its characterization is performed using

TABLE 3.9 Various treatments of weighting materials in OBDFs (Siddig et al., 2022)

Treatment	Acids	Weighting materials	References
Water or diesel + mutual solvent + acid	Formic acid precursor (FAP)	Calcium carbonate	Davidson et al. (2006)
	Ethylene glycol monobutyl ether (EGMBE)		Ali et al. (2007)
	Glycol acid precursor (GAP)		Mohamed et al. (2019)
	HCl	Ilmenite	Xiao et al. (2015)
Brine + surfactant + acid	Formic acid precursor (FAP)	Calcium carbonate	Binmoqbil et al. (2009)
	Acetic Acid		Al-Otaibi et al. (2010)
Water + mutual solvent + surfactant + acid	Organic acid precursor (OAP)	Calcium carbonate	Al-Kuait et al. (2012)
	Acetic acid		Siddig et al. (2020)
	EDTA + K_2CO_3	Barite	Mahmoud and Elkatatny (2019)

TABLE 3.9 Various treatments of weighting materials in OBDFs (Siddig et al., 2022)—Cont'd

Treatment	Acids	Weighting materials	References
Microemulsions + acid	HCl/acetic acid	Calcium Carbonate/ Barite	Quintero et al. (2005)
	Formic acid precursor (FAP)		Quintero et al. (2007)
	Acetic acid	Calcium carbonate	Lavoix et al. (2007)
	Inorganic acid	Barite	Addagalla et al. (2016)

several techniques, including scanning electron microscope (SEM), CT scan, X-ray fluorescence (XRF), and nuclear magnetic resonance (NMR) (Siddig et al., 2022). Filter cake properties such as permeability, porosity, hardness, thickness, structure, and particle size influence formation productivity. The weighting agents and polymers in the drilling fluid and the clay, salinity, and sand content of the mud cake layer all affect the quality of the filter cake and its removal (Siddig et al., 2022).

Further formation damage caused by clay swelling can be controlled by incorporating shale swelling inhibitors in drilling and completion fluids. Several chemicals, including nanomaterials, graphene, polymers, ionic liquids, surfactants, natural products, and composite materials, of shale inhibitors have been developed (Aftab et al., 2017; Ahmed Khan et al., 2020; Arfaj et al., 2020; Chu et al., 2020; Kazempour et al., 2013; Murtaza et al., 2020a; Olayiwola and Dejam, 2020; Rana et al., 2020; Xie et al., 2020a,b). KCl is the most used shale swelling inhibitor. None of the developed shale inhibitor compounds has successfully stabilized the borehole due to various properties such as the clay shape, various particle size, different charges types, the flexibility of the layers, and the ion exchange capacity. As a result, developing a water-based shale inhibitor with multiple functionalities, chemical, and thermal stability and improved efficiency is still required in all environments (Roldan et al., 2019).

3.6 Interactions of drilling fluid filtrate with the formation rocks and fluids

During drilling operations, some of the drilling fluids invade into the formation around the wellbore, and the fluid invasion can cause several issues, especially in tight reservoirs (Bialy et al., 2011; Elkatatny et al., 2019).

This section discusses the problems associated with the drilling fluid invasion. The interactions between drilling fluid filtrates and formation rocks and fluids are discussed. The different quantification techniques that are used to understand the impact of fluid filtration on the formation properties are presented. Also, the drilling fluids used in unconventional formations are discussed.

The interactions between drilling filtrate and formation systems are complex and require a range of experimental approaches to understand the induced interactions (Barbati et al., 2016). The type and components of drilling fluids play significant roles in determining the filtrate/rock interactions. Also, the formation type and composition showed different responses during the drilling operations (Mahmoud et al., 2020). Unconventional formations such as shale are more challenging than conventional sandstone and carbonate rocks; therfore, more consideration is needed to design the drilling and completion fluids (Ali et al., 2020; Mohammadi et al., 2021). Pore impairments and clay swelling are the common results of filtrate/rock interactions (Ali et al., 2020; Kazempour et al., 2013).

Different quantification techniques can be utilized to investigate the interactions of drilling fluid filtrate and formation rock and fluids (Murtaza et al., 2021). Core flooding is the common approach used to evaluate the impacts of drilling filtrate on the formation properties (Ahmed et al., 2021; Siddig et al., 2020). The pressure drop profiles before and after mud filtration can be used to assess the reduction in the formation permeability. Also, imaging techniques such as atomic force microscopy (AFM) can generate high-resolution images for the pore spaces. However, AFM would require much time to carry out the measurements. Therefore, faster techniques such as nuclear magnetic resonance (NMR) are introduced to the petroleum industry as an effective and accurate technique to study the changes in pore spaces under different conditions (Ali et al., 2020; Siddig et al., 2020). Overall, integrating various techniques such as coreflooding, AFM, and NMR would result in a better understating for the fluid filtrates and formation rock/fluid interactions.

Water-based mud (WBM) is the common fluid used during drilling operations; however, using WBM in drilling unconventional formations such as shale can lead to severe formation damages (Siddig et al., 2020). WBM will increase the possibility of clay swelling due to the interactions between the water and active clay minerals (Ahmed Khan et al., 2020; Kazempour et al., 2013). Shale swelling will lead to various problems such as wellbore instability and reduction in the borehole diameter (Rana et al., 2020; Zhang et al., 2017). Therefore, chemical additives are used to inhibit shale swelling and improve the drilling fluid's rheological and filtration properties (Ahmed Khan et al., 2020; Xie et al., 2020a). The application of conventional mud additives such as inorganic and organic chemicals is not recommended in extreme drilling conditions (Moajil et al., 2013; Murtaza et al., 2021). Advanced materials such as nanomaterials have been proposed as promising alternatives to maximize the drilling performance and minimize formation damage (Ali et al., 2020; Cheraghian, 2021).

In tight reservoirs, several types of formation blockages can be induced by the drilling and completion fluids (Bialy et al., 2011). The extremely low permeability and super high capillary forces in tight formations lead to complicating the issue of formation blockages; therefore, more attention is required to ensure that the used fluids during drilling and completion operations will not induce formation blockages. Water blockage is a regular type of formation damage that can be induced in horizontal or multilateral wells. The drilling fluid filtrates into the formation due to the overbalance operation, leading to accumulating the liquid and fine particles inside the formation around the wellbore. The liquid blockage can kill the formation productivity, especially in tight gas reservoirs (Hassan et al., 2019). Therefore, avoiding such problems is very important, which can be performed by ensuring the compatibility between the drill fluids and the formation characteristics.

Van Zanten et al. (2011) studied the impact of using different surfactant treatments on the rock wettability of different tight formations. They found out that using microemulsion has the best performance and can lead to 100% permeability retain. Bialy et al. (2011) proposed a drill-in fluid utilizing a combination of manganese tetroxide (Mn_3O_4) and potassium formate (KCOOH). Using manganese tetroxide leads to inducing acidic environment and thereby maintaining a high level of permeability retain; 99% of the initial permeability was achieved using the proposed Mn_3O_4 and KCOOH fluids. Han et al. (2012) used sodium chloride (NaCl) and potassium chloride (KCl) to prepare drilling fluids; 3% glycol additives were also used to improve the fluid rheological properties. The developed drilling mud was tested in drilling several horizontal wells in a Brazilian field. The studied formations are reactive shale, and the proposed drilling mud was able to maintain around 70% of the initial shale permeability. Moreover, Elkatatny et al. (2019) proposed a new water-based drilling mud that contains sodium silicate (Na_2SiO_3). The introduced mud showed high performance in drilling tight gas formations; a thin and impermeable filter cake was formed, which leads to avoiding the water blockage and minimizing the formation damage.

Jiang et al. (2021) developed a novel water-based fluid that can be used for drilling and completion operations. The developed fluid can improve the borehole quality for conventional and unconventional formations. Microscopic and macroscopic measurements were carried out to assess the performance of the proposed fluid. They reported that the developed fluid outperforms the typical drilling fluids in terms of formation protection, lubricity, and wellbore stability. The proposed fluid was tested in around 1000 unconventional gas and oil wells in China, where minimum formation damage and high wellbore quality were obtained. Also, they mentioned that the developed drilling fluid led to reduce the wellbore collapse by around 83% and increase the rate of penetration by 33% compared to the conventional WBM and OBM.

Hassan et al. (2019) applied the thermochemical treatment approach to remove water blocking from tight formations. Several treatment cycles were implemented using thermochemical solutions, and a considerable increase in core conductivity was obtained. Also, the capillary pressure was reduced by around 54%, the formation productivity was increased by 1.79, and the required pore pressure was improved by 78%. Additionally, the pore pressure and heat generated by the exothermic reaction may cause microfractures in the sandstone core sample, decreasing the core's capillary forces. Alade et al. (2021) applied thermochemical treatment approach in removing emulsion blockage. By injecting one pore volume of thermochemical fluids into the core sample, 72% of the W/O emulsion was removed. Furthermore, the pressure profile observed during the flooding experiment demonstrates that the thermochemical flooding caused no precipitation or damage. As a result, it is anticipated that in situ heat generation will alleviate the emulsion blockage problem and provide advantages over existing methods regarding environmental friendliness and damage removal efficiency.

Overall, there is a significant shift in conventional ways by utilizing environment-friendly approaches, such as natural resources to minimize the drilling cost and mitigate the formation damages. Also, advanced technology such as ultrasonic methods has been introduced as a safe and efficient technique for removing the formation damage such as paraffin deposition, asphaltene precipitation, polymer, and other clogs that prevent oil and gas inflow (Shi et al., 2017; Towler et al., 2007; Tunio et al., 2011).

3.7 Concluding remarks

Drilling fluids are essentially complex fluids that comprise fluid and solid phases and are thus sensitive to variation in mechanical and thermal forces and changes in composition. At HT/HP conditions of wells, drilling and completion fluid systems showed different types of interactions. Also, they experience instability in the composition due to density variations and degradation in the rheological and filtration properties. To ensure successful drilling operations, particularly in harsh conditions, instability issues must be avoided. The stability of water-based mud has been commonly affected by high temperatures and high-pressure conditions. Various investigations have been conducted on the development of drilling fluid formulations using specially designed polymeric materials and nanoparticles as additives to enhance the rheological and filtration properties of the fluids. The outcomes from these researches have shown promising results on the thermal stability of the drilling fluids. However, future research directions should focus on technological advancement which promotes the sustainability of the process. This includes improvement in the development of environmentally friendly green additives such as bio-based polymers and nanoparticles. Moreover, economic sustainability should receive more attention from the research through process optimization.

References

Abbaszadeh, M., Nasiri, M., Riazi, M., 2016. Experimental investigation of the impact of rock dissolution on carbonate rock properties in the presence of carbonated water. Environ. Earth Sci. Springer. https://doi.org/10.1007/s12665-016-5624-3.

Addagalla, A.K.V., Kosandar, B.A., Lawal, I.G., Jadhav, P.B., Imran, A., Al Saqer, Q.R., El Sherbeny, W., Ansari, A., Pino, R., Gad-Alla, A.E., Olivaresantunez, T., 2016. Overcoming OBM filter cake damage using micro-emulsion remediation technology across a high-temperature formation. In: SPE International Conference and Exhibition on Formation Damage Control. Society of Petroleum Engineers. https://doi.org/10.2118/178942-MS.

Aftab, A., Ismail, A.R., Ibupoto, Z.H., 2017. Enhancing the rheological properties and shale inhibition behavior of water-based mud using nano-silica, multi-walled carbon nanotube, and graphene nanoplatelet. Egypt. J. Pet. 26 (2), 291–299.

Agarwal, S., Phuoc, T., Yee, S., Donald, M., Gupta, R.K., 2011. Flow Behavior of Nanoparticle Stabilized Drilling Fluids and Effect of High-Temperature Aging. In: AADE National Technical Conference and Exhibition, Houston.

Ahmad, H.M., Kamal, M.S., Al-Harthi, M.A., 2018. High molecular weight copolymers as a rheology modifier and fluid loss additive for water-based drilling fluids. J. Mol. Liq. 252, 133–143.

Ahmed Khan, R., Murtaza, M., Abdulraheem, A., Kamal, M.S., Mahmoud, M., 2020. Imidazolium-based ionic liquids as clay swelling inhibitors: mechanism, performance evaluation, and effect of different anions. ACS Omega 5 (41), 26682–26696. https://doi.org/10.1021/acsomega.0c03560.

Ahmed, A., Alsaihati, A., Elkatatny, S., 2021. An overview of the common water-based formulations used for drilling onshore gas wells in the Middle East. Arab. J. Sci. Eng. 46 (7), 6867–6877.

Akpan, E.U., Enyi, G.C., Nasr, G., Yahaya, A.A., Ahmadu, A.A., Saidu, B., 2019. Water-based drilling fluids for high-temperature applications and water-sensitive and dispersible shale formations. J. Petrol. Sci. Eng. 175, 1028–1038.

Akpan, E., Enyi, G., Nasr, G., 2020. (2020). Enhancing the performance of xanthan gum in water-based mud systems using an environmentally friendly biopolymer. J. Pet. Explor. Prod. Technol. 10, 1933–1948. https://doi.org/10.1007/s13202-020-00837-0.

Al Jaberi, J., Bageri, B.S., Adebayo, A.R., Patil, S., Barri, A., Salin, R.B., 2021. Evaluation of formation damages during filter cake deposition and removal process: the effect of primary damage on secondary damage. Pet. Sci. 18, 1153–1162. https://doi.org/10.1016/J.PETSCI.2021.07.004.

Al Moajil, A.M., Nasr-El-Din, H.A., Al-Yami, A.S., Al-Aamri, A.D., AlAgil, A.K., 2008. Removal of filter cake formed by manganese tetraoxide-based drilling fluids. SPE International Conference and Exhibition on Formation Damage Control. https://doi.org/10.2118/112450-MS.

Al Moajil, A.M., Nasr-El-Din, H.A., 2013. Removal of manganese tetraoxide filter cake using combination of HCl and organic acid. In: Proceeding of the SPE Heavy Oil Conference Canada Held in Calgary, Alberta, Canada, 11–13 June. https://doi.org/10.2118/165551-MS. SPE-165551-MS.

Alade, O., Mahmoud, M., Hassan, A., Murtaza, M., Al Shehri, D., Al-Nakhli, A., Bataweel, M., 2021. A novel method of removing emulsion blockage after drilling operations using thermochemical fluid. SPE Drill. Complet. 36, 88–100. https://doi.org/10.2118/199315-PA.

Albdiry, M.T., Almensory, M.F., 2016. Failure analysis of drillstring in petroleum industry: a review. Eng. Fail. Anal. 65, 74–85.

Ali, S.A., Needham, W.J., Luyster, M.R., Jones, A.G.K., 2007. Novel Chemistry Eliminates Post-Cleanup of OBM Filter Cakes. In: European Formation Damage Conference. Society of Petroleum Engineers. https://doi.org/10.2118/106703-MS.

Ali, I., Ahmad, M., Ganat, T., 2021. Biopolymeric formulations for filtrate control applications in water-based drilling muds: a review. J. Pet. Sci. Eng. 110021.

Ali, M., Jarni, H.H., Aftab, A., Ismail, A.R., Saady, N.M.C., Sahito, M.F., Keshavarz, A., Iglauer, S., Sarmadivaleh, M., 2020. Nanomaterial-based drilling fluids for exploitation of unconventional reservoirs: a review. Energies 13 (13), 3417.

Al-Kuait, A.M.S., Al-yateem, K.S., Olivares, T., Zubail, M.A., El Bialy, M., Ezell, R.G., Maghrabi, S., 2012. Improved producibility after delayed filter cake breaker treatment in the safaniya offshore field in Saudi Arabia. In: SPE Kuwait International Petroleum Conference and Exhibition. Society of Petroleum Engineers. https://doi.org/10.2118/163357-MS.

Al-Otaibi, M.B., Nasr-El-Din, H.A., Siddiqui, M.A., 2004. Chemical treatments to enhance productivity of horizontal and multilateral Wells: lab studies and case histories. In: SPE/DOE Symposium on Improved Oil Recovery Tulsa, Oklahoma. https://doi.org/10.2118/89467-MS.

Al-Otaibi, M.B., Nasr-El-Din, H.A., Al Moajil, A.M., 2006. In-situ acid system to clean up drill-in fluid damage in high-temperature gas wells. In: IADC/SPE Asia Pacific Drilling Technology Conference and Exhibition, Bangkok, Thailand. https://doi.org/10.2118/103846-MS.

Al-Otaibi, M.A., Binmoqbil, K.H., Rabba, A., Abitrabi, A.N., 2010. Single-stage chemical treatment for oil-based mud cake cleanup: lab studies and field case. In: SPE International Symposium and Exhibition on Formation Damage Control. Society of Petroleum Engineers. https://doi.org/10.2118/127795-MS.

Amanullah, M., AlArfaj, M.K., Al-abdullatif, Z.A., 2011. Preliminary test results of nano-based drilling fluids for oil and gas field application. In: Proceedings of the 2011 SPE/IADC Drilling Conference and Exhibition, 1–3 March, Amsterdam, The Netherlands, pp. 1–9. https://doi.org/10.2118/139534-MS.

Anderson, R.L., Ratcliffe, I., Greenwell, H.C., Williams, P.A., Cliffe, S., Coveney, P.V., 2010. Clay swelling—a challenge in the oilfield. Earth-Sci. Rev. 98 (3–4), 201–216.

Arfaj, M.K., Rana, A., Saleh, T.A., 2020. Highly efficient modified activated carbon as shale inhibitor for water based drilling mud modification. In: Abu Dhabi International Petroleum Exhibition & Conference, Society of Petroleum Engineers; 2020. https://doi.org/10.2118/203410-MS.

Ba Geri, B.S., Mahmoud, M., Al-Majed, A.A., Al-Mutairi, S.H., Abdulazeez, A., Shawabkeh, R., 2017a. Water Base barite filter cake removal using non-corrosive agents. In: Paper presented at the SPE Middle East Oil & Gas Show and Conference, Manama, Kingdom of Bahrain. https://doi.org/10.2118/183653-MS.

Ba Geri, B.S., Mahmoud, M., Abdulraheem, A., Al-Mutairi, S.H., Elkatatny, S., Shawabkeh, R.A., 2017b. Single-stage filter cake removal of barite weighted water-based drilling fluid. J. Petrol. Sci. Eng. 149, 476–484. https://doi.org/10.1016/j.petrol.2016.10.059.

Barbati, A.C., Desroches, J., Robisson, A., McKinley, G.H., 2016. Complex fluids and hydraulic fracturing. Annu. Rev. Chem. Biomol. Eng. 7, 415–453.

Barri, A., Mahmoud, M., Elkatatny, S., 2016. Evaluation of rock mechanical properties alteration during matrix stimulation with chelating agents. ASME. J. Energy Resour. Technol. 138 (3). 032907 https://doi.org/10.1115/1.4032546.

Barry, M.M., Jung, Y., Lee, J.K., Phuoc, T.X., Chyu, M.K., 2015. Fluid filtration and rheological properties of nanoparticle additive and intercalated clay hybrid bentonite drilling fluids. J. Petrol. Sci. Eng. 127, 338–346.

Belayneh, M., Aadnøy, B., Strømø, S.M., 2021. MoS2 nanoparticle effects on 80 °C thermally stable water-based drilling fluid. Matererials 2021 (14), 7195. https://doi.org/10.3390/ma14237195.

Bialy, M.E., Mohsen, M., Ezell, R.G., Abdulaziz, M.E., Kompantsev, A., Khakimov, A., Ashoor, A., 2011. Utilization of non-damaging drilling fluid composed of potassium Formate brine and manganese tetra oxide to Drill sandstone formation in tight gas reservoir. In: SPE/IADC Middle East Drilling Technology Conference and Exhibition, Muscat,Oman, 24-26.

Binmoqbil, K.H., Al-Otaibi, M.A., Al-Faif, M.G., Al-Khudair, W.S., AlAamri, A.D., 2009. Cleanup of oil-based mud filter cake using an in-situ acid generator system by a single-stage treatment. In: SPE Saudi Arabia Section Technical Symposium. Society of Petroleum Engineers. https://doi.org/10.2118/126065-MS.

Caenn, R., Chillingar, G.V., 1996. Drilling fluids: state of the art. J. Petrol. Sci. Eng. 14 (3-4), 221–230.

Cheraghian, G., 2021. Nanoparticles in drilling fluid: a review of the state-of-the-art. J. Mater. Res. Technol. 13, 737–753.

Chu, Q., Su, J., Lin, L., 2020. Inhibition performance of amidocyanogen silanol in water-based drilling fluid. Appl. Clay Sci. https://doi.org/10.1016/j.clay.2019.105315.

Contreras, O., Hareland, G., Husein, M., Nygaard, R., Al-Saba, M., 2014. Application of in-house prepared nanoparticles as filtration control additive to reduce formation damage. In: Proceedings of the 2014 SPE International Symposium and Exhibition on Formation Damage Control, 26–28 February, Lafayette, LA, USA, pp. 1–11. https://doi.org/10.2118/168116-MS.

Dahab, A.S., 1991. Thermal stability of drilling fluids prepared from Saudi Palygorskite. J. Can. Pet. Technol. 30. https://doi.org/10.2118/91-03-04.

Davidson, E., McMillan, D.N., Martin, F., Morton, K., Lenz, R., 2006. Successful deployment of a new stimulation chemical, post horizontal open-hole gravel pack in wells drilled with both water-based and oil-based drill-in-fluids. In: SPE/IADC Indian Drilling Technology Conference and Exhibition. Society of Petroleum Engineers. https://doi.org/10.2118/101964-MS.

Davis, E., Hodge, R., Dick, M., Mason, S., Martens, H., 2004. Optimizing drill-in fluid & completion cleanup processes for open-hole horizontal gravel packed completions in low-temperature environments. In: SPE International Symposium and Exhibition on Formation Damage Control, Lafayette, Louisiana. https://doi.org/10.2118/86510-MS.

Dye, W., Hemphill, T., Gusler, W., Mullen, G., 2001. Correlation of ultralow-shear-rate viscosity and dynamic barite sag. SPE Drill. Comple. 16 (01), 27–34.

Dye, W., Mullen, G., Gusler, W., 2003. Drilling processes: the other half of the barite sag equation. In: Proceedings of the SPE Asia Pacific Oil and Gas Conference and Exhibition, Jakarta, Indonesia, 9–11 September 2003.

Elkatatny, S., 2018. Enhancing the stability of invert emulsion drilling fluid for Drilling in High-Pressure High-Temperature Conditions. Energies 11 (9), 2393. https://doi.org/10.3390/en11092393.

Elkatatny, S., Jafarov, T., Al-Majed, A., Mahmoud, M., 2019. Formation damage avoidance by reducing invasion with sodium silicate-modified water-based drilling fluid. Energies. https://doi.org/10.3390/en12081485.

Elkatatny, S., Ahmed, A., Abughaban, M., Patil, S., 2020. Deep illustration for loss of circulation while drilling. Arab. J. Sci. Eng. 45 (2), 483–499.

Elkatatny, S.M., Xiao, J., Nasr-El-Din, H.A., Al-Bagoury, M., 2013. Using hydrochloric acid to remove ilmenite water-based filter cake in HPHT applications. In: Paper presented at the

SPE European Formation Damage Conference & Exhibition, Noordwijk, The Netherlands. https://doi.org/10.2118/165181-MS.
Fink, J., 2015. Chapter 1 - Drilling muds. Petroleum Engineering Guide to Oil Field Chemical Fluids, Second Ed. Gulf Professional Publishing, Boston, pp. 1–61. https://doi.org/10.1016/B978-0-12-803734-8.00001-1.
Fink, J., 2021. *Petroleum engineer's guide to oil field chemicals and fluids.* Gulf Professional Publishing.
Gautam, S., Guria, C., 2020. Optimal synthesis, characterization, and performance evaluation of high-pressure high-temperature polymer-based drilling fluid: the effect of viscoelasticity on cutting transport, filtration loss, and lubricity. SPE J. 25, 1333–1350. https://doi.org/10.2118/200487-PA.
Gautam, R., Sahai, M., Kumar, S., 2020. Recent advances in application of nanomaterials as additives for drilling fluids. Energy Sources, Part A: Recovery, Utilization, and Environmental Effects 1–24.
Han, L., Clinch, D., van der Zwaag, C., Galletti, E., Fornasier, F., Estevez, F., Patey, I., 2012. Customizing drill-in fluid for peregrino project in Brazil: laboratory development and field experience. In: SPE International Symposium and Exhibition on Formation Damage Control, Lafayette, LA, USA.
Hassan, A.M., Mahmoud, M.A., Al-Majed, A.A., Al-Nakhli, A.R., Bataweel, M.A., 2019. Water blockage removal and productivity index enhancement by injecting thermochemical fluids in tight sandstone formations. J. Petrol. Sci. Eng. 182, 106298. https://doi.org/10.1016/j.petrol.2019.106298.
Hoelscher, K.P., De Stefano, G., Riley, M., Young, S., 2012. Application of nanotechnology in drilling fluids. In: Proceedings of the 2012 SPE International Oilfield Nanotechnology Conference and Exhibition, 12–14 June, Noordwijk, The Netherlands, pp. 1–7. https://doi.org/10.2118/157031-MS.
Hossain, M.E., Al-Majed, A.A., 2015. Fundamentals of sustainable drilling engineering. John Wiley & Sons. Scrivener Publishing LLC, pp. 73–139.
Ikram, R., Jan, B.M., Vejpravova, J., 2021. Towards recent tendencies in drilling fluids: application of carbon-based nanomaterials. J. Mater. Res. Technol. 15, 3733–3758.
Javeri, S.M., Haindade, Z.M.W., Jere, C.B., 2011. Mitigating loss circulation and differential sticking problems using silicon nanoparticles. In: Proceedings of the 2011 SPE/IADC Middle East Drilling Technology Conference and Exhibition, 24–26 October, Muscat, Oman, pp. 1–4. https://doi.org/10.2118/145840-MS.
Jia, H., Niu, C., Dai, C., 2021. Solid-free flexible colloidal completion fluid with variable density for gas well completion in high-temperature and high-pressure reservoirs: experimental study and pilot test. SPE J. https://doi.org/10.2118/208590-PA.
Jiang, G., Sun, J., He, Y., Cui, K., Dong, T., Yang, L., Wang, X., 2021. Novel water-based drilling and completion fluid technology to improve wellbore quality during drilling and protect unconventional reservoirs. Engineering, 1–14.
Jose, A., Imqam, A., 2020. Silica and Graphene Oxide Nanoparticle Formulation to Improve Thermal Stability and Inhibition Capabilities of Water-Based Drilling Fluid Applied to Woodford Shale. SPE Drill. Complet. 35 (02), 164–179. https://doi.org/10.2118/193567-PA.
Kazempour, M., Manrique, E.J., Alvarado, V., Zhang, J., Lantz, M., 2013. Role of active clays on alkaline–surfactant–polymer formulation performance in sandstone formations. Fuel 104, 593–606. https://doi.org/10.1016/J.FUEL.2012.04.034.
Khodja, M., Khodja-Saber, M., Canselier, J.P., Cohaut, N., Bergaya, F., 2010. Drilling fluid technology: performances and environmental considerations. In: Products and services; from R&D to final solutions, pp. 227–256.

Khosravanian, R., Aadnøy, B.S., 2021. *Methods for Petroleum Well Optimization: Automation and Data Solutions*. Gulf Professional Publishing.

Knight, R., Pyrak-Nolte, L.J., Slater, L., Atekwana, E., Endres, A., Geller, J., Lesmes, D., Nakagawa, S., Revil, A., Sharma, M.M., Straley, C., 2010. Geophysics at the interface: response of geophysical properties to solid-fluid, fluid-fluid, and solid-solid interfaces. Rev. Geophys. 48 (4), 1–30.

Lakatos, I., Lakatos-Szabo, J., Kosztin, B., 2002. Comparative study of different barite dissolvers: technical and economic aspects. In: Paper Presented at the International Symposium and Exhibition on Formation Damage Control, Lafayette, Louisiana. https://doi.org/10.2118/73719-MS.

Larson, R.G., 1999. The Structure and Rheology of Complex Fluids. Oxford University Press, New York.

Lavoix, F., Jones, T.A., Quintero, L., Le Prat, X., Leschi, P., Aubry, E., 2007. First application of novel microemulsion technology for sand control remediation operations - a successful case history from the rosa field, a deepwater development project in Angola. In: European Formation Damage Conference, Scheveningen, The Netherlands. https://doi.org/10.2118/107341-MS.

Li, G., Zhang, J., Zhao, H., Hou, Y., 2012. Nanotechnology to improve sealing ability of drilling fluids for shale with micro-cracks during drilling. In: Proceedings of the 2012 SPE International Oilfield Nanotechnology Conference and Exhibition, 12–14 June, Noordwijk, The Netherlands, pp. 1–7. https://doi.org/10.2118/156997-MS.

Li, M.C., Wu, Q., Song, K., Qing, Y., Wu, Y., 2015. Cellulose nanoparticles as modifiers for rheology and fluid loss in bentonite water-based fluids. ACS Appl. Mater. Interfaces 7, 5006–5016.

Li, W., Zhao, X., Ji, Y., Peng, H., Chen, B., Liu, L., Han, X., 2016. Investigation of biodiesel-based drilling fluid, part 1: biodiesel evaluation, invert-emulsion properties, and development of a novel emulsifier package. SPE J. 21 (05), 1755–1766. https://doi.org/10.2118/180918-PA.

Liu, H., Papangelakis, V.G., Alam, M.S., Singh, G., 2003. Solubility of hematite in H2SO4 solutions at 230–270 °c. Can. Metall. Quart. 42, 199–207. https://doi.org/10.1179/cmq.2003.42.2.199.

Mahmoud, M., Elkatatny, S., 2017. Towards a complete removal of barite weighted water and oil based-drilling fluids in a single stage. In: SPE Annual Technical Conference and Exhibition. San Antonio, Texas, USA, October 2017. https://doi.org/10.2118/187122-MS.

Mahmoud, M.A.N.E.-D., Elkatatny, S., 2019. Removal of barite-scale and barite-weighted water- or oil-based-drilling-fluid residue in a single stage. SPE Drill Complet. 34, 016–026. https://doi.org/10.2118/187122-PA.

Mahmoud, M., Hussein, I.A., Sultan, A., Saad, M.A., Buijs, W., Vlugt, T.J.H., 2018. Development of efficient formulation for the removal of iron sulfide scale in sour production wells. Can. J. Chem. Eng. 96, 2526–2533. https://doi.org/10.1002/cjce.23241.

Mahmoud, H., Hamza, A., Nasser, M.S., Hussein, I.A., Ahmed, R., Karami, H., 2020. Hole cleaning and drilling fluid sweeps in horizontal and deviated wells: comprehensive review. J. Petrol. Sci. Eng. 186, 106748.

Mahmoud, O., Nasr-El-Din, H.A., Vryzas, Z., Kelessidis, V.C., 2016. February. Nanoparticle-based drilling fluids for minimizing formation damage in HP/HT applications. In: *SPE international conference and exhibition on formation damage control*. OnePetro.

Matsumoto, K., Taniguchi, S., Kikuchi, A., 1994. Leaching rate of hematite in nitric acid solutions in agitated vessel. J. Japan Inst. Met. 58, 291–296. https://doi.org/10.2320/jinstmet1952.58.3_291.

Moajil, A., Abdullah, M., Nasr-El-Din, H.A., 2013. Removal of manganese Tetraoxide filter cake using combination of HCl and organic acid. In: Paper presented at the SPE Heavy Oil Conference-Canada, Calgary, Alberta, Canada. https://doi.org/10.2118/165551-MS.

Mohamed, A., Elkatatny, S., Basfar, S., 2019. One-stage calcium carbonate oil-based filter cake removal using a new biodegradable acid system. In: SPE Kuwait Oil & Gas Show and Conference. Society of Petroleum Engineers. https://doi.org/10.2118/198045-MS.

Mohamed, A., Al-Afnan, S., Elkatatny, S., Hussein, I., 2020. Prevention of barite sag in water-based drilling fluids by a urea-based additive for drilling deep formations. Sustain. 12 (7), 2719.

Mohammadi, S., Kord, S., Mohammadzadeh, O., Moghadasi, J., 2021. An experimental study into rock dissolution mechanism during diluted seawater injection in carbonate rocks. Upstream Oil Gas Technol. 6, 100031. https://doi.org/10.1016/J.UPSTRE.2021.100031.

Murtaza, M., Ahmed Khan, R., Kamal, M.S., Hussain, S.M.S., Mahmoud, M., 2020a. Poly(oxyethylene)-amidoamine based Gemini cationic surfactants with hydrophilic spacers as clay stabilizers. Energy Fuel 34, 10619–10630. https://doi.org/10.1021/acs.energyfuels.0c01010.

Murtaza, M., Alarifi, S.A., Kamal, M.S., Onaizi, S.A., Al-Ajmi, M., Mahmoud, M., 2021. Experimental investigation of the rheological behavior of an oil-based drilling fluid with rheology modifier and oil wetter additives. Molecules 26 (16), 4877. https://doi.org/10.3390/molecules26164877. PMID: 34443465; PMCID: PMC8398167.

Nasr-El-Din, H.A., Al Moajil, A.M., 2007. Evaluation of in-situ generated acids for filter cake clean up. In: Paper Presented at the European Formation Damage Conference, Scheveningen, The Netherlands. https://doi.org/10.2118/107537-MS.

Nazar, S., Yang, J., Thomas, B.S., Azim, I., Rehman, S.K.U., 2020. Rheological properties of cementitious composites with and without nano-materials: a comprehensive review. J. Clean. Prod. 272, 122701.

Nguyen, T., Miska, S., Yu, M., Takach, N., Ahmed, R., Saasen, A., Maxey, J., 2011. Experimental study of dynamic barite sag in oil-based drilling fluids using a modified rotational viscometer and a flow loop. J. Petrol. Sci. Eng. 78 (1), 160–165.

Ofei, T.N., Bavoh, C.B., Rashidi, A.B., 2017. Insight into ionic liquid as potential drilling mud additive for high-temperature wells. J. Mol. Liq. 242, 931–939.

Olayiwola, S.O., Dejam, M., 2020. Experimental study on the viscosity behavior of silica Nanofluids with different ions of electrolytes. Ind. Eng. Chem. Res. https://doi.org/10.1021/acs.iecr.9b06275.

Parizad, A., Shahbazi, K., Tanha, A.A., 2018. SiO2 nanoparticle and KCl salt effects on filtration and thixotropical behavior of polymeric water-based drilling fluid: with zeta potential and size analysis. Results Phys. 9, 1656–1665.

Pedrosa, C., Saasen, A., Ytrehus, J.D., 2021. Fundamentals and physical principles for drilled cuttings transport—cuttings bed sedimentation and Erosion. Energies 14 (3), 545.

Perween, S., Beg, M., Shankar, R., Sharma, S., Ranjan, A., 2018. Effect of zinc titanate nanoparticles on rheological and filtration properties of water-based drilling fluids. J. Petrol. Sci. Eng. 170, 844–857.

Pinto, F., Meo, M., 2017. Design and manufacturing of a novel shear thickening fluid composite (STFC) with enhanced out-of-plane properties and damage suppression. Appl. Compos. Mater. 24, 643–660. https://doi.org/10.1007/s10443-016-9532-1.

Quintero, L., Jones, T.A., Clark, D.E., 2005. One step acid removal of an invert emulsion. In: The SPE European Formation Damage Conference. Society of Petroleum Engineers. https://doi.org/10.2118/94604-MS.

Quintero, L., Jones, T.A., Clark, D., Twynam, A.J., 2007. NAF filter cake removal using micro-emulsion technology. In: European Formation Damage Conference. Society of Petroleum Engineers. https://doi.org/10.2118/107499-MS.

Rabia, H., 2001. Chapter 7. Well Engineering and Construction. Entrac Petroleum, London, United Kingdom, pp. 197–234.

Rafati, R., Smith, S.R., Haddad, A.S., Novara, R., Hamidi, H., 2018. Effect of nanoparticles on the modifications of drilling fluids properties: a review of recent advances. J. Petrol. Sci. Eng. 161, 61–76.

Rana, A., Arfaj, M.K., Saleh, T.A., 2020. Graphene grafted with glucopyranose as a shale swelling inhibitor in water-based drilling mud. Appl. Clay Sci. 199. https://doi.org/10.1016/j.clay.2020.105806.

Rasool, K., Nasrallah, G.K., Younes, N., Pandey, R.P., Abdul Rasheed, P., Mahmoud, K.A., 2018. Green ZnO-interlinked chitosan nanoparticles for the efficient inhibition of sulfate-reducing Bacteria in inject seawater. ACS Sustain. Chem. Eng. 6, 3896–3906.

Roldan, V.J., Johanna, L., Hernando, L.S., Carlo, C.J., Disney, Z.R., Andrés, F.C., Cortés, F.B., 2019. Effect of silica nanoparticles on thermal stability in bentonite free water-based drilling fluids to improve its rheological and filtration properties after aging process. In: Paper presented at the Offshore Technology Conference Brasil, Rio de Janeiro, Brazil, October 2019. https://doi.org/10.4043/29901-MS.

Salehi, S., Madani, S.A., Kiran, R., 2016. Characterization of drilling fluids filtration through integrated laboratory experiments and CFD modeling. J. Nat. Gas Sci. Eng. 29, 462–468.

Shi, X., Xu, H., Yang, L., 2017. Removal of formation damage induced by drilling and completion fluids with a combination of ultrasonic and chemical technology. J. Nat. Gas Sci. Eng. 37, 471–478. https://doi.org/10.1016/j.jngse.2016.11.062.

Siddig, O., Mahmoud, A.A., Elkatatny, S., 2020. A review of different approaches for water-based drilling fluid filter cake removal. J. Petrol. Sci. Eng. 192, 107346.

Siddig, O., Mahmoud, A.A., Elkatatny, S., 2022. A review of the various treatments of oil-based drilling fluids filter cakes. J. Pet. Explor. Prod. Technol. 12, 365–381.

Tariq, Z., Kamal, M.S., Mahmoud, M., Alade, O., Al-Nakhli, A., 2021. Self-destructive barite filter cake in water-based and oil-based drilling fluids. J. Petrol. Sci. Eng. 197, 107963. https://doi.org/10.1016/J.PETROL.2020.107963.

Temple, C., Aterson, A. F., Leith, C. D. (2004). Method for Reducing Sag in Drilling, Completion, and Workover Fluids. U.S. Patent WO2004113467A1, 29 December 2004.

Towler, B.F., Chejara, A., Mokhatab, S., 2007. Experimental investigations of ultrasonic waves effects on wax deposition during crude oil production. In: SPE Annual Technical Conference and Exhibition, Anaheim, California, U.S.A. https://doi.org/10.2118/109505-MS.

Tunio, S.Q., Tunio, A.H., Ghirano, N.A., El Adawy, Z.M., 2011. Comparison of different enhanced oil recovery techniques for better oil productivity. Int. J. Appl. Sci. Technol. 1 (5), 143–153.

Van Zanten, R., Horton, D., Tanche-Larsen, P., 2011. Engineering drill-in fluids to improve reservoir producibility. In: Presented at the SPE European Formation Damage Conference, Noordwijk, The Netherlands, pp. 7–10.

Vryzas, Z., Kelessidis, V.C., 2017. Nano-based drilling fluids: a review. Energies 10 (4), 540.

Wagner, N.J., Brady, J.F., 2009. Shear thickening in colloidal dispersions. Phys. Today 62, 27–32. https://doi.org/10.1063/1.3248476.

Wei, M., Lin, K., Liu, H., 2019. Experimental investigation on hysteretic behavior of a shear thickening fluid damper. Struct. Control Health Monit. 26, e2389. https://doi.org/10.1002/stc.2389.

William, J.K.M., Ponmani, S., Samuel, R., Nagarajan, R., Sangwai, J.S., 2014. Effect of CuO and ZnO nanofluids in xanthan gum on thermal, electrical and high-pressure rheology of water-based drilling fluids. J. Petrol. Sci. Eng. 117, 15–27. https://doi.org/10.1016/j.petrol.2014.03.005.

Xiao, J., Nasr-El-Din, H.A., Al-Bagoury, M., 2015. Removal of ilmenite oil-based filter cake under HP/HT conditions using hydrochloric acid. In: SPE North Africa Technical Conference and Exhibition. Society of Petroleum Engineers. https://doi.org/10.2118/175728-MS.

Xie, G., Xiao, Y., Deng, M., Luo, Y., An, W., Yu, H., Luo, P., 2020a. Investigation on the inhibition mechanism of alkyl diammonium as montmorillonite swelling inhibitor: experimental and molecular dynamics simulations. Fuel 282, 118841.

Xie, G., Xiao, Y., Deng, M., Luo, Y., Luo, P., 2020b. Low molecular weight branched polyamine as a clay swelling inhibitor and its inhibition mechanism: experiment and density functional theory simulation. Energy Fuels 34 (2), 2169–2177.

Xindi, S.U.N., Baojun, B.A.I., 2017. Comprehensive review of water shutoff methods for horizontal wells. Pet. Explor. Dev. 44 (6), 1022–1029.

Zhang, X., Jiang, G., Xuan, Y., Wang, L., Huang, X., 2017. Associating copolymer acrylamide/Diallyldimethylammonium chloride/butyl acrylate/2-Acrylamido-2-methylpropanesulfonic acid as a Tackifier in clay-free and water-based drilling fluids. Energy Fuel 31, 4655–4662.

Chapter 4

Interactions of fluids during sandstone acidizing operations

Amjed Hassan[a], Murtada Saleh Aljawad[a], Mohamed Mahmoud[a], and Muhammad Shahzad Kamal[b]

[a]*Department of Petroleum Engineering, College of Petroleum Engineering & Geosciences, King Fahd University of Petroleum & Minerals, Dhahran, Saudi Arabia*

[b]*Center for Integrative Petroleum Research, King Fahd University of Petroleum and Minerals, Dhahran, Saudi Arabia*

4.1 Basic chemistry of sandstone acidizing

Sandstone formations are types of clastic sedimentary rocks that consist mainly of silicate minerals such as quartz, feldspar, and different types of clay (Etemad-Saeed et al., 2011; Hill et al., 1981; Miall, 1988; Shaughnessy and Kunze, 1981). Different treatments are used to improve the production from sandstone reservoirs as well as to remove the formation damage (Economides et al., 2013; Ponce da Motta et al., 1992). The acid injection is considered one of the most effective techniques in stimulating sandstone reservoirs. Different types of acids can be injected to interact with the sandstone rocks, leading to the dissolving part of the formation matrix (Andotra, 2014; Aneto, 2012; Garcia et al., 2016; Ji et al., 2014; Legemah et al., 2015; Pituckchon, 2014; Shafiq et al., 2018; Zhou and Nasr-El-Din, 2016; Zhou et al., 2016). As a result, the formation permeability will be increased and the fluid flow rate from the reservoir to the wellbore will be improved (Kalfayan and Metcalf, 2000). Hydrochloric acid (HCl) and hydrofluoric acid (HF) are the common acids used for sandstone acidizing (Shafiq and Mahmud, 2017; Shafiq et al., 2018; Smith and Hendrickson, 1965).

Different types of interactions take place during the acid injections in the sandstone formations. The main parameters that affect the rock–acid interactions are the sandstone mineralogy and the acid type (Mahmoud and Gomaa, 2022; Shafiq and Mahmud, 2017). In general, sandstone formation composes of two major mineral types: silicate minerals and carbonate minerals (Mahmoud and Gomaa, 2022; Shafiq et al., 2018). The first group includes minerals such

Developments in Petroleum Science, Vol. 78. https://doi.org/10.1016/B978-0-323-99285-5.00004-1

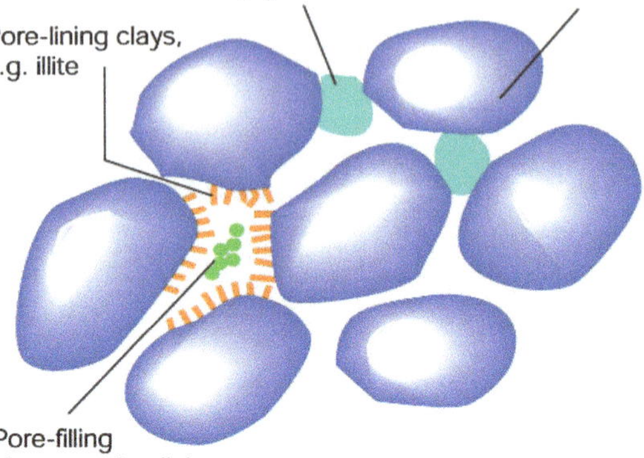

FIG. 4.1 The typical composition of sandstone rocks. *After Crowe, C., Masmonteil, J., & Thomas, R., 1992. Trends in matrix acidizing. Oilf. Rev. 4 (4), 24–40.*

as quartz, kaolinite, bentonite, orthoclase, and albite (Leong et al., 2018; Leong and Ben Mahmud, 2019), and the second group covers minerals such as calcite, siderite, and dolomite (Crowe et al., 1992; Muecke, 1982). Fig. 4.1 shows the typical composition of sandstone rocks. Moreover, the acid types and concentrations showed significant impacts on the sandstone interactions. Strong acids such as HF or HCl can dissolve considerable portions of the sandstone rocks, while weak acids such as organic or chelating agents can lessen the degree of sandstone dissolutions (Leong et al., 2018; Shafiq and Mahmud, 2017; Shafiq et al., 2018; Smith and Hendrickson, 1965). For all types of acids, increasing the acid concentration will lead to increasing the reaction rate between the sandstone rocks and the injected acids (Leong et al., 2018).

During sandstone acidizing, the interaction between the injected acids and sandstone rocks can lead to significant changes in the petrophysical properties. Dissolving part of the sandstone matrix results in increasing the rock permeability and porosity (Crowe et al., 1992). However, considerable reductions in the sandstone permeability can be induced due to the behavior of clays, and some minerals generally are defined as pre-lining or pore-filling minerals (see Fig. 4.1). Using the mixture of HF and HCl acids, which is commonly named mud acid, can minimize the sandstone damage by dissolving almost all types of sandstone minerals including the pre-lining or pore-filling clays (Smith and Hendrickson, 1965). However, the main issues of using mud acids for sandstone acidizing are corrosion and environmental impacts (Gidley, 1985; Gomaa et al., 2013; Wang et al., 2020).

Two types of chemical reactions are occurring due to the injection of acids in the sandstone formations; mutual reactions between the injected fluids and heterogeneous reactions between the acids and sandstone minerals (Al-Shaalan and Nasr-El-Din, 2000). The mutual reactions depend on the component of the acid system; all chemical constituents should be precisely selected to ensure that the components are compatible with each other to avoid any type of precipitations. Moreover, the heterogeneous reactions between the injected acids and sandstone minerals are affected mainly by the concentration and type of acid system (Al-Shaalan and Nasr-El-Din, 2000). For example, HF injection in the sandstone formations can lead to dissolving all rock minerals, however, at different rates based on the exposed rock areas and the intrinsic reaction rates. Fast or slow reaction rates can be induced based on the rock mineralogy; clay and feldspar usually show a faster reaction rate compared to quartz minerals (Martin, 2004). Furthermore, the chemical reactions during sandstone acidizing can induce some precipitations such as the depositions of hydrated silica, sodium, potassium, and calcium minerals, as explained by the following equations (Al-Shaalan and Nasr-El-Din, 2000; Leong and Ben Mahmud, 2019; Shafiq et al., 2018; Shaughnessy and Kunze, 1981):

$$4HF + SiO_2 \rightarrow SiF_4 + 2H_2O \tag{4.1}$$

$$6HF + SiO_2 \rightarrow H_2SiF_6 + 2H_2O \tag{4.2}$$

$$H_2SiF_6 + 4H_2O \rightarrow Si(OH)_4 + 6HF \tag{4.3}$$

$$H_2SiF_6 + 2Na^+ \rightarrow Na_2SiF_6 + H_2 \tag{4.4}$$

$$H_2SiF_6 + 2K^+ \rightarrow K_2SiF_6 + H_2 \tag{4.5}$$

$$2HF + Ca+ \rightarrow CaF_2 + H_2 \tag{4.6}$$

Overall, different types of reactions can take place during sandstone acidizing. The acid system and the rock compositions are the main parameters that affect the rock–acid interactions. The acid type and concentration should be selected based on the rock composition to stimulate the sandstone formation without inducing minerals participation. Table 4.1 summarizes the main chemical reactions during sandstone acidizing.

4.2 Composition of stimulations fluids during sandstone acidizing

In the oil and gas industry, sandstone acidizing has been widely implemented using different types of acid systems (Al-Shaalan and Nasr-El-Din, 2000; Leong et al., 2018; Leong and Ben Mahmud, 2019). The most common acid system is the mud acid which is a mixture of different percentages of HF and HCl acids. Mud acid showed efficient performance in increasing the permeability and porosity of different sandstone formations due to the effectiveness of HF acid in dissolving a wide range of sandstone minerals while HCl is used mainly to control the mineral precipitation. Gomaa et al. (2013) investigated the performance of mud acid in stimulating sandstone formations. The impact of acid percentages on the permeability improvements was studied at

TABLE 4.1 The common chemical reactions during sandstone acidizing (Guo, 2011; Leong and Ben Mahmud, 2019)

Acid type	Rock mineralogy	Chemical reactions
HF/HCl	Quartz	$SiO_2+4HF \leftrightarrow SiF_4+2H_2O$ $SiF_4+2HF \leftrightarrow H_2SiF_6$
	Kaolinite	$Al_4Si_8O_{10}(OH)_8+40HF$ $+4H^+ \leftrightarrow 4AlF_2^++8SiF_4+18H_2O$
	Montmorillonite (Bentonite)	$Al_4Si_8O_{20}(OH)_4+40HF$ $+4H^+ \leftrightarrow 4AlF_2^++8SiF_4+24H_2O$
	Orthoclase	$KAlSi_3O_8+14HF+2H+\leftrightarrow K++AlF2^++3SiF4+8H2O$
	Albite	$NaAlSi_3O_8+14HF$ $+2H^+ \leftrightarrow Na^++AlF_2^++3SiF_4+8H_2O$
HCl	Calcite	$CaCO_3+2HCl \rightarrow CaCl_2+CO_2+H_2O$
	Dolomite	$CaMg(CO3)_2+4HCl \rightarrow CaCl_2+MgCl_2+2CO_2+2H_2O$
	Siderite	$FeCO_3+2HCl \rightarrow FeCl_2+CO_2+H_2O$

a temperature of 180°F. They concluded that using HF concentrations of 1.9–2.8% and HCl concentrations between 3% and 15% can successfully improve the sandstone permeability. Also, they mentioned that increasing the acid concentrations can lead to higher permeability improvement and fewer acid volumes.

In deep or high-temperature reservoirs, mud acid showed poor performance due to the fast reaction rates, leading to consuming the acid volume without achieving deep penetration within the treated formations. Also, using mud acid can lead to several problems such as corrosion and hazardous issues. Moreover, phosphoric (H_3PO_4) and fluoroboric (HBF_4) acids were proposed as effective substitutions for the mud acids. Mixing H_3PO_4 and HBF_4 showed deeper penetrations compared to the mud acid (Al-Harthy et al., 2008). At downhole conditions, HBF_4 can hydrolyze and generate HF while H_3PO_4 can stabilize the rock minerals to avoid participation. Consequently, the H_3PO_4/HBF_4 acid system leads to higher permeability and fewer corrosion issues compared to mud acid (Van Hong and Mahmud, 2017).

In addition, chelating agents have been introduced as effective fluids to stimulate sandstone formations (Leong et al., 2018; Leong and Ben Mahmud, 2019; Mahmoud et al., 2011, 2015; Shafiq et al., 2018). The common types of chelating agents that can be used for sandstone acidizing are ethylenediaminetetraacetic acid (EDTA), hydroxyethylethylenediaminetriacetic acid

(HEDTA), ethanoldiglycine acid (EDG), nitrilotriacetic Acid (NTA), and glutamic acid *N,N*-diacetic acid (GLDA) (Hassan et al., 2020). Those chemicals are soluble in water resulting in stable solutions at different pH values. Injecting chelating agents' solution into sandstone rocks leads to considerable dissolution in the sandstone matrix, resulting in high rock permeability. Ali et al. (2008) investigated the use of sodium hydroxyethylethylenediaminetriacetic acid (Na_3HEDTA) as sandstone acidizing fluid for high-temperature wells. The used acid showed high performance and the formation permeability was significantly increased. LePage et al. (2011) evaluated the performance of different types of chelating agents such as GLDA, EDTA, NTA, EDG, and HEDTA. Among all tested chemicals, GLDA showed the best acidizing performance where less volume of acid was used without inducing corrosion problems.

Mahmoud et al. (2011) studied the effectiveness of GLDA solutions in stimulating sandstone rocks. The impact of several variables such as acid volume, injection rate, system temperature, and original acid pH on the treatment performance was examined. They concluded that GLDA is very stable at temperatures up to 300°F and can effectively increase the sandstone permeability. The main mechanism for increasing the permeability is the sequestration of calcium, magnesium, and iron from the rock surface. Nasr-El-Din (2013) reported that GLDA has less corrosion affinity and high stability in high-temperature conditions. Also, increasing the temperature showed higher increases in the rock permeability during GLDA stimulation. The permeability improvement was increased from 21% to 84% when the system temperature was raised from 200 to 300°F, without inducing any minerals precipitation.

Retarded and organic acids were used to stimulate the sandstone formations (Alhamad et al., 2020; Farley et al., 1970; Shafiq and Mahmud, 2017). Initially, a mixture of organic acids and HF acids were utilized to increase the sandstone permeability (Al-Harbi et al., 2012). Different types of organic acids can be used including formic (HCOOH), acetic (CH_3COOH), and citric ($C_6H_8O_7$). The treatment performance depends mostly on the organic acid type, solution pH, and acid concentrations (Al-Harbi et al., 2012). These chemicals showed minimum precipitations. However, the presence of a high percentage of aluminum-based minerals in the rock can lead to considerable precipitations (Al-Harbi et al., 2012; Yang et al., 2012).

Zhou and Nasr-El-Din (2013) compared the performance of GLDA and HEDTA with the efficiency of HF-phosphonic acid in stimulating sandstone formations, at a temperature of 300°F. The mixture of HF showed higher permeability improvements compared to HEDTA and GLDA. However, HEDTA and GLDA showed better performance compared to HF-based fluid in removing the carbonate minerals from the sandstone rocks. Moreover, Shafiq et al. (2015) and Shafiq and Mahmud (2016) proposed new acid systems that can be replaced the conventional mud acid. Mixtures of fluoboric (HBF_4) acid-HCOOH, orthophosphoric (H_3PO_4) acid-HF, and HCOOH-HF were used. The impact of the acid system on permeability, porosity, and mineral dissolution was studied. Among all tested acid systems, H_3PO_4-HF showed the best acidizing performance. The proposed acid system showed permeability improvement of more than 130% while using conventional mud acid (HF/HCl) increased the permeability by around 100%, under the same conditions. However, the major limitation is that all experimental measurements were conducted at room temperature and pressure (Shafiq and Mahmud (2016)).

Furthermore, Zhou and Nasr-El-Din (2016) studied the effectiveness of a phosphonic-based HF fluid system at a temperature of 300°F. The impact of system temperature, acid concentration, and treatment time on the treatment efficiency was determined. The phosphonic-based HF acid system showed permeability improvement of higher than 177% while around 130% permeability enhancement was achieved using the conventional mud acid, under the same conditions. Overall, a wide range of chemical mixtures was proposed to replace the conventional mud acid. Organic and retarded acids showed considerable enhancement in the sandstone permeability. The main advantage of using organic and retarded acids is that these chemicals have fewer correction issues and they can penetrate deeper within the treated formations leading to better acidizing performance (Alhamad et al., 2020; Shafiq and Mahmud, 2017).

Recently, Mahmoud et al. (2020) proposed a new retarded fluid system by using Gemini surfactants. The proposed acid system can stabilize the reactive minerals and avoid clay damage, by adsorbing on the clay surfaces and minimizing the chemical reactions between the injected acids and clay minerals. Also, the developed acid system improves the reaction between the mud acid and the silicate minerals, leading to deeper penetrations and higher permeability improvements. The new acid system was evaluated using clay-rich sandstone rocks. Moreover, numerical techniques were utilized to assess the reaction between mud acid and sandstone minerals. The permeability improvement and acid penetration were estimated for core and field scales using a calibrated numerical-based model. Injecting 0.5 wt.% of the new system as pre-flush stage showed considerable impacts on stabilizing the reactive clay minerals as indicated by the pressure drop profiles and ions concentration analysis. Overall, the proposed treatment can improve the permeability enhancements by reducing the minerals' surface area by more than fivefolds.

In summary, each acid system has its advantages and limitations. Mud acid is the most common fluid used for sandstone acidizing; however, the corrosion and fast reaction rate are the main limitations of mud acid, especially in high-temperature reservoirs. At temperatures above 200°F, retarded and organic acids showed minimum corrosion issues, but they provided lower permeability enhancements compared to mud acid. The selection of the sandstone acidizing system depends mainly on the required permeability improvements and the treatment conditions. Table 4.2 lists the advantages and disadvantages of several acid systems that are commonly used in sandstone acidizing (Van Hong and Mahmud, 2017).

TABLE 4.2 Advantages and disadvantages of the common acid systems

Acid system	Formula	Advantages	Disadvantages	References
Conventional mud acid	HF and HCl	– HF dissolves almost all sandstone minerals – HCl minimizes the minerals precipitation – Different chemical concentrations can be used	– Fast reaction rates at a high temperature above 200°F – Acid consumed faster, at high temperature – Very corrosive and hazardous	Al-Harthy et al. (2008), Gomaa et al. (2013)
Chelating agents	GLDA EDTA HEDTA NTA HEIDA HTDP APCA	– Stable at temperature up to 300°F. – Minimum corrosion and hazardous – Effective performance in clay-rich rocks	– Less performance in cleaning uniform sandstone – Expensive compared to common acids	Ali et al. (2008), Parkinson et al. (2010), LePage et al. (2011), Mahmoud et al. (2011), Hassan et al. (2020), Nasr-El-Din (2013)
Retarded and organic acids	HCOOH CH_3COOH HBF_4 H_3PO_4 $C_6H_8O_7$	– Deeper penetration within the formation – Less corrosion – Slow reaction rate – Stable at different temperatures	– Chemical reaction not well cleared – Less investigation at high temperatures – Fewer field applications	Al-Harbi et al. (2012), Shafiq and Mahmud (2016), Yang et al. (2012), Zhou and Nasr-El-Din (2013), Zhou and Nasr-El-Din (2016)

4.3 Compatibility of the stimulation fluid ingredients

Sandstone stimulation is conducted through several stages. The first stage is called pre-flush, the second stage is called the main flush, and the third stage is called the post-flush. These stages can be repeated when required to reduce the skin damage in the near-wellbore area and improve the well productivity based on the real-time monitoring of the stimulation process. Table 4.3 shows the different stages of the sandstone acidizing and the fluids used in each stage.

The main flush consists of different fluids such as HF acid and another acid to maintain the reaction product soluble such as HCl, organic acid, or chelating agents. Retarders may be added to this stage to have deep penetration of the acid system and remove deeper damage. A low concentration of HF is compatible with other acids (Mahmoud et al., 2015). The concentration of HF should be designed based on the clay and feldspar content in addition to the type of the other acid. For example, if the acid is HCl, up to 3 wt.% HF can be used, and if the acid was chelating-based such as GLDA, an HF concentration of 1 wt.% is recommended to avoid fluid computability issues. Higher content of clays and feldspars requires low HF concentrations as well to prevent the subsequent formation damage (Abdelmoneim and Nasr-El-Din, 2015).

4.4 Acid/sandstone interactions during sandstone acidizing

Sandstone formations consist of different types of minerals such as clays, feldspars, carbonate, sulfides, sulfates, etc. Clay minerals such as smectite, illite, chlorite, and kaolinite may exist in sandstone rocks in different quantities. The composition of the three common clay types in sandstone reservoirs is listed in Table 4.4 (Al-Zain et al., 2009; Thomas et al., 2002).

The surface charge of clay minerals is considered the most critical parameter that affects the fluid/rock interactions during sandstone acidizing. The pH of the stimulation fluid will affect the value of the surface charge. Rock minerals' surface charge in a specific solution can be evaluated using zeta potential. In an acidic environment, chlorite clay minerals will possess a negative surface charge. Chlorite clay minerals in an iron-rich mineral and their surface charge are mainly controlled by the abundance of ferrous ions. Compared to illite and kaolinite, chlorite has complex behavior and complex surface charge. Chlorite will have a negative surface charge due to silicon and aluminum ions, a positive charge on the surface of the magnesium hydroxide group, and a positive charge at low pH at the edge of silicon and magnesium hydroxide.

Iron-rich minerals exist in sandstone reservoirs in different quantities and will have a major impact on the fluid–rock interactions. Iron is the most

TABLE 4.3 Sandstone acidizing stages

Stage	Fluids	Function	Comments	References
Pre-flush	– Brines such as NaCl, KCl, $CaCl_2$, NH_4Cl, and clay stabilizers – HCl	– To stabilize clays – Remove carbonate minerals	The type of salt should be selected carefully to prevent subsequent interactions with the following stages	Hill et al. (1994), Gdanski and Shuchart (1998), Mahmoud and Gomaa (2022)
Main flush	Mud acid (HCl/HF) HF/organic acids (formic, acetic, citric) HF/chelating agents (GLDA, HEDTA) In situ generated HF acid Retarded acid system Single-stage acid system	To remove silica and silicate damage HF concentration depends on the clay and feldspar content in the rock	This stage is very critical and may cause severe damage if not properly designed Compatibility between the fluids should be tested and with the formation minerals as well	Shafiq and Mahmud (2017), Leong et al. (2018)
Post-flush	– Brines such as NaCl, KCl, $CaCl_2$, NH_4Cl, and clay stabilizers – HCl	To remove the damage caused by the main flush and stabilize the clays		Gdanski and Shuchart (1998), Leong and Ben Mahmud (2019), Rozo et al. (2007)

TABLE 4.4 Chemical composition of clays

Clay type	Chemical composition	Surface area, m^2/g
Kaolinite	Al_2O_3. $2SiO_2$. $2H_2O$	20
Chlorite	$(Mg, Al, Fe^{2+}, Fe^{3+})_{12}[(Si, Al)_8O_{20}](OH)_{16}$	100
Illite	$K_{1-1.5}Al_4[Si_{6.5-7}Al_{1-1.5}O_{20}](OH)_4$	100

TABLE 4.5 Mineral composition of actual sandstone reservoirs

	Concentration, wt.%		
Mineral	*Rock A*	*Rock B*	*Rock C*
Chlorite	0–2	0–3.7	0–0.3
Ankerite	0–7	0–8	—
Siderite	—	0–0.5	0–2.3
Illite	0–13	0–3.8	0–7.2
Quartz	68–97	68–99	89.4–99
Feldspar	0–19	—	0–0.6
Kaolinite	0–8	—	0.1–2.2
Calcite	0–1	—	—
Anhydrite	0–4	—	0–2.9

impacting element when it comes to the surface charge evaluation (Bailey, 1986; Caenn et al., 2011; Ezzat, 1990; Greenwood et al., 2007; Jones, 1981; Nasser and James, 2006; Welton, 1984; Wilson, 2013). Table 4.5 shows the mineral composition in different sandstone reservoir rocks. Iron-containing minerals are highlighted such as chlorite, ankerite, siderite, and illite. Table 4.6 shows the surface charge based on the pH value for different clay minerals.

Clay minerals are very fine grains and are characterized by a particle size of fewer than 5 μm. This very low particle size gives a larger surface area of the particles as shown in Table 4.4. The main source of damage in sandstone reservoirs can be attributed to the interactions between clay minerals and stimulation fluids. Table 4.7 shows the different damage potential due to the interactions of fluids with clay minerals during sandstone stimulation.

TABLE 4.6 Surface charge of different clays

Clay mineral	Surface charge	pH-dependent charge	References
Chlorite	Permanent negative charge	The positive charge at low pH The negative charge at high pH Contains Mg^{2+}, Fe^{2+}, Fe^{3+} (the main cause of oil wetness)	Nasser and James (2006) Denekas and Mattax (1959)
Kaolinite	Permanent negative charge	Net positive surface charge at the edges	Michaels and Bolger (1962) Olphen (1963)
Illite	Mixed surface charge	Net negative surface charge at moderate and high pH	Mahmoud (2018)

In addition to clay minerals, other minerals such as calcite, dolomite, feldspars, iron oxides, and iron carbonate can introduce severe damage to sandstone reservoirs during the stimulation operations. Mud acids (a combination of HCl and HF acids at different concentrations) have different and complex reactions with different minerals in the sandstone rocks. Table 4.8 shows the different reactions and potential damage sources during these reactions with different sandstone minerals.

HF acid in the mud acid dissolves quartz, feldspars, clays, and micas. The primary objective of sandstone acidizing is to remove silica and silicate damage; therefore, the mud acid should be retarded to avoid its consumption by unnecessary reactions. The presence of carbonate minerals will consume the HF acid and produce hard calcium fluoride that will cause permanent damage; it is necessary to pre-flush the formation with carbonate-dissolving acid before the sandstone stimulation process. All clay minerals have very high solubility in mud acid compared to quartz because of the larger surface area. Calcite and dolomite have very high reactivity with mud acid and will produce calcium and magnesium fluoride precipitate. Feldspars have moderate solubility in mud acid (Allen and Roberts, 1993).

Several researchers recommended replacing HF acid with other acids in case the formation has other reactive minerals with HCl, organic acids, or chelating agents. HCl and chelating agents can be used to stimulate sandstone reservoirs with high calcite content. Table 4.9 shows the recommended alternatives for mud acid to stimulate sandstone formations.

TABLE 4.7 Different clay damage types and potential during sandstone acidizing

Source of damage	Damage type	Damage severity	Damaging fluids	References
Illite	Clay swelling Fines migration due to dispersion Production of amorphous silica gel due to reaction with HCl Attack of the alumina layer	Worst damage type Can plug the pore throats and reduce permeability Very difficult to remove and stabilize	HCl can cause fines migration Drilling fluid invasion may cause swelling Mud acid Recommendation to use low concentrated HCl or organic acids At very high temperatures, chelating agents only can be used	Amaefule et al. (1988), Mahmoud et al. (2015), Simon and Anderson (1990)
Kaolinite	Swelling	Permanent damage and plug to the pore space and pore throat	Fresh water-based fluids	Mahmoud et al. (2015)
Chlorite	Production of amorphous silica gel due to reaction with HCl	Severe damage, very difficult to remove	HCl can cause fines migration Drilling fluid invasion may cause swelling Mud acid Recommendation to use low concentrated HCl or organic acids At very high temperatures, chelating agents only can be used	Thomas et al. (2001), Simon and Anderson (1990)

TABLE 4.8 Fluid/clay minerals interactions during sandstone acidizing

Reacting fluid	Reacting minerals	Damage type	Damaging severity	References
HF	Quartz Alumina silicates (feldspars, clays, mica) Carbonate	Amorphous silica gel Silica gel Calcium fluoride	Severe damage difficult to remove Calcium fluoride precipitation can be avoided by pre-flushing the rock with HCl to remove the calcium source before contacting HF acid	Gdanski (1997), Hill et al. (1981), Fogler and McCune (1976), Li et al. (1998)
H_2SiF_6 (Fluosilicic acid)	Minerals that have sources of calcium, potassium, and sodium	Sodium fluosilicate (Na_2SiF_6) Potassium fluosilicate (K_2SiF_6) Calcium fluosilicate ($CaSiF_6$)	Solid precipitation that will cause permanent damage The solution is to prevent the contact between the HF acid and the Ca-, K-, Na-containing minerals such as using ammonium chlorite as a spacer	Gdanski (1997), Mahmoud et al. (2015)
HCl	Clays Iron-rich minerals Carbonates	Break clays cause fine migration Reacts with chlorite to produce residue silica gel Sand production	Iron-based damage is very difficult to remove	Mahmoud et al. (2015) Mahmoud (2017)
Organic acids	Iron-rich minerals and carbonates	Iron hydroxides Organic acid salts	Iron-based damage is very difficult to remove	Mahmoud et al. (2015)
Chelating agents	Iron-rich minerals Carbonate Chelate multivalent cations from all minerals	Gentle to clays and other minerals	Low possibility of damage	Mahmoud et al. (2011a), Mahmoud et al. (2011b)

TABLE 4.9 Mud acid alternatives for sandstone stimulation

Alternative	Pros	Cons histories	Case histories	References
HCl	Remove carbonate	May cause sand production Cannot be used in case clay content is high	High illite content sandstone with 10% carbonate was stimulated with 15 wt.% HCl HCl caused clay breakage and subsequent fines migration	Mahmoud et al. (2015)
Organic acids	Remove carbonate Chelate irons	Weak reactants at low temperatures	Berea sandstone cores were stimulated with several organic acids and showed good permeability improvement	Mahmoud et al. (2011c)
Chelating agents	Remove carbonate Chelate irons Enhanced the permeability	Can be used only at high pH to avoid damage Slow reaction	Berea sandstone cores treated with HEDTA GLDA chelating agent was used to stimulate different sandstone rocks EDTA used to stimulate Berea sandstone	Parkinson et al. (2010) Ali et al. (2002) Mahmoud et al. (2015) Gdanski and Shuchart (1998)

4.5 Adsorption and retention of stimulation fluids (ingredients) in the formation rocks

Injection of acids in the sandstone formations can lead to different types of rock–fluids interactions. The injected acids can dissolve part of the sandstone matrix, but the acids can be adsorbed on the rock surfaces leading to alter the rock properties. Also, fluid retention can be induced during the sandstone acidizing (Curbelo et al., 2007). Acid adsorption on the rock surface will result in reducing the acid efficiency; shallow acid penetration will be achieved due to the consumption of the injected acids. In some cases, the adsorption of acid on the sandstone surface can lead to different changes in the flow conditions by altering the rock surface wettability (Amirmoshiri et al., 2020). NMR, CT scan, SEM, and AFM techniques can be employed to study acid adsorption and retention (Rezaeizadeh et al., 2021).

Chemical retention is usually used to describe the reduction of acid concentration from the injected fluids. During chemical injection, fluid retention can lead to acid adsorption, phase trapping, and hydrodynamic retention (Amirmoshiri et al., 2020). Acid trapping usually refers to the mechanical entrapment of the injected chemicals where the large molecules will be trapped within the small pores or inside the narrow flow channels. Hydrodynamic retention indicates that some of the acid molecules are trapped in stagnant flow regions. All types of acid retention would reduce the acidizing efficiency by consuming the acid or inducing some precipitation. Also, the trapped fluids within the sandstone formations will lead to blocking the hydrocarbon flow. During well treatment, the injected acids can be trapped around the wellbore, especially in tight formations characterized by high capillary pressure. The trapped acids block will reduce the hydrocarbon production significantly by restricting the flow from the reservoir to the wellbore (Bennion, 2002).

Several mechanisms can contribute to chemical adsorption such as ion exchange and electrostatic interactions (Zhang and Somasundaran, 2006). The rock composition plays a significant role in the acid–rock interactions. The presence of divalent ions such as calcium and magnesium can lead to minerals precipitation due to chemical retention (Mannhardt et al., 1993, 1994). The precipitated minerals reduce the pore throat and decrease the formation permeability. Also, the rock surface charges showed conservable impacts on the acid–rock interactions. Injection of acids that have similar charges as the rock surface would result in repulsive forces between the injected fluids and rock surfaces. However, attraction forces could be induced when the rock and the injected fluids have different surface charges. Attraction forces would lead to acid adsorption on the rock surface, thereby less acid performance (Mannhardt et al., 1993). The amount of the adsorbed chemicals on the rock surfaces depends on the chemical type, structure as well as rock type, and mineralogy. Also, parameters such as solution pH and system temperatures can affect chemical adsorption. In general, increasing the temperature or pH can lead to decreasing the chemical adsorption, due to the fast reaction between the injected acids and rock surfaces (Zhang et al., 2016).

In situations where rock surface and surfactants are charged, electrostatic interactions dominate the adsorption mechanism. Sandstone reservoirs are typically composed of silica, feldspar, and a substantial amount of clay minerals (usually illite, kaolinite, chlorite, and montmorillonite), carbonates that serve as cementing materials (with divalent ions such as Ca^{2+}, Mg^{2+}, Fe^{2+}, and Mn^{2+}), and iron oxides such as siderite ($FeCO_3$) (Amirmoshiri et al., 2020; Levitt et al., 2015). Since silica is negatively charged over a wide range of pH (greater than 1.7–3.5), sandstone reservoirs are generally considered negatively charged (Kosmulski, 1998). Thus, using chemicals of negative charges is preferentially employed for sandstone acidizing (Mannhardt et al., 1993). However, the adsorption of several chemicals on sandstone was reported due to the presence of clay (Iglauer et al., 2010); therefore, in such cases, the adsorption will depend on the mode of spread of the clay mineral in the formation (Ma et al., 2013).

The chemical solubility has an important impact on the rock–fluid interactions; using less soluble chemicals can induce different types of precipitations. Almost all acids used during sandstone acidizing are soluble; however, the presence of formation brine within the treated formations may interrupt the chemical stability. Reducing the chemical solubility can result in deposition or fluid trapping due to incompatibility issues. Therefore, the pre-flush stage is very important to ensure that the treated formation is cleaned and ready for acid injections. The pre-flush fluids should be precisely designed to stabilize the reactive minerals and minimize acid adsorptions. The selection of pre-flush fluids depends on the rock mineralogy and the type of acid system that will be used during the sandstone acidizing.

Overall, chemical adsorption or retention can lead to severe types of formation damages due to the reduction of pore throat or the trapping of injected acids. The rock mineralogy and surface charges are the main parameters that affect acid adsorption. Carbonate rocks have a higher possibility of inducing chemical adsorption or retention compared to sandstone rocks (Amirmoshiri et al., 2020). The formation damages due to acid retention or adsorption can be minimized by the proper design of the acid fluid system.

4.6 Fines migration and clay swelling during sandstone acidizing

Fine migration and clay swelling are the common issues during sandstone acidizing (AlMubarak et al., 2015; Gdanski and Shuchart, 1998). Injection of fluids into sandstone formations can lead to detachments of several minerals such as clay, feldspar, and amorphous silica (Khilar and Fogler, 1998; Zheng et al., 2018). The detachments of fine minerals from the rock surface depend on several factors such as the rock morphology, mineralogical composition, and the injected chemicals (You et al., 2015). The detached fines can migrate within the rock porosity system due to the pressure gradient; higher pressure drop is usually associated with severe fine migrations. Pore throat plugging and production reduction are the main problems induced by fine migrations (Zheng et al., 2018; Zou et al., 2014).

Sandstone acidizing can induce different types of formation damage if the acid system is not well selected. Releasing of fine particles is one of the main issues during sandstone acidizing. The released particles can precipitate and plug the pore throat, leading to significant reductions in the formation permeability (Dehgani, 2010; O'Driscoll et al., 2005; Rabie and Nasr-El-Din, 2015). The interactions between clay minerals and the injected acid are reported to be the main cause of fine migration (Hanafy and Nasr-El-Din, 2016; Hanafy et al., 2015). Several parameters can increase the possibility of fine migrations; the common parameters are acid type, concentration, and formation temperature. Increasing any of these parameters will lead to an increase in the severity of fine migration. Injection of hydrochloric acid at high-temperature conditions (150–250 °F) showed a severe reduction in the sandstone permeability due to fine migration issues. Several additives are used during sandstone acidizing to stabilize the reactive minerals and minimize the problems of fine migrations (Bybee, 2003; Leong and Ben Mahmud, 2019). Table 4.10 summarizes the common additives and their main objectives during sandstone acidizing.

TABLE 4.10 The common additives during sandstone acidizing treatments

Acid additives	Functions
Clay stabilizer	To keep clay and fines in suspension To prevent clay migration and swelling
Corrosion inhibitors	To reduce the corrosion rate To create a thin film on the metal surface To protect downhole and surface equipment from corrosion
Diverting agents	To place the reactive fluid evenly To improve the performance in multilayers
Iron control agents	To dissolve the corrosion products To minimize iron predication
Surfactants	To reduce the surface and interfacial tensions To improve the wettability conditions To minimize emulsion blockages

Clay detachment and migration can be affected by the rock wettability conditions. Inducing water wet status can lead to interrupting the stability of some clay minerals such as illite, leading to illite migrations. While kaolinite is the most sensitive clay type, inducing oil-wet or water-wet surfaces can result in kaolinite movements (Kudrashou and Nasr-El-Din, 2019). The kaolinite edges showed different electrical charges depending on the solution pH. Negative surface charges were reported for the octahedral layer (Al) side of kaolinite at pH 6.5, while other kaolinite edges (such as the tetrahedral layer) showed neutral surface charges (White and Zelazny, 1988). The imbalance in the surface charges of kaolinite edges will increase the possibility of kaolinite migrations (Rueslatten et al., 1994). Overall, altering the rock wettability can cause clay detachments and migrations, leading to severe formation damages (Kudrashou and Nasr-El-Din, 2019).

Kalfayan and Watkins (1990) proposed an acid system to stabilize the clay minerals and minimize clay migration and swelling during sandstone acidizing. The reported approach showed good performance in stabilizing clay as well as no clay minerals such as siliceous fines. Organosilane chemical was added to the conventional mud acid (HCl-HF) to stabilize the reactive sandstone minerals. The acid additive was able to improve the mud acid performance by reducing the severe dissolution of sandstone minerals, which led to higher permeability and less clay damage. The proposed acid system was tested on a lab and field scale, where reasonable acidizing efficiency was obtained.

Rozo et al. (2007) proposed a nonacid-based system to minimize the fine migrations. The proposed fluid system was prepared using chelating agents and different additives. The fluid system showed excellent compatibility with sandstone minerals, leading to minimum fine migrations. They mentioned that the proposed acid system can be implemented as a single stage, minimizing the treatment time and cost. Also, demulsifiers or surfactants were needed as the fluid system showed low corrosion issues as well as no acid trapping. Extensive laboratory measurements were carried out to assess the efficiency of the chelating agent-based system, and successful field implementations were reported.

Hanafy et al. (2015) assessed the influence of fines migration, during sandstone acidizing, on the rock petrophysical properties. Core flooding experiments were carried out using Bandera rock samples that have a wide range of clay percentages. Deionized water was injected into the rock samples to induce fine migration. X-ray diffraction (XRD) and computational tomography (CT) were used to capture the changes in rock samples due to fine migrations. The core permeability and porosity were reduced significantly due to the fine migration. Increasing the temperature showed high permeability impairments due to the increase in reaction rates between the injected acid (HCl) and sandstone rocks. Moreover, Hanafy and Nasr-El-Din (2016) used the NMR technique to evaluate the generation and removal of fine migration during mud acid injection into sandstone rocks. The pore size was reduced by one order of magnitude due to the fine migrations. Once the fine migration was induced, the mud acid was not able to dissolve the precipitated fines.

Kamal et al. (2019) quantify the formation damage due to clay minerals during sandstone acidizing. The influence of clay type and percentages on the sandstone damage was examined utilizing SEM and NMR techniques. The changes in clay distribution within the treated rocks due to HCl injection were determined using SEM. Fine migration and clay swelling led to severe formation damage during HCl injection. Among all clay types, chlorite and illite have shown the highest contribution to clay damage. Illite clay was broken down into fine particles and then migrates within the porosity system, while chlorite mineral led to precipitation of iron hydroxide. Consequently, the sandstone permeability was 84% due to pore throat plugging induced by clay swelling and fine migrations.

The fine migration can be mitigated using several approaches such as oil coating, surface charge alterations, acidification, and stabilizers injections (Byrne and Waggoner, 2009; Hibbeler et al., 2003; Huang et al., 2002; Rozo et al., 2007). Organo-silane additives are the most common clay stabilizer that can effectively prevent clay detachments from rock surfaces. However, it showed poor efficiency in controlling the fine movements, once they are already free or detached from rock surfaces. Also, organo-silane was not able to stabilize nonclay minerals such as silicates (Kalfayan and Watkins, 1990; Stanley et al., 1995). Mitigation approaches should be applied before conducting sandstone acidizing because it is very challenging to remove the induced clay damage (Ahmadi et al., 2013; El-Monier and Nasr-El-Din, 2011; Hibbeler et al., 2003). Therefore, it is very important to inject clay stabilizers during the pre-flush stage. Coating the clay mineral using nanoparticles is one of the effective approaches to minimize the fine migrations (Ahmadi et al., 2013; Zheng et al., 2018). Nanoparticles can effectively prevent fine migration as well as improve the flow conditions by creating favorable wettability situations (Ahmadi et al., 2013; Zheng et al., 2018).

In summary, clay swelling and fine migration can lead to severe formation damages due to the pore throat plugging and fine precipitations. Injection of strong acids can result in mineral detachments, thereby increasing the possibility of fine migrations. Clay stabilizers should be injected before conducting the acidizing treatments to minimize the clay damage. NMR and imaging techniques can be utilized to capture the fine migrations and clay swelling. Nanoparticles showed effective performance in preventing the formation damages by coating the reactive minerals and avoiding clay swelling or fine migrations.

4.7 Effect of corrosion inhibitors on sandstone wettability

Corrosion inhibitors are commonly utilized for sandstone acidizing to avoid acid-related corrosion on metal surfaces, downhole completion, and tubing. Corrosion inhibitors (CI) primarily work by adsorbing and making a barrier against acid attacks on the metal surface. Acid corrosion inhibitors are a mixture of several chemicals to achieve effective performance. A typical corrosion inhibitor formulation consists of surfactant to wet the metallic surfaces, disperse the inhibitor in the acid and solvents to enhance the stability, and solubility, and improve the viscosity. Intensifiers are also part of acid corrosion inhibitors formulation for high-temperature applications. In the acid stimulation treatments of underground formation with temperatures higher than 80°C, the protective film formed by a typical alcohol-amine-based CI becomes thermally sensitive and becomes less and less effective to protect the metal from corrosion. In these cases, most formulations of corrosion inhibitors required the addition of a corrosion inhibitor intensifier, to boost its usefulness and to keep corrosion rates below the specified safety levels (less than 24.4 mg/cm^2/operation time). The major criteria for any corrosion inhibitor are its ability to contain acid-related corrosion and cost-effectiveness. However, for oil and gas applications such as sandstone acidizing, compatibility with other materials, high-temperature stability, pressure stability, cleanup, and potential formation damage are other important criteria.

Formation damage may occur due to the insoluble materials in the inhibitor formulation, and adsorption of inhibitor on clays and other minerals present in the matrix of the rock. The acid-insoluble materials could result in surface plugging of the formation face. Corrosion inhibitors have strong interactions with other acidizing additives which can cause severe operational issues and alter the performance requirement. For example, corrosion inhibitors can interact with viscoelastic surfactants and can impact the viscosity of the solution (Li et al., 2012; Nasr-El-Din et al., 2008). Acrylamide-based polymeric viscosifier and gel systems can be severely affected by corrosion inhibitors (Cassidy et al., 2011). The interactions of corrosion inhibitors can result in premature or delayed cross-linking which can impact the diversion efficiency. Corrosion inhibitors interacting with the acid could result in a change in the dissolution rate. However, the use of mutual solvents can minimize this effect. Corrosion inhibitors can also change the physical properties of the acids. In field operations, a major fraction of the inhibitor is lost due to adsorption in the formation (Dabbousi et al., 1999). Adsorption of inhibitors and plugging could influence the well-clean-up rate and effective permeability (Crowe and Minor, 1985). Adsorbed inhibitors can alter the wettability or can result in severe emulsion problems resulting in slow or incomplete clean-up after the treatment.

Wettability alteration is a major problem related to formation damage during sandstone acidizing. The wettability alteration took place due to the invasion of corrosion inhibitors and other additives. The corrosion inhibitors can invade the formation and change the wettability from water-wet to oil-wet. This change in wettability results in the excessive production of water. The change in the wettability is associated with the interactions between rock and corrosion inhibitors. Most of the minerals present in the sandstone rock have a net negative charge. On the other hand, most corrosion inhibitors are positively charged. Cationic corrosion inhibitors having nitrogen compounds strongly adsorb on clay surfaces. These adsorbed materials can have a major influence on the clean-up rate and, in some cases, the long-term capacity of the well. The difference in the charge results in higher adsorption of cationic species on negatively charged mineral surfaces. In addition to corrosion inhibitors, the cationic surfactants present in the corrosion inhibitors have a very strong tendency to adsorb on mineral surfaces (Kamal et al., 2017). To prevent the problem of plugging and adsorption, mutual solvents should be added to the acid to solubilize the damaging inhibitor residues and prevent the formation of plugging during injection of the acid. Also, the selection criteria of corrosion inhibitors should not be only based on their performance to avoid corrosion but also on their formation damage potential.

Change in the wettability also results in enhancing the fine migration. The change in wettability could result in detaching the clay particles and transferring them into the oil phase. Fig. 4.2 illustrates the fine migration toward the oil phase. Initially, the clay particles were stable in the water phase. However, as soon as the wettability changes, the clay particles also move toward the oil phase. This could result in the formation damage due to pore plugging. The fines can also trap a significant portion of corrosion inhibitors due to the adsorption (Dabbousi et al., 1999).

Corrosion inhibitors can also cause sand aggregation due to wettability alteration to more oil-wet. This aggregation could result in the formation of stick deposits. A study conducted by Gulbrandsen et al. suggested that corrosion inhibitors can be adsorbed on sand surface and make it less hydrophilic (Gulbrandsen and Pedersen, 2007). The presence of small particles due to fine migration could result in the formation of Pickering emulsions if the particles are intermediate wet to strongly water wet. The emulsions could be either oil-in-water or water-in-oil depending on the exact wettability.

In summary, corrosion inhibitor formulations consist of several additives such as surfactants, intensifiers, biocides, and solvents to achieve corrosion inhibition. However, fluid–fluid and rock–fluid interactions due to the presence of corrosion inhibitors and other additives could cause wettability alteration and formation damage during sandstone acidizing. Therefore, it is extremely important to understand these interactions to design the corrosion inhibitor formulation with formation damage.

4.8 Modeling of sandstone acidizing

Sandstone acidizing involves a complex reaction between HF/HCl system and sandstone minerals such as quartz, feldspar, mica, clay, carbonate, sulfate, etc.

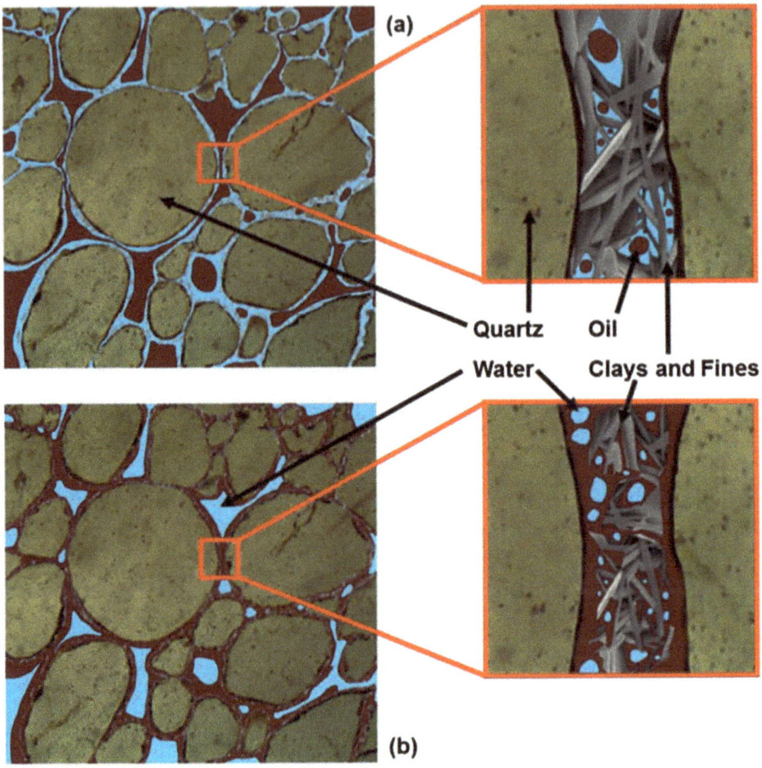

FIG. 4.2 Effect of wettability on formation damage: (A) water wet case; (B) oil-wet case (Dabbousi et al., 1999).

(Economides and Nolte, 2000). The reaction with HF usually contains many steps that could lead to the precipitation of other minerals. Other types of acids such as chelating agents (Shafiq et al., 2017), retarded acids (Aneto, 2012), organic acids (Akanni and Nasr-El-Din, 2016), and fluoroboric (HBF4) have been utilized. The literature contains mainly models that predict the interaction between HF and sandstone and is lacking when it comes to the other acid types (Leong et al., 2018).

Modeling sandstone acidizing usually targets predicting the enhancement/impairment of the formation permeability after stimulation. Fluid flow models assuming Darcy's law are the first building block of the model. They could be linear to represent the acid flow within the core samples, simulating lab outcomes, or radial to simulate field treatments. The second building block represents the acid mass balance to track the concentration distribution. The reaction kinetics between the acid system and rock minerals should be understood in this step. Also, the mineral mass balance is implemented to track the change of minerals as the reaction progresses. Finally, other constitutive equations that relate, for example, porosity to permeability and surface area are utilized. The models in the literature could be divided based on reaction understanding into the lumped-parameter model (Labrid, 1975; Lund et al., 1973; McCune et al., 1975; Schechter and Gidley, 1969; Williams and Whiteley, 1971), two-parameter model (Bartko et al., 1997; Da Motta et al., 1993; Guimares et al., 2017; Hekim et al., 1982; Lea et al., 1992; Perthuis et al., 1989; Rahman, 2000; Taha et al., 1986), four-parameter model (Bryant, 1991; Li et al., 2004; etc.), and the detailed chemistry model (Murtaza et al., 1999; Sevougian et al., 1995; Ying-Hsiao et al., 1998). We will divide the models into lumped and detailed and discussed them below for simplifications.

4.9 Lumped-parameter model

The lumped model considers the reaction between one acid type (i.e., HF) and one mineral. Due to its simplicity, analytical solutions exist for the lumped model and can be tuned with experimental data on sandstone core flooding (Ali et al., 2016). The lumped model predicts the effluent acid concentration and core permeability change. It is a one-dimensional (direction of flow) model that does not have a flow model (i.e., utilizing an average velocity). The reactive transport equation can be written as

$$\phi\frac{\partial(C_A)}{\partial t} + U\frac{\partial(C_A)}{\partial z} + \sum_{j=1}^{N_m} v_j k_{rj} C_A = 0 \tag{4.7}$$

where C_A is the HF acid concentration, U is the average velocity, v is the stoichiometric coefficient, N_m is the number of minerals, k_r is the reaction rate constant of a mineral (j) with HF. The above equation represents a mass balance for the acid and contains three components; accumulation, convection, and reaction. For the sake of simplicity, the reaction is assumed to be first order in this model. Similarly, a mass balance for the solid is used to indicate the change of mineral mass as follows:

$$(1-\phi)\frac{\partial(W_j)}{\partial t} = k_{rj} C_A W_j \tag{4.8}$$

where W_j is the concentration of reacting solids (mol/m^3 of solid). The above equations can be converted into dimensionless form utilizing the dimensionless parameters in Table 4.11 and assuming one reacting mineral. The solutions to dimensionless equations are as follows:

$$\psi = \left(1 + \exp\left(\text{Da.Ac}\left(\left(1+\frac{1}{\text{Ac}}\right)\varepsilon - \Theta\right) - \exp\left(\text{Da.Ac}(\varepsilon - \Theta)\right)\right)\right)^{-1} \tag{4.9}$$

$$\eta = \left(1 + \exp\left(-\text{Da.Ac}\left(\left(1+\frac{1}{\text{Ac}}\right)\varepsilon - \Theta\right) - \exp\left(-\text{Da}\varepsilon\right)\right)\right)^{-1} \tag{4.10}$$

TABLE 4.11 Definition of the dimensionless parameters

Symbol	Definition	Dimensionless parameter
ψ	C_A/C_{Ao}	Acid concentration
η	W/W_o	Mineral concentration
Θ	$\frac{tU}{\phi L}$	Time
ε	x/L	Position
Ac	$\frac{\phi C_{Ao}}{v(1-\phi)(\Delta W)}$	Acid capacity
Da	$\frac{k_f \Delta WL}{U}$	Damköhler number

where ψ represents the dimensionless acid concentration and η is the dimensionless mineral concentration. The solution is a function of two dimensionless parameters: the acid capacity (Ac) and Damköhler numbers (Da). The solution also depends on dimensionless time and position. The acid capacity represents the potential of acid to dissolve a certain volume of minerals within a rock. It is defined as the amount of mineral dissolved by pore volume of acid to the mineral amount in a unit rock volume. The Damköhler number, on the other hand, represents the relation between acid reaction and acid convention. Table 4.11 lists the definitions of the dimensionless parameters. Fig. 4.3 shows the solution results of high and low Damköhler numbers. At a high Da number, the acid moves like a shock front due to its high reactivity and leaves no reactive minerals beyond its front. At low Da, the acid reaction is slow compared to its convection, and hence the acid concentration declines gradually.

Once dissolution is obtained, the change in porosity can be calculated analytically as

$$\frac{\Delta\phi}{\Delta\phi_{\max}} = \frac{\exp\left(-\mathrm{Da}\cdot\mathrm{Ac}\left(1+\frac{1}{\mathrm{Ac}}\varepsilon-\Theta\right)\right)-\exp\left(-\mathrm{DA}\varepsilon\right)}{1+\exp\left(-\mathrm{Da}\cdot\mathrm{Ac}\left(1+\frac{1}{\mathrm{Ac}}\varepsilon-\Theta\right)\right)-\exp\left(-\mathrm{DA}\varepsilon\right)} \tag{4.11}$$

Then, a correlation (Fogler and McCune, 1976) is used to relate the change in porosity to the change in permeability as follows:

$$\frac{k}{k_o} = \exp\left(B\frac{\Delta\phi}{\Delta\phi_{\max}}\right) \tag{4.12}$$

where k is the rock permeability and B is a matching parameter.

Fig. 4.4 shows a typical outcome of the lumped model, which shows the change of rock permeability with acid injection time. Lund and Fogler (1976) used the model to match experimental data, which provided a decent match.

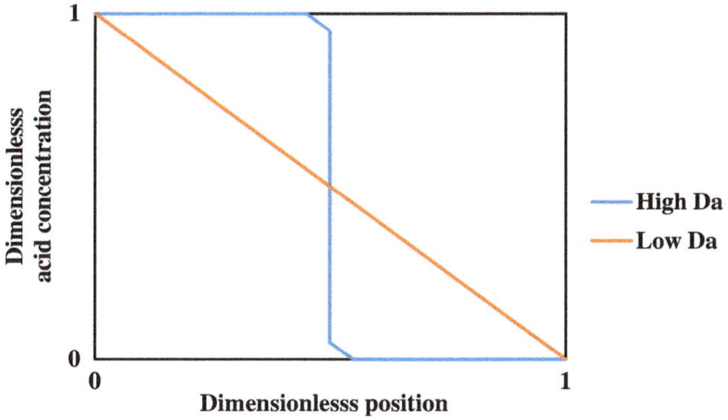

FIG. 4.3 An acid solution as a function of Damköhler number.

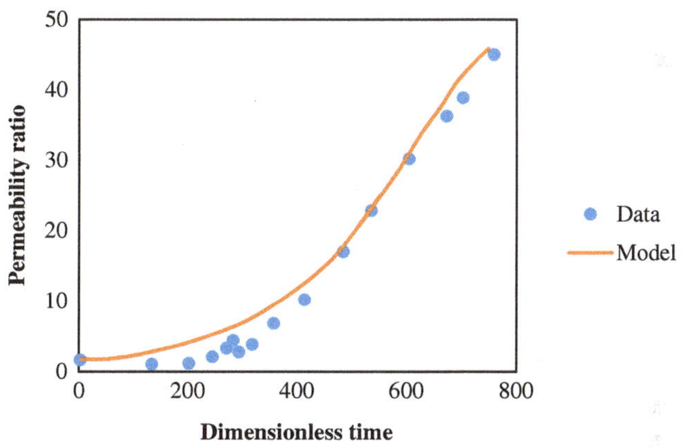

FIG. 4.4 Matching between experimental data and the lumped model. *Modified after Lund, K., Fogler, H.S., 1976. Acidization—V: the prediction of the movement of acid and permeability fronts in sandstone. Chem. Eng. Sci. 31 (5), 381–392.*

4.10 Detailed reaction models

We included the two-parameter, four-parameter, and detailed chemistry models in this category. These models utilize the same algorithm and differ in details of the considered acid and minerals. For instance, the two-parameter model assumes that only mud acid reacts with two types of minerals: fast-reacting minerals such as feldspar and clay and slow-reacting minerals such as quartz. On the other hand, the four-parameter model considers both HF, the primary acid, and H_2SiF_6, a byproduct acid. The two acids react with both fast and slow reacting minerals, hence the name four-parameter model. The detailed chemistry model considers individual acids and minerals' kinetics and thermodynamics that control the dissolution precipitation reactions.

All the above-mentioned models consist of the same algorithm and solution methodology. The acid injection in sandstone formations is a reactive transport process, and hence flow models should be resolved to estimate the fluid velocity and pressure. For fluid flow in a porous medium, the diffusivity equation is implemented to estimate the pressure profile, which is written as

$$\nabla \cdot (k \cdot \nabla p) = \phi \mu_f c_t \frac{\partial p}{\partial t} \tag{4.13}$$

where p is the pressure, k is the permeability tensor, μ_f is the viscosity of the fluid, and c_t is the total compressibility. The equation above can be used either for linear or radial flow by changing the boundary conditions. The pressure profile can be converted into a velocity profile by applying the following equation:

$$u = -\frac{k}{\mu_f} . \nabla p \tag{4.14}$$

where $\boldsymbol{u}$ is the velocity. The second model considers the acid transport with the formation, and hence, the acid concentration profile should be resolved according to the following equation:

$$\frac{\partial(\phi C_i)}{\partial t} + \nabla.(\mathbf{u}C_i) = R(C_i) \tag{4.15}$$

where C_i is the acid concentration, $R(C_i)$ is the reaction rate of the acid with minerals, and i represents the acid type. The equation is a convective-reactive equation where the first term represents the accumulation of acid, the second term is the convention of acid, and the last term is the reaction. The reaction term can be written as

$$R(C_j) = \rho_{ss} \sum_{j=1}^{N_m} v_{i,j} k_{ri,j} C_i^{n_{i,j}} \left(1 + K_j C_{HCl}{}^{ej}\right) MF_j \alpha S_j (1 - \phi) \tag{4.16}$$

where ρ_{ss} is the density of sandstone rock, n is the exponent of reaction, MF is the mass fraction of a mineral, S is the minerals specific surface area, and α is an empirical parameter for tuning the surface area. Note that HCl acid does not react with sandstone minerals but acts as a catalyst for the reaction between HF and feldspars. In Eq. (4.16), C_{HCl} is the HCl concentration, K is the reaction rate constant, and ε is the exponent of reaction. Notice that the subscript i represents the acid type (i.e., HF and H_2SiF_6) and j represents the mineral type.

Detailed chemistry models should include equilibrium calculation which requires a database of species thermodynamic data such as enthalpy, Gibbs free energy, heat capacity, solubility, etc. For example, according to Ying-Hsiao et al. (1998), the three groups of species that require equilibrium calculations are aluminum fluorides, aluminum hydroxides, and silica fluorides.

Tracking the change of mineralogy and porosity is an essential component of a sandstone acidizing model. The following equation can be used to estimate the change in mineral fraction:

$$\frac{\partial(\mathrm{MF}_j)}{\partial t} = \sum_{i=1}^{N_a} \mathrm{MW}_i v_{i,j} k_{ri,j} C_i^{ni,j} \left(1 + K_j C_{\mathrm{HCl}}{}^{\varepsilon j}\right) \mathrm{MF}_j \alpha S_j \beta_{i,j} \tag{4.17}$$

where N_a is the number of reacting acids, MW is the acid molecular weight, and β is the acid dissolving power of a certain mineral. Also, the porosity change can be estimated through the following:

$$\frac{\partial(\phi)}{\partial t} = (1-\phi)\rho_{\mathrm{ss}} \sum_{j=1}^{N_m} \sum_{i=1}^{N_a} k_{ri,j} C_i^{ni,j} \left(1 + K_j C_{\mathrm{HCl}}{}^{\varepsilon j}\right) \mathrm{MF}_j \alpha S_j \tag{4.18}$$

Usually, constitutive equations are used to update the petrophysical properties such as permeability and specific area as a function of the porosity. For instance, Mahmoud et al. (2020) utilized the following correlations:

$$k = k_{\mathrm{o}} \left[\left(\frac{\phi}{\phi_{\mathrm{o}}}\right)^3 \left(\frac{1-\phi_{\mathrm{o}}}{1-\phi}\right)^2 \right]^{\gamma} \tag{4.19}$$

where the subscript o refers to the initial property and γ is a tuning parameter. Similarly, another correlation can be used for the specific area such as

$$S = S_{\mathrm{o}} \left[\left(\frac{\phi}{\phi_{\mathrm{o}}}\right) \left(\frac{1-\phi_{\mathrm{o}}}{1-\phi}\right) \right]^{\eta} \tag{4.20}$$

where η is an empirical parameter. Notice that both Eq. 4.13 and 4.15 are partial differential equations and they need both require initial and boundary conditions to be numerically solved. These boundary conditions depend on the scale of simulation (i.e., core or filed) and the flow pattern (radial or linear). The algorithm of a typical sandstone model follows the flow shown in Fig. 4.5. First, a heterogeneous domain in terms of mineralogy, porosity, and permeability should be populated. Then, at each time step, the fluid

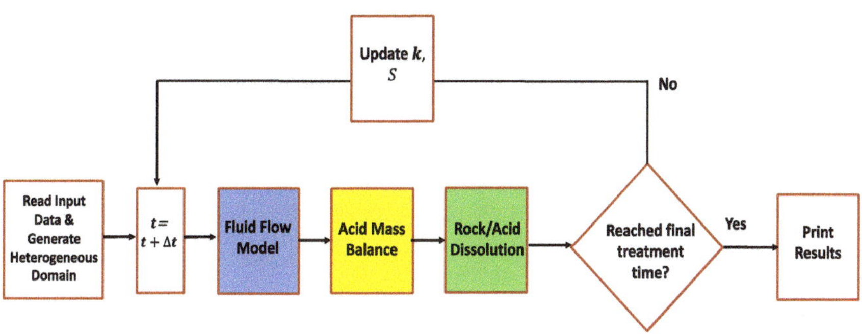

FIG. 4.5 Model algorithm of sandstone stimulation.

velocity and pressure are resolved (see Eqs. 4.13 and 4.14). That enables solving the concentration distribution of acids in the domain (see Eqs. 4.15 and 4.16). Then, the rock/acid dissolution can be estimated using Eqs. 4.17 and 4.18, where both mass fraction and porosity are calculated. Finally, the permeability and other petrophysical properties should be updated using empirical correlations as shown in equations.

More detailed models might include the thermodynamics of dissolution and precipitation of different scales and minerals. Such a model was applied by Mahmoud et al. (2020) to predict the concentration of mud acid when injected into Scioto sandstone core sample containing quartz, illite, plagioclase, chlorite, kaolinite, and potassium feldspar. Fig. 4.6 shows that the model could match the experimental outcome by modifying the tuning parameter. Once the model could match the experimental data, it could be used to estimate the field stimulation outcomes in terms of acid penetration distance and permeability enhancement, as shown in Fig. 4.7.

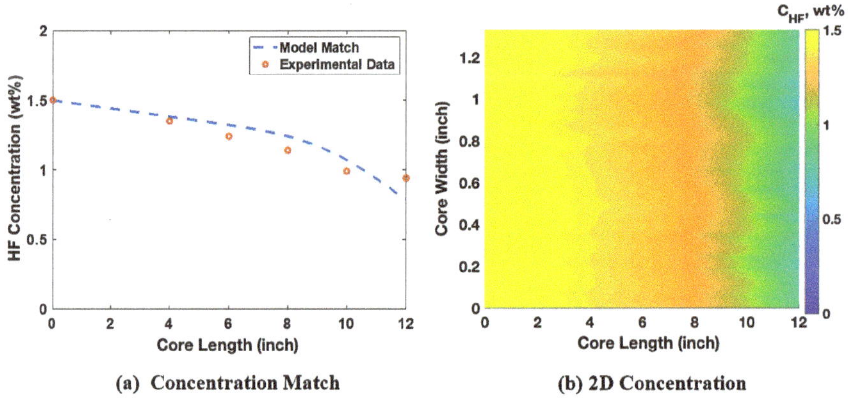

FIG. 4.6 Match between the theoretical model and experimental data of mud acid reaction with Scioto sandstone (Mahmoud et al., 2020).

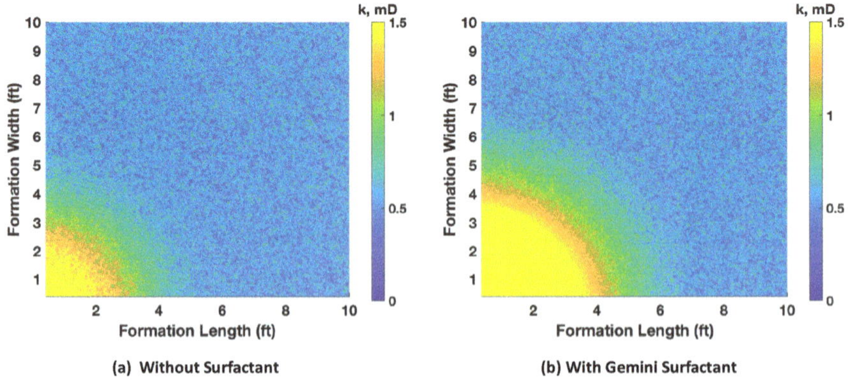

FIG. 4.7 Radial mud stimulation is based on a model tuned with experimental data sandstone (Mahmoud et al., 2020).

4.11 Conclusions

Sandstone formations consist of a wide range of minerals with different properties, making the treatment of sandstone reservoirs very challenging. Sandstone acidizing is conducted in three main stages: pre-flush, main flush, and post-flush. This chapter covers the interactions of different fluids and additives with sandstone minerals. The basic chemistry of sandstone acidizing and the composition and additives of sandstone acidizing fluids are discussed. Different types of sandstone acid systems and additives such as HCl, HF, retarders, corrosion inhibitors, and clay stabilizers are explained. This chapter sheds a light on the compatibility between different fluid stages and the fluid in the same stage. Fluid/rock interactions during sandstone acidizing may cause fines migration, clay swelling, fluid retention, and fluid adsorption. These interactions will affect both the rock porosity and permeability in addition to the rock surface properties such as wettability and interfacial tension (IFT). Finally, the modeling techniques of sandstone acidizing are discussed including the common models such as the lumped-parameter model and detailed reaction models.

Abbreviations

AFM	atomic force microscopy
CI	corrosion inhibitors
CT	computed tomography
EDG	ethanoldiglycine acid
EDTA	ethylenediaminetetraacetic acid
GLDA	glutamic acid *N,N*-diacetic
HCl	hydrochloric acid
HEDTA	hydroxyethylethylenediaminetriacetic acid
HF	hydrofluoric acid
NMR	nuclear magnetic resonance
NTA	nitrilotriacetic acid
SEM	scanning electron microscope
XRD	X-ray diffraction

References

Abdelmoneim, S.S., Nasr-El-Din, H.A., 2015. Determining the optimum HF concentration for stimulation of high temperature sandstone formations. In: Paper Presented at the SPE European Formation Damage Conference and Exhibition, Budapest, Hungary, 3–5 June. SPE-174203-MS. https://doi.org/10.2118/174203-MS.

Ahmadi, M., Habibi, A., Pourafshary, P., Ayatollahi, S., 2013. An experimental study of interaction between Nanoparticles' deposition on a sintered porous medium and migratory fines. J. Porous Media 16 (5), 459–467.

Akanni, O.O., Nasr-El-Din, H.A., 2016, September. Modeling of wormhole propagation during matrix acidizing of carbonate reservoirs by organic acids and chelating agents. In SPE Annual Technical Conference and Exhibition. OnePetro.

Alhamad, L., Alrashed, A., Al Munif, E., Miskimins, J., 2020, February. A review of organic acids roles in acidizing operations for carbonate and sandstone formations. In: SPE International Conference and Exhibition on Formation Damage Control. OnePetro.

Al-Harbi, B.G., Al-Dahlan, M.N., Al-Khaldi, M.H., 2012, February. Aluminum and iron precipitation during sandstone acidizing using organic-HF acids. In: SPE International Symposium and Exhibition on Formation Damage Control. OnePetro.

Al-Harthy, S., Bustos, O.A., Samuel, M., Still, J., Fuller, M.J., Hamzah, N.E., Parapat, A., 2008. Options for high-temperature well stimulation. Oilf. Rev. 20 (4), 52–62.

Ali, A.H.A., Frenier, W.W., Xiao, Z., et al., 2002. Chelating agent-based fluids for optimal stimulation of high-temperature wells. In: Paper Presented at the SPE Annual Technical Conference and Exhibition, San Antonio, Texas.

Ali, S.A., Ermel, E., Clarke, J., Fuller, M.J., Xiao, Z., Malone, B., 2008. Stimulation of high-temperature sandstone formations from West Africa with chelating agent-based fluids. SPE Prod. Oper. 23 (01), 32–38.

Ali, S.A., Kalfayan, L., Montgomery, C., 2016. Acid Stimulation. Society of Petroleum Engineers, Richardson, Texas, USA.

Allen, O.T., Roberts, A.P., 1993. Production Operation: Well Completions, Workover, and Stimulation, fourth ed. vol. 2 OGCI, Tulsa, Oklahoma, U.S.A.

AlMubarak, T., AlDajani, O., AlMubarak, M., 2015, December. A collective clay stabilizers review. In: International Petroleum Technology Conference. OnePetro.

Al-Shaalan, T.M., Nasr-El-Din, H.A., 2000, March. Mathematical modeling of sandstone stimulation: a critical review of available models. In: CORROSION 2000. OnePetro.

Al-Zain, A.K., Abitrabi, A.N., Al-Jandal, A.A., Said, R., 2009. An innovative stimulation approach for horizontal power injectors in sandstone reservoir, Saudi Arabia: field experience. In: Paper SPE 119997 Presented at SPE Middle East Oil and Gas Show and Conference, 15-18 March, Bahrain.

Amaefule, J.O., Kersey, D.G., Norman, D.L., Shannon, P.M., 1988. Advances in formation damage assessment and control strategies. In: CIM Paper No. 88-39-65, Proceedings of the 39th Annual Technical Meeting of Petroleum Society of CIM and Canadian Gas Processors Association, Calgary, Alberta, June 12-16.

Amirmoshiri, M., Zhang, L., Puerto, M.C., Tewari, R.D., Bahrim, R.Z.B.K., Farajzadeh, R., Biswal, S.L., 2020. Role of wettability on the adsorption of an anionic surfactant on sandstone cores. Langmuir 36 (36), 10725–10738.

Andotra, G., 2014. Investigating the Use of Chelating Agents for Clay Dissolution and Sandstone Acidizing Purposes. (Doctoral dissertation).

Aneto, N., 2012. An In-Depth Investigation of an Aluminum Chloride Retarded Mud Acid System on Sandstone Reservoirs (Master's Thesis). Texas A&M University.

Bailey, S.W., 1986. In: Brindley, G.W., Brown, G. (Eds.), Crystal Structure of Clay Minerals and Their X-ray Identification. London Mineralogical Society (Chapter 1).

Bartko, K.M., Acock, A.M., Robert, J.A., Thomas, R.L., 1997. A Field Validated Matrix Acidizing Simulator for Production Enhancement in Sandstone and Carbonates. Society of Petroleum Engineers. https://doi.org/10.2118/38170-MS.

Bennion, D.B., 2002. An overview of formation damage mechanisms causing a reduction in the productivity and injectivity of oil and gas producing formations. J. Can. Pet. Technol. 41 (11), 29–36.

Bryant, S.L., 1991. An Improved Model of Mud Acid/Sandstone Chemistry. Society of Petroleum Engineers.

Bybee, K., 2003. Hydraulic fracturing in the Ghawar field carbonate and sandstone reservoirs. J. Petrol. Tech. 55 (3), 47–48.

Byrne, M.T., Waggoner, S.M., 2009, May. Fines migration in a high temperature gas reservoir-laboratory simulation and implications for completion design. In: 8th European Formation Damage Conference. OnePetro.

Caenn, R., Darley, H.C., Gray, G.R., 2011. Composition and Properties of Drilling and Completion Fluids, sixth ed. Elsevier Inc., ISBN: 978-0-12-383858-2.

Cassidy, J., Kiser, C., Smith, C., 2011. Corrosion-inhibitor design for compatibility with acid-gelling polymers. Proc. SPE Int. Symp. Oilfield Chem. 2, 809–829. https://doi.org/10.2118/141435-MS.

Crowe, C.W., Minor, S.S., 1985. Effect of acid corrosion inhibitors on matrix stimulation results. J. Petrol. Tech. 37, 1853–1860. https://doi.org/10.2118/11119-PA.

Crowe, C., Masmonteil, J., Thomas, R., 1992. Trends in matrix acidizing. Oilf. Rev. 4 (4), 24–40.

Curbelo, F.D., Santanna, V.C., Neto, E.L.B., Dutra Jr., T.V., Dantas, T.N.C., Neto, A.A.D., Garnica, A.I., 2007. Adsorption of nonionic surfactants in sandstones. Colloids Surf. A Physicochem. Eng. Asp. 293 (1–3), 1–4.

Da Motta, E.P., Plavnik, B., Schecter, R.S., Hill, A.D., 1993. Accounting for silica precipitation in the design of sandstone acidizing. SPE Prod. Facil. 8 (02), 138–144.

Dabbousi, B.O., Nasr-El-Din, H.A., Al-Muhaish, A.S., 1999. Influence of oilfield chemicals on the surface tension of stimulating fluids. Proc. SPE Int. Symp. Oilfield Chem., 313–324. https://doi.org/10.2118/50732-MS.

Dehgani, M., 2010, April. Matrix acid stimulation. In: The First International Geological Congress. Department of Geology, Islamic Azad University.

Denekas, M.O., Mattax, C.C., 1959. Effects of crude oil components on rock wettability. Trans. AIME 216, 330–333.

Economides, M.J., Nolte, K.G., 2000. Reservoir Stimulation. Wiley, Chichester, England.

Economides, M.J., Ehlig-Economides, C., Zhu, D., Hill, A.D., 2013. Petroleum Production Systems. Pearson Education (US).

El-Monier, I.A., Nasr-El-Din, H.A., 2011, June. A New Al-Based Stabilizer for High pH Applications. Brasil Offshore. OnePetro.

Etemad-Saeed, N.A., Hosseini-Barzi, M.A., Armstrong-Altrin, J.S., 2011. Petrography and geochemistry of clastic sedimentary rocks as evidences for provenance of the lower Cambrian Lalun formation, Posht-e-badam block, Central Iran. J. Afr. Earth Sci. 61 (2), 142–159.

Ezzat, A.M., 1990. Completion fluids design criteria and current technology weaknesses. In: Paper SPE 19434 Presented at the SPE Formation Damage Control Symposium, Lafayette, Louisiana, February 22-23.

Farley, J.T., Miller, B.M., Schoettle, V., 1970. Design criteria for matrix stimulation with hydrochloric-hydrofluoric acid. J. Petrol. Tech. 22 (04), 433–440.

Fogler, H.S., McCune, C.C., 1976. On the extension of the model of matrix acid stimulation to different sandstones. AIChE J. 22 (4), 799–805.

Garcia, E.A., LaBlanc, A., Beuterbaugh, A., Calabrese, T., 2016, February. Developments in sandstone HF acidizing: HF fluid compatible with Na or K brines and carbonate-laden mineralogy for high temperatures (360 F). In: SPE International Conference and Exhibition on Formation Damage Control. OnePetro.

Gdanski, R.D., 1997. Kinetics of the primary reaction of HF aluminosilicates. In: Paper SPE 37459 Presented at the 1997 SPE Production Operations Symposium, Oklahoma City, Oklahoma, 9-11 May.

Gdanski, R.D., Shuchart, C.E., 1998. Advanced sandstone-acidizing designs with improved radial models. SPE Prod. Facil. 13 (04), 272–278.
Gidley, J.L., 1985, September. Acidizing sandstone formations: a detailed examination of recent experience. In: SPE Annual Technical Conference and Exhibition. OnePetro.
Gomaa, A.M., Cutler, J., Qu, Q., Boles, J., Wang, X., 2013, March. Matrix stimulation: an effective one-step sandstone acid system. In: SPE Production and Operations Symposium. OnePetro.
Greenwood, R., Lapčíková, B., Surýnek, M., Waters, K., Lapčík, L., 2007. The zeta potential of kaolin suspensions measured by electrophoresis and electroacoustics. Chem. Pap. 61 (2), 83–92.
Guimares, M.S., de A Grana, T., Matos Sarkis, S.R., 2017. Study and Simulation of Matrix Acidizing in Sandstones Reservoirs Applied to Injection Wells of the Mature Fields. Society of Petroleum Engineers. https://doi.org/10.2118/184928-MS.
Gulbrandsen, E., Pedersen, A., 2007. Alteration of sand wettability by corrosion inhibitors and its effect on formation of sand deposits. In: Proc - SPE Int Symp Oilf Chem, pp. 635–640. https://doi.org/10.2118/106492-MS.
Guo, B., 2011. Petroleum Production Engineering, a Computer-Assisted Approach. Elsevier.
Hanafy, A.M., Nasr-El-Din, H.A., 2016, November. NMR study to assess fines migration damage and its removal using regular mud acid in different sandstone cores. In: Abu Dhabi International Petroleum Exhibition & Conference. OnePetro.
Hanafy, A., Ali, A., Nasr-El-Din, H.A., Heidari, Z., 2015, December. Evaluating the effects of acid stimulation treatment before and after fines migration on petrophysical properties in sandstone reservoirs. In: International Petroleum Technology Conference. OnePetro.
Hassan, A., Mahmoud, M., Bageri, B.S., Aljawad, M.S., Kamal, M.S., Barri, A.A., Hussein, I.A., 2020. Applications of chelating agents in the upstream oil and gas industry: a review. Energy Fuel 34 (12), 15593–15613.
Hekim, Y., Fogler, H.S., McCune, C.C., 1982. The Radial Movement of Permeability Fronts and Multiple Reaction Zones in Porous Media. Society of Petroleum Engineers.
Hibbeler, J., Garcia, T., Chavez, N., 2003, April. An integrated long-term solution for migratory fines damage. In: SPE Latin American and Caribbean Petroleum Engineering Conference. OnePetro.
Hill, A.D., Lindsay, D.M., Silberberg, I.H., Schechter, R.S., 1981. Theoretical and experimental studies of sandstone acidizing. Soc. Pet. Eng. J. 21 (01), 30–42.
Hill, A.D., Sepehrnoori, K., Wu, P.Y., 1994. Design of the HCl preflush in sandstone acidizing. SPE Prod. Facil. 9 (02), 115–120.
Huang, T., McElfresh, P.M., Gabrysch, A., 2002, February. High temperature acidization to prevent fines migration. In: International Symposium and Exhibition on Formation Damage Control. OnePetro.
Iglauer, S., Wu, Y., Shuler, P., Tang, Y., Goddard III, W.A., 2010. New surfactant classes for enhanced oil recovery and their tertiary oil recovery potential. J. Petrol. Sci. Eng. 71 (1–2), 23–29.
Ji, Q., Zhou, L., Nasr-El-Din, H.A., 2014, May. Acidizing sandstone reservoirs using fines control acid. In: SPE Latin America and Caribbean Petroleum Engineering Conference. OnePetro.
Jones, A.A., 1981. Charges on the surfaces of two chlorites. J. Clay Miner. 16, 347–359.
Kalfayan, L.J., Metcalf, A.S., 2000, October. Successful sandstone acid design case histories: exceptions to conventional wisdom. In: SPE Annual Technical Conference and Exhibition. OnePetro.
Kalfayan, L.J., Watkins, D.R., 1990, April. A new method for stabilizing fines and controlling dissolution during sandstone acidizing. In: SPE California Regional Meeting. OnePetro.

Kamal, M.S., Hussein, I.A., Sultan, A.S., 2017. Review on surfactant flooding: phase behavior, retention, IFT, and field applications. Energy Fuel 31, 7701–7720.

Kamal, M.S., Mahmoud, M., Hanfi, M., Elkatatny, S., Hussein, I., 2019. Clay minerals damage quantification in sandstone rocks using core flooding and NMR. J. Pet. Explor. Prod. Technol. 9 (1), 593–603.

Khilar, K.C., Fogler, H.S., 1998. Migrations of Fines in Porous Media. vol. 12 Springer Science & Business Media.

Kosmulski, M., 1998. Positive electrokinetic charge of silica in the presence of chlorides. J. Colloid Interface Sci. 208 (2), 543–545.

Kudrashou, V.Y., Nasr-El-Din, H.A., 2019, March. Formation damage and compatibility issues associated with use of corrosion inhibitors in well acidizing—A review. In: SPE/ICoTA Well Intervention Conference and Exhibition. OnePetro.

Labrid, J.C., 1975. Thermodynamic and Kinetic Aspects of Argillaceous Sandstone Acidizing. Society of Petroleum Engineers.

Lea, C.-M., Hill, A.K., Sepehrnoori, K., 1992. Simulation of Sandstone Acidizing of a Damaged Perforation. Society of Petroleum Engineers.

Legemah, M., De Benedictis, F., Workneh, N., Smith, D., Le, H., Sun, H., Moulton, M., 2015. Pushing the limit on guar loading: treatment of a Green River sandstone formation by use of a novel low-polymer crosslinked fluid. SPE Prod. Oper. 30 (01), 59–68.

Leong, V.H., Ben Mahmud, H., 2019. A preliminary screening and characterization of suitable acids for sandstone matrix acidizing technique: a comprehensive review. J. Pet. Explor. Prod. Technol. 9 (1), 753–778.

Leong, V.H., Mahmud, H.B., Law, M.C., Foo, H.C.Y., Tan, I.S., 2018. A comparison and assessment of the modelling and simulation of the sandstone matrix acidizing process: a critical methodology study. J. Nat. Gas Sci. Eng. 57, 52–67.

LePage, J.N., De Wolf, C.A., Bemelaar, J.H., Nasr-El-Din, H.A., 2011. An environmentally friendly stimulation fluid for high-temperature applications. SPE J. 16 (01), 104–110.

Levitt, D.B., Weatherl, R.K., Harris, H.W., McNeil, R.I., Didier, M., Loriau, M., Bourrel, M., 2015, April. Adsorption of EOR chemicals under laboratory and reservoir conditions, part 1-iron abundance and oxidation state. In: IOR 2015–18th European Symposium on Improved Oil Recovery. European Association of Geoscientists & Engineers, p. cp-445.

Li, Y.H., Fambrough, J.D., Montgomery, C.T., 1998. Mathematical modeling of secondary precipitation from sandstone acidizing. SPEJ 3 (4), 393–401.

Li, C., Xie, T., Pournik, M., Zhu, D., Hill, A.D., 2004. Fine-Scale Simulation of Sandstone Acidizing. Society of Petroleum Engineers.

Li, L., Nasr-El-Din, H.A., Zhou, J., Holt, S., 2012. Compatibility and phase behavior studies between corrosion inhibitors and surfactants-based acids. In: Proc - SPE Int Symp Form Damage Control. vol. 1, pp. 533–552. https://doi.org/10.2118/151641-MS.

Lund, K., Fogler, H.S., 1976. Acidization—V: the prediction of the movement of acid and permeability fronts in sandstone. Chem. Eng. Sci. 31 (5), 381–392.

Lund, K., Fogler, H.S., McCune, C.C., Ault, J.W., 1973. Kinetic Rate Expressions for Reactions of Selected Minerals with HCl and HF Mixtures. Society of Petroleum Engineers.

Ma, K., Cui, L., Dong, Y., Wang, T., Da, C., Hirasaki, G.J., Biswal, S.L., 2013. Adsorption of cationic and anionic surfactants on natural and synthetic carbonate materials. J. Colloid Interface Sci. 408, 164–172.

Mahmoud, M., 2017. New formulation for sandstone acidizing that eliminates sand production problems in oil and gas sandstone reservoirs. J. Energy Resour. Technol. 139 (4), 042902.

Mahmoud, M., Gomaa, I., 2022. Sandstone matrix stimulation. In: Fluid Chemistry, Drilling and Completion. Gulf professional publishing, pp. 341–386.

Mahmoud, M.A., Nasr-El-Din, H.A., De Wolf, C.A., Alex, A.K., 2011, April. Sandstone acidizing using a new class of chelating agents. In: SPE International Symposium on Oilfield Chemistry. OnePetro.
Mahmoud, M.A., Nasr-El-Din, H.A., De Wolf, C.A., LePage, J.N., 2011a. Optimum injection rate of A new chelate that can be used to stimulate deep carbonate reservoirs. SPEJ 16 (4), 968–980.
Mahmoud, M.A., Nasr-El-din, H.A., De Wolf, C.A., LePage, J.N., Bemelaar, J.H., 2011b. Evaluation of a new environmentally friendly chelating agent for high temperature applications. SPEJ 16 (3), 559–574.
Mahmoud, M., Nasr-El-Din, H., De Wolf, C., Alex, A., 2011c. Sandstone acidizing using a new class of chelating agents. In: Paper SPE-139815-MS Presented at the Paper Presented at the SPE International Symposium on Oilfield Chemistry, The Woodlands, Texas, USA, April 2011.
Mahmoud, M.A., Nasr-El-Din, H.A., De Wolf, C.A., 2015. High-temperature laboratory testing of illitic sandstone outcrop cores with HCl-alternative fluids. SPE Prod. Oper. 30 (01), 43–51.
Mahmoud, M.A., 2018. Effect of chlorite clay-mineral dissolution on the improved oil recovery from sandstone rocks during diethylenetriaminepentaacetic acid chelating-agent flooding. SPE J. 23 (05), 1880–1898.
Mahmoud, M., Aljawad, M.S., Kamal, M.S., Hussain, S.M.S., 2020. Sandstone acidizing using a new retarded acid system based on gemini surfactants. J. Petrol. Sci. Eng. 194, 107459.
Mannhardt, K., Schramm, L.L., Novosad, J.J., 1993. Effect of rock type and brine composition on adsorption of two foam-forming surfactants. SPE Adv. Technol. Ser. 1 (01), 212–218.
Mannhardt, K., Novosad, J.J., Jha, K.N.N., 1994. Adsorption of foam-forming surfactants in Berea sandstone. J. Can. Pet. Technol. 33 (02), 34–43.
Martin, A.N., 2004, September. Stimulating sandstone formations with non-HF treatment systems. In: SPE Annual Technical Conference and Exhibition. OnePetro.
McCune, C.C., Fogler, H.S., Cunningham, J.R., 1975. A new model of the physical and chemical changes in sandstone during acidizing. Soc. Pet. Eng. 15 (5), 361–370.
Miall, A.D., 1988. Facies architecture in clastic sedimentary basins. In: New Perspectives in Basin Analysis. Springer, New York, NY, pp. 67–81.
Michaels, A.S., Bolger, J.C., 1962. Settling rates and sediment volumes of flocculated kaolin suspensions. Ind. Eng. Chem. Fundam. 1 (1), 24–33.
Muecke, T.W., 1982, March. Principles of acid stimulation. In: International Petroleum Exhibition and Technical Symposium. OnePetro.
Murtaza, Z., Berndt, O., Robert, J., 1999. An Improved Sandstone Acidizing Model: The Importance of Secondary and Tertiary Reactions. Society of Petroleum Engineers.
Nasr-El-Din, H.A., 2013. HT stimulation fluid based on GLDA meets productivity, environmental need. J. Petrol. Tech. 65 (01), 28–31.
Nasr-El-Din, H.A., Al-Ghamdl, A.W.H., Al-Qahtani, A.A., Samuel, M.M., 2008. Impact of acid additives on the rheological properties of a viscoelastic surfactant and their influence on field application. SPE J. 13, 35–47. https://doi.org/10.2118/89418-PA.
Nasser, M.S., James, A.E., 2006. Settling and sediment bed behavior of kaolinite in aqueous media. J. Sep. Purif. Technol. 51 (1), 10–17. http://10.1016/j.seppur.2005.12.017.
O'Driscoll, K.P., Stolyarov, S., Kalfayan, L., 2005, May. A review of matrix aciding sandstones in western Siberia Russia. In: SPE European Formation Damage Conference. OnePetro.
Olphen, H.V., 1963. Introduction to Clay Colloid Chemistry. Inter science, New York.
Parkinson, M., Munk, T., Brookley, J., Caetano, A., Albuquerque, M., Cohen, D., Reekie, M., 2010, February. Stimulation of multilayered high-carbonate-content sandstone formations in West Africa using chelant-based fluids and mechanical diversion. In: SPE International Symposium and Exhibition on Formation Damage Control. OnePetro.

Perthuis, H., Touboul, E., Piot, B., 1989. Acid Reactions and Damage Removal in Sandstones: A Model for Selecting the Acid Formulation. Society of Petroleum Engineers.

Pituckchon, A., 2014. Further Investigation of Fluoboric Acid in Sandstone Acidizing Using^(11) B and^(19) F NMR. (Doctoral Dissertation).

Ponce da Motta, E., Plavnik, B., Schechter, R.S., 1992. Optimizing sandstone acidization. SPE Reserv. Eng. 7 (01), 149–153.

Rabie, A.I., Nasr-El-Din, H.A., 2015, September. Effect of acid additives on the reaction of stimulating fluids during acidizing treatments. In: SPE North Africa Technical Conference and Exhibition. OnePetro.

Rahman, S.S., 2000. A 3-D Numerical Model for Designing and Planning of Matrix Acid Stimulation in Low-Permeability Rocks. Society of Petroleum Engineers.

Rezaeizadeh, M., Hajiabadi, S.H., Aghaei, H., Blunt, M.J., 2021. Pore-scale analysis of formation damage; A review of existing digital and analytical approaches. Adv. Colloid Interface Sci. 288, 102345.

Rozo, R.E., Paez, J., Mendoza Rojas, A., Milne, A.W., Soler, D.F., Abuseif, H., 2007, November. An alternative solution to sandstone acidizing using a nonacid based fluid system with fines-migration control. In: SPE Annual Technical Conference and Exhibition. OnePetro.

Rueslatten, H., Oren, P.E., Robin, M., Rosenberg, E., 1994, April. A combined use of Cryo-SEM and NMR-Spectroscopy for studying the distribution of oil and brine in sandstones. In: SPE/DOE Improved Oil Recovery Symposium. OnePetro.

Schechter, R.S., Gidley, J.L., 1969. The change in pore size distribution from surface reactions in porous media. AIChE J. 15 (2), 339–350.

Sevougian, S.D., Lake, L.W., Schechter, R.S., 1995. A new geochemical simulator to design more effective sandstone acidizing treatments. Soc. Pet. Eng. 10 (1), 13–19.

Shafiq, M.U., Mahmud, H.K.B., 2016, March. An effective acid combination for enhanced properties and corrosion control of acidizing sandstone formation. In: IOP Conference Series: Materials Science and Engineering. Vol. 121 (1). IOP publishing, p. 012002.

Shafiq, M.U., Mahmud, H.B., 2017. Sandstone matrix acidizing knowledge and future development. J. Pet. Explor. Prod. Technol. 7 (4), 1205–1216.

Shafiq, M.U., Mahmud, H.K.B., Hamid, M.A., 2015, March. Comparison of buffer effect of different acids during sandstone acidizing. In: IOP Conference Series: Materials Science and Engineering. 78 (1). IOP Publishing, p. 012008.

Shafiq, M.U., Ben Mahmud, H., Rezaee, R., Testamandi, N., 2017. Investigation of changing pore topology and porosity during matrix acidizing using different chelating agents. IOP Conf. Ser. Mater. Sci. Eng. 217, 012023. https://doi.org/10.1088/1757-899X/217/1/012023.

Shafiq, M.U., Mahmud, H.K.B., Arif, M., 2018. Mineralogy and pore topology analysis during matrix acidizing of tight sandstone and dolomite formations using chelating agents. J. Petrol. Sci. Eng. 167, 869–876.

Shaughnessy, C.M., Kunze, K.R., 1981. Understanding sandstone acidizing leads to improved field practices. J. Petrol. Tech. 33 (07), 1196–1202.

Simon, D.E., Anderson, M.S., 1990. Stability of clay minerals in acid. In: Paper SPE 20624 Presented at the Formation Damage Control Symposium, Lafayette, Louisiana, 22-23 February.

Smith, C.F., Hendrickson, A.R., 1965. Hydrofluoric acid stimulation of sandstone reservoirs. J. Petrol. Tech. 17 (02), 215–222.

Stanley, F.O., Ali, S.A., Boles, J.L., 1995, April. Laboratory and field evaluation of organosilane as a formation fines stabilizer. In: SPE Production Operations Symposium. OnePetro.

Taha, R., Hill, A.D., Sepehrnoori, K., 1986. Simulation of Sandstone-Matrix Acidizing in Heterogeneous Reservoirs. Society of Petroleum Engineers.

Thomas, R.L., Nasr-El-Din, H.A., Lynn, J.D., Mehta, S., Zaidi, S.R., 2001. Precipitation during the acidizing of a HT/HP illitic sandstone reservoir in eastern Saudi Arabia: a laboratory study. In: Paper SPE 71690 Presented at the SPE ATCE, New Orleans, Louisiana, 30 September-3 October.

Thomas, R.L., et al., 2002. Channels vs. matrix sandstone acidizing of a HPHT reservoir in Saudi Arabia. In: SPE 73702 Presented at the SPE International Symposium on Formation Damage Control Held in Lafayette, Louisiana, 20-21 February.

Van Hong, L., Mahmud, H.B., 2017, July. A comparative study of different acids used for sandstone acid stimulation: a literature review. In: IOP Conference Series: Materials Science and Engineering. Vol. 217 (1). IOP Publishing, p. 012018.

Wang, J., Huang, Y., Zhou, F., Song, Z., Liang, X., 2020. Study on reservoir damage during acidizing for high-temperature and ultra-deep tight sandstone. J. Petrol. Sci. Eng. 191, 107231.

Welton, J.E., 1984. SEM Petrology Atlas. American Association of Petroleum Geologists, Tulsa, Oklahoma, p. 237.

White, G.N., Zelazny, L.W., 1988. Analysis and implications of the edge structure of dioctahedral phyllosilicates. Clays Clay Miner. 36 (2), 141–146.

Williams, B.B., Whiteley, M.E., 1971. Hydrofluoric Acid Reaction with a Porous Sandstone. Society of Petroleum Engineers.

Wilson, M.J., 2013. Rock-Forming Minerals Volume 3C-Sheet Silicates: Clay Minerals, second ed. Geological Society of London, London.

Yang, F., Nasr-El-Din, H.A., Al-Harbi, B., 2012, May. Acidizing sandstone reservoirs using HF and organic acids. In: SPE International Production and Operations Conference & Exhibition. OnePetro.

Ying-Hsiao, L., Fambrough, J.D., Montgomery, C.T., 1998. Mathematical Modeling of Secondary Precipitation From Sandstone Acidizing. Society of Petroleum Engineers.

You, Z., Bedrikovetsky, P., Badalyan, A., Hand, M., 2015. Particle mobilization in porous media: temperature effects on competing electrostatic and drag forces. Geophys. Res. Lett. 42 (8), 2852–2860.

Zhang, R., Somasundaran, P., 2006. Advances in adsorption of surfactants and their mixtures at solid/solution interfaces. Adv. Colloid Interface Sci. 123, 213–229.

Zhang, J., Wang, D., Olatunji, K., 2016, May. Surfactant adsorption investigation in ultra-lower permeable rocks. SPE Low Perm Symposium. OnePetro.

Zheng, X., Perreault, F., Jang, J., 2018. Fines adsorption on nanoparticle-coated surface. Acta Geotech. 13 (1), 219–226.

Zhou, L., Nasr-El-Din, H.A., 2013, June. Acidizing sandstone formations using a sandstone acid system for high temperatures. In: SPE European Formation Damage Conference & Exhibition. OnePetro.

Zhou, L., Nasr-El-Din, H.A., 2016. Phosphonic-based hydrofluoric acid: interactions with clay minerals and flow in sandstone cores. SPE J. 21 (01), 264–279.

Zhou, H., Liu, H., Hu, D., Yang, F., Lu, J., Zhang, F., 2016. Anisotropies in mechanical behaviour, thermal expansion and P-wave velocity of sandstone with bedding planes. Rock Mech. Rock Eng. 49 (11), 4497–4504.

Zou, Y.S., Zhang, S.C., Zhang, J., 2014. Experimental method to simulate coal fines migration and coal fines aggregation prevention in the hydraulic fracture. Transp. Porous Media 101 (1), 17–34.

Chapter 5

Interactions of fluids during hydraulic and acid fracturing operations

Satyajit Chowdhury[a,b], Mayank Rakesh[c], and Jitendra S. Sangwai[a,d,e]
[a]*Gas Hydrate and Flow Assurance Laboratory, Petroleum Engineering Program, Department of Ocean Engineering, Indian Institute of Technology Madras, Chennai, India*
[b]*Assam Energy Institute, A Centre of Rajiv Gandhi Institute of Petroleum Technology, Sivasagar, Assam, India*
[c]*Department of Petroleum Engineering and Earth Sciences, University of Petroleum and Energy Studies, Dehradun, India*
[d]*Department of Chemical Engineering, Indian Institute of Technology Madras, Chennai, India*
[e]*Center of Excellence on Subsurface Mechanics and Geo-Energy, Indian Institute of Technology Madras, Chennai, India*

5.1 Introduction

Hydraulic fracturing has boosted hydrocarbon production and raised its current worth tremendously. Apart from improving production, hydraulic fracturing is also a means to maintain a well and ensure productivity. This is accomplished by introducing/placing a channel that is conductive through the damaged formation. This formation reaches further depths that increase productivity. In some cases, these channels are placed such that it contributes to altering the nature of fluid flow from the formation to the wellbore. A hydraulic fracturing job can be incorporated due to the following reasons:

- To restore "natural" productivity by circumventing the damaged wellbore zone.
- To induce a conductive path into the formation, pay-zone, and increases permeability past the natural level.
- Alteration of preferential fluid flow within the formation.

Thousands of fracture treatments have been conducted since Stanolind (Amoco) launched fracturing in 1947 (Al-Ameri and Gamadi, 2020). Fracture stimulation procedures have been credited for producing billions of barrels of oil and trillions of cubic feet of gas that would otherwise be uneconomic to

Developments in Petroleum Science, Vol. 78. https://doi.org/10.1016/B978-0-323-99285-5.00006-5

produce. Conventionally, the fracturing job has often rested with the drilling engineering department as field application requires the use of packers, high-pressure pumps, etc. Yet, the outcome of a fracturing job is predominantly of a production engineer's interest. Also, a reservoir engineer who plays a role at a macrolevel must be well informed about the job. Expertise in rock mechanics is of great importance, as designing treatments eventually dominate the fracture geometry (Osiptsov, 2017). Furthermore, knowledge in the domain of fluid mechanics (that controls proppant fluid interaction inside fractures) and chemistry (that involves designing fluid properties and other materials) is significant. These operations, at all times, are restricted by well and field environments.

In a well, if a fluid of a certain density is pumped then it can leak off or escape, and the formation around the wellbore breaks. The constant rise in pressure generates tensile hoop stresses that eventually split the wellbore. The compressive "principal stresses" always act on a rock at any given point among which two are horizontal and the remaining two are vertical. Apart from exceptional cases of reverse faults or thrust, the minimum horizontal stress (σ_{Hmin}) is the minimum stress (σ_{min}) while the overburden stress (σ_v) is the maximum stress (σ_{max}). On the other hand, once a fracture opens up, a closer pressure (P_{CL}) always tends to push the fracture close. The stress associated with this pressure is often called "closure stress" (σ_{CL}) (Zheng et al., 2020). Fig. 5.1 depicts the in-situ principal stress orientation during hydraulic fracturing. Generally, a rock fractures owing to the hydraulic pressure exerted by the fracturing fluid in a direction orthogonal to the minimum stress. The least stress for a vertical well is the minimum horizontal stress. Hence, the fracture in this scenario initiates in the vertical plane. The fracture created exposes the fracturing fluid to a new formation and the rate of leakage for these fluids constantly increases. However, if the volumetric rate of

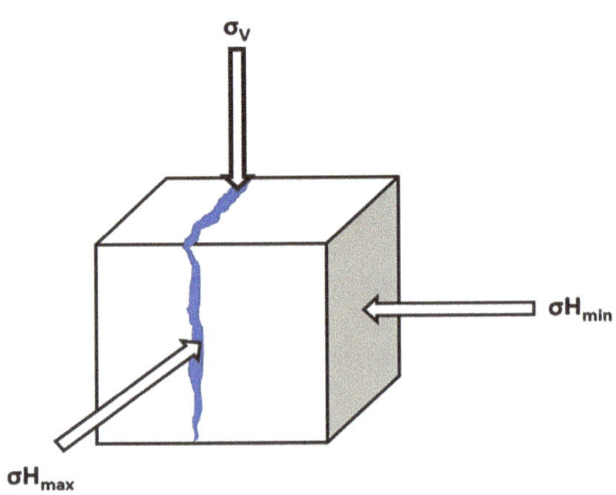

FIG. 5.1 In-situ principle stresses during fracture creation and propagation.

injection is more than that of the fluid being lost, the fracture grows and propagates. The void that has been created is usually large in two dimensions, height and length which are of relatively higher magnitude than that width. The fracture closes as soon as the pumping of fluids stops. To maintain a conductive path, or to avoid fracture closers, propping agents are pushed inside the fracture, so when the pumping of fluid comes to a halt, the proppants are trapped inside the fracture.

This further keeps the fracture open and enhances the formation flow area during the production phase. Proppants are usually sand-based or high-strength materials.

Acidizing treatments are routinely used in the oilfield to reduce near-wellbore degradation and boost formation productivity. Following the injection pressure and whether or not fractures are generated, acidizing treatments are of two types: matrix acidizing and acid fracturing. Acid matrix treatment is another name for matrix acidizing. It is a method of stimulating wells by pumping acid as well as other pads and additional fluids below the fracture pressure of the reservoir to improve well inflow performance. In matrix acidizing, an acid solution is injected through the wellbore to dissolve certain minerals, allowing sandstones to permeabilize again (removing skin) or treating carbonates around the wellbore to permeabilize more (creating wormholes). Acid fracturing involves first creating cracks in the formation and then injecting acid into those fractures. Because acids unevenly scrape the faces of fractures, some tiny channels will remain open when the cracks close that eventually increasing the reservoir's permeability. To accomplish reservoir stimulation, the acid fracturing treatment should create a conductive flow channel that is long enough to change the fluid flow in the reservoir from radial to linear. However, before the commencement of the job, fracture geometry during treatment, communicative fracture length, and fracture conductivity caused by acid fracturing start must all be predicted and analyzed (Abass et al., 2006).

5.2 Reaction kinetics of different fracturing fluids with carbonates

Fracturing complex reservoirs bearing low permeability with pressured fracturing fluid is a common method of extracting hydrocarbons. Fracturing fluid is a combination of water/brine and chemical additives that comes into contact with the targeted formation. This gives the fracturing fluid plenty of time to interact with the formation and reservoir fluid at reservoir temperature and pressure. Precipitation and dissolution, creation of organometallic complexes, organic matter degradation, and ion adsorption on rock and clay minerals are all possible effects of the resting fluid in the formation. Mineral dissolution and precipitation processes, for example, have a considerable impact on reservoir rock characteristics. Differences between modeled and actual mature stage hydrocarbon production in hydraulically fractured wells have been linked to changes in flow dynamics caused by precipitation processes. Mineral dissolution and organic matter degradation can undoubtedly improve reservoir rock permeability and porosity, but they raise the toxicity of generated water and the danger of spills toward the ground-water table. One of the most vital factors affecting reservoir geochemistry is carbonate mineral reactions. Carbonate minerals play a key role in rock–fracturing fluid interactions by buffering the pH, which has a substantial influence on most reactions in the formation and influences the equilibrium of other reactive constituents. Calcite ($CaCO_3$) ranging from 5% to 50%, dolomite ($CaMg(CO_3)_2$) varying between 0% and 10%, and limited quantities of other carbonate compounds such as siderite make up the fundamental components of carbonate minerals. Acids such as hydrochloric acid (HCl) and hydrofluoric acid (HF) present in the fracturing fluid undergo a chemical reaction with carbonates to liberate calcium, magnesium (for dolomite), bicarbonate (HCO_3^-), and metals like strontium (Sr^{+2}), which are either dissolved or ionized (Ferguson et al., 2021; Jew et al., 2017). The reaction neutralizes acid by forming bicarbonate (HCO_3^-), which also serves to buffer the chemical system and regulates pH. The degree to which fracturing fluid reacts with carbonate minerals is determined by the carbonate concentration of the formations as well as the acid supplied to the fracturing fluid. Studies have found a clear link between changes in pH value, calcium, magnesium, and dissolved inorganic carbon (DIC) proportional to the quantity of carbonate in the matrix. Depending on the well's physical dynamics, carbonate mineral dissolution can either increase pore size, which is good for hydrocarbon extraction, or create instability and collapse microfractures, which on the other hand is unfavorable. Acidic components present in the fracturing fluids are consumed more rapidly during matrix treatments in carbonates than in sandstone reservoirs for the following reasons:

The carbonate dissolves at a rapid rate as a result of limited mass transfer even at a lower temperature of around 0 °C, but the dissolution of dolomite (calcium magnesium carbonate) mass transfer is limited above 150° C (Hoefner and Fogler, 1989).

$$CaCO_3 + 2H^+ = CO_2 + H_2O + Ca^{++}$$

$$CaMg[CO_3]_2 + 4H^+ = 2CO_2 + 2H_2O + Ca^{++} + Mg^{++}$$

In carbonate rock, the amount of material accessible for dissolution is substantially larger, while only interstitial material is dissolved in sandstone. However, quartz that makes up for the larger part remains inert. Additionally, low fluid velocities result in lower acid penetration when the injection pressure of the fracture fluid is maintained below the fracture pressure.

In carbonate-rich formations, the pH of the solution differs dramatically. In cases of contact between the shale-fracturing fluid and carbonate-rich rock samples, a considerable rise in pH post reaction is seen. A carbonate-free rock, on the other hand, maintains a steady acidic pH. The pH of the rock-fracturing fluid solutions is primarily controlled by carbonate minerals. Carbonate dissolution, according to some research, raises the calcium ion saturation in the solution. However, investigations using fracturing fluid with a significant quantity of calcium added to the brine component of the aqueous mixture have shown that calcium chloride swamps the growing calcium trend caused by carbonate dissolution.

Hydrochloric acid (HCl) as a fracturing fluid for carbonate formation is widely used in the industry for acid-fracturing operations. The downsides of utilizing HCl acid during acid treatment in carbonates include rapid tube corrosion, formation surface dissolution, poor acid pumping rates, induced formation of skin, and damage due to emulsion formation after mixing with crude oil with a higher asphaltene concentration. In the context of multilateral and extended reach drilling (ERD) wells, the acid has a severe impact; therefore, chelating agents and organic acids are now utilized (Elsarawy et al., 2016).

5.3 Components and compatibility of fracturing fluids properties

Hydraulic fracturing, as mentioned in earlier sections, is a technique for increasing a well's production. In hydrocarbon-bearing zones, fracturing fluids are introduced into the well to generate conductive cracks and bypass near-wellbore damage. Consequently, an increase in the reservoir's productive surface area is attained. Chemical additives are used to provide the fluid with a set of desired properties, such as viscosity, friction, formation compatibility, fluid loss control, etc. (Barati and Liang, 2014). Fracturing fluids must fulfill several criteria at the same time out of which stability at high pumping rates and high temperature is of vital significance. These extreme circumstances might cause the fluids to disintegrate which drives the proppants to settle out prematurely before the fracturing is finished. The majority of fracturing fluids in the market are aqueous mixtures in either gelled or foamed states (Xiong et al., 1996).

Acids are one of the functional classes of fracturing fluids. They are used to increase injection capability or penetration distance and then to dissolve minerals and clays to minimize clogging and enable reservoir fluid to flow to the surface. Bioadditives are used to prevent bacteria from the corrosion of pipes and fittings and to break down gelants that maintain fluid viscosity and proppant movement. Breakers like ammonium persulfate and magnesium peroxide are used to degrade gelling which transports the proppant and improves the flowback after the fracturing process. Cross-linkers often with metal salts are used to thicken fluids, enhance viscosity, and proppant transport. Clay stabilizers are used to form a fluid barrier to prevent clay mobilization. Corrosion inhibitors are used to protect pipes and casing. Proppants, in addition to the previously listed compounds, make up a significant portion of the fracture fluid composition. Proppants such as styrene, crystalline silica, ceramic, and graphite are commonly utilized to hold the fractures open, also enabling hydrocarbon to escape from the broken formation. Foamers and de-foamers are used to change the carrying capacity of proppants and let trapped gas escape during transport. Friction reducers in fracturing fluid reduce friction caused under high pressure, allowing the fluid to travel more seamlessly. To suspend heavy proppant particles during transit, propylene glycol, ethylene glycol, and guar gum are employed to improve viscosity and gel strength. pH controllers and scale inhibitors are also utilized to keep the pH buffer stable at different stages and avoid scale build-up (Harris, 1988).

Slick water, liner gel, cross-linked gel, and viscoelastic surfactant solutions are all water-based fracturing fluids. Water with a gelling ingredient like guar, hydroxypropyl guar (HPG), hydroxyethyl cellulose (HEC), carboxymethyl hydroxypropyl guar (CMHPG), or carboxymethyl hydroxyethyl cellulose (CMHEC) is known as linear gel fluid. Polymers expand and produce a thick gel with a high viscosity. When compared to water fracture fluid, these fluids have a dynamic viscosity of 10^{-30} mPas, which allows for better proppant transport and a broader fracture. A similar function of polymer-based compounds can also be seen in designing drilling fluids during the exploration phase (Medhi et al., 2020, 2021). Low-viscosity linear gels and friction reducers, on the other hand, are unable to provide effective suspension properties within the fractures under static conditions. This eventually causes the proppant to settle prematurely. Hence, high-viscosity-laden fluids have been the preferred option for improved proppant suspension and transport. The following table consists of common conditions for the use of different fracturing fluids (Table 5.1).

5.4 Fracturing fluids and formation damage (fracture face skin)

Fracture stimulation usually leads to a negative skin value that indicates an enhancement in permeability. However, various postfracture treatment effects might cause further pressure drop and prevent the fractured well from delivering to its full potential. Gel dehydration, proppant embedment or crushing within the fracture, clogging the fracture by overflushing (which usually causes screen-out), and fluid leakage through fracture faces are only a few of them that are responsible for limiting production rates after stimulation jobs. The increased pressure loss also might restrict its effective wellbore radius even further. The degree of this effect is determined by fracture geometry, reservoir characteristics, fracturing fluid filtrate viscosity, and the amount of fluid leakage into the reservoir. The fracture face skin effect refers to the obstruction of fluid flow along the fracture surface in the reservoir (Li et al., 2016). Choked fracture is a term used to represent flow limitation caused by lower permeability throughout the fracture near the wellbore

TABLE 5.1 Types of commonly used fluid systems for fracturing operations and their applicability.

Fluid base	Fluid type	Main composition	Generally used for
Water-based fluid	Linear fluids	Gelled water, HPG, HEC, CMHPG, CMHEC	Short fractures, low temperatures
	Cross-linked fluids	Cross-linker+ HPG, HEC	Long fractures, high temperatures
Oil-based fluid	Linear fluids	Oil, gelled oil	Water-sensitive formations, short fractures
	Cross-linked fluids	Cross-linked+oil	Water-sensitive formations, long fractures
	Water external poly-emulsion	Emulsifier+oil+water	Good for fluid loss control
Foam-based fluid	Acid-based foam	Acid+foamer+N_2	Low-pressure, water-sensitive formations
	Water-based foam	Water+foamer+CO_2 or N_2	Low-pressure formations
	Alcohol-based foam	Methanol +foamer+N_2	Low-pressure formations with water-blocking problems
Alcohol-based fluid	Linear system	Gelled water+alcohol	Removal of water blocks
	Cross-linked system	Cross-linked system+ alcohol	

region. In fractured wells, both of these forms of flow impairments lead to poor production. For a long time, the notion of "fracture face skin" has piqued people's attention. With the continuous exploration and production of low permeability reserves, it has been a hotspot of discussion in recent years owing to its criticality concerning fracturing and acidizing design. Based on their findings, Adegbola and Boney (2002) proposed the following correlation (Fig. 5.2) for assessing fracture damage skin effect S_{fs} in terms of fracture half-length X_f, reservoir penetration breadth b_s, and intact to affected permeability ratio $\frac{K}{K_s}$ as seen in Eq. (5.1)

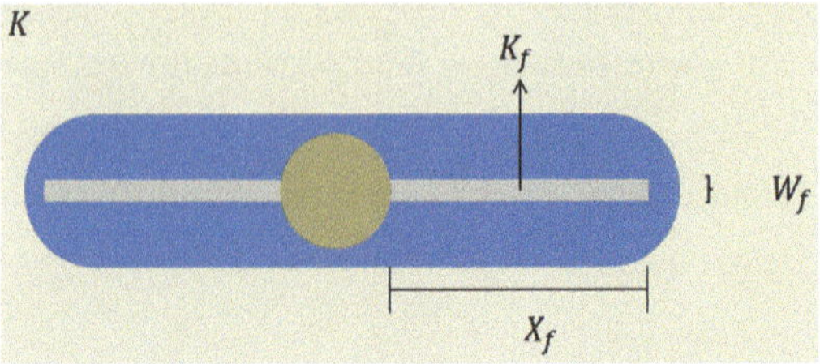

FIG. 5.2 Cinco-Ley model for fracture skin effect modeling where K is reservoir permeability.

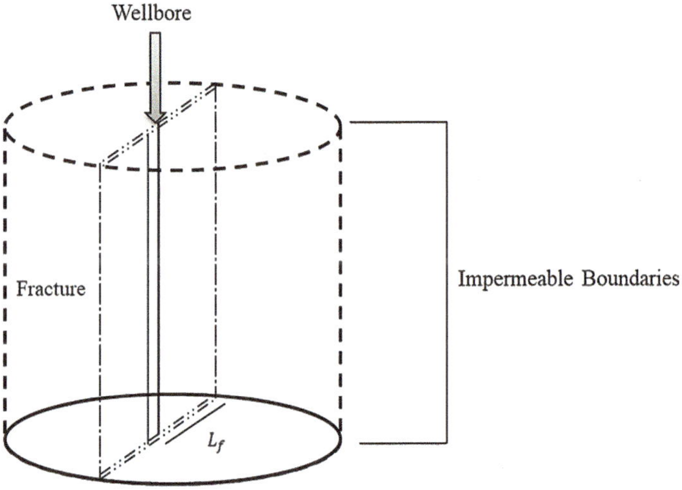

FIG. 5.3 A schematic diagram for hydraulically induced fracture.

$$S_{fs} = \frac{\pi b_s}{2X_f}\left(\frac{K}{K_s} - 1\right) \tag{5.1}$$

Fig. 5.3 represents a 3D schematic for induced fractures in formations. Traditional fracture modeling approaches such as PKN (Perkins and Kern) and KGD (Khristinaovic–Geertsma–de Klerk) (Fig. 5.4) consider the skin of the fracture face as a uniformly distributed leak-off region. The PKN model assumes that the fracture height is substantially less than the fracture length and that the energy delivered by the fluid injection is likewise consumed by dissipation of energy from fluid flow (viscosity-dominated regime), neglecting fracture toughness. Unlike PKN, KGD assumes a constant fracture aperture that is higher than the fracture length. In addition, unlike PKN, rock stiffness is only included in the model.

Increases in fracture half-length result in more oil recovery, but fracture face skin reduces flow rate, implying that there is a best fracture half-length for long-term production. High fracture skin can be caused by a fluid loss

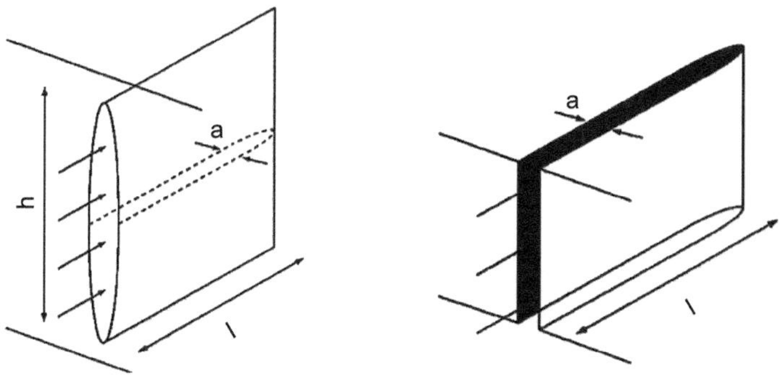

FIG. 5.4 Fracture propagation models: (A) PKN model and (B) KGD propagation model.

in extremely high-pressure oil and gas wells; however, the effect on production is obvious early on, and the fracture face skin has minimal impact on productivity later on.

5.5 Proppant embedment and its effect on fracture conductivity

Proppants are suspended in fracturing fluids while injection and are propagated into the fracture cracks to prevent them from closing after the injection pressure is released. This provides conducive pathways inside the formation through which hydrocarbons flow. The fracturing fluid viscosity may be lowered once at least one crack has been produced and at least a portion of the proppant is substantially in place. Proppants in the fracture are subjected to a variety of stresses, and the interaction between the proppant-pack and the rock formation has a significant impact on proppant-pack permeability. Proppant may be embedded in the rock or crushed into little pieces if the proppant size and strength characteristics are not adjusted appropriately. An inferior design can result in a loss in proppant-pack permeability and fracture aperture, which subsequently results in reduced production output. The most significant metric to evaluate the efficacy of hydraulic fracturing is fracture conductivity (F_{cd}) of a propped hydraulic fracture, which is calculated as the product of aperture width and permeability of the proppant pack (Eq. 5.2).

$$F_{cd} = \frac{k_f w}{k x_f} \tag{5.2}$$

where F_{cd} represents fracture conductivity, k_f denotes proppant-pack permeability, w denotes fracture aperture width, k denotes permeability, and x_f represents fracture half-length. It can be demonstrated theoretically that for pseudo-radial and pseudo-steady-state circumstances, a fracture conductivity of roughly 2 is the ideal value for well deliverability. A short wide fracture can be made with a greater magnitude of $k_f w$ for a given amount of proppant. Furthermore, a long, thin fracture can be formed with a lower value of $k_f w$ for a given amount of proppant. As seen in Fig. 5.5, increasing effective stress

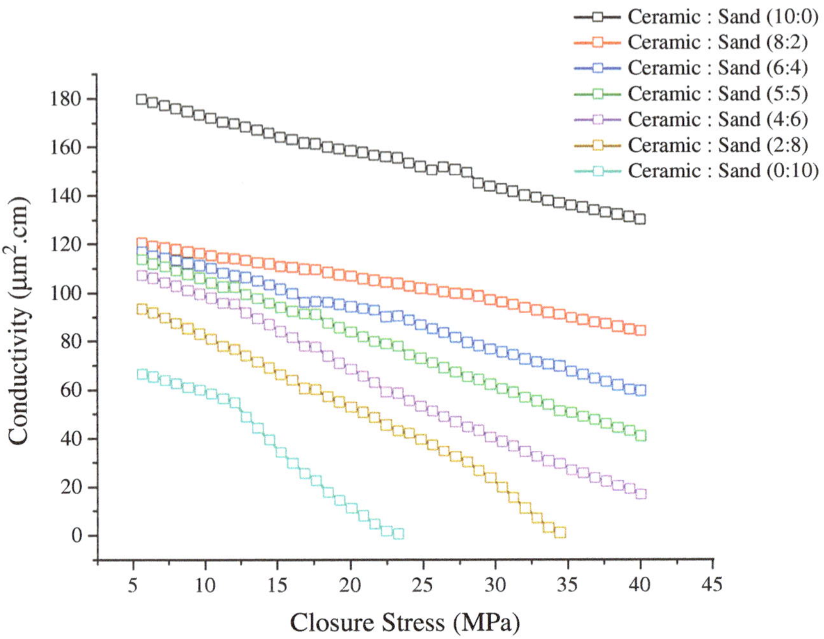

FIG. 5.5 Decreasing fracture conductivity trends with a change in closure stress for ceramic sands. *The figure is redrawn and adapted from Liang, X., Zhou, F., Liang, T., Huang, Y., Wei, D., Ma, S., 2020. The effect of combined proppants upon the fracture conductivity in tight gas reservoirs. Energy Rep. 6, 879–884. doi:10.1016/j.egyr.2020.04.007 with permission from Elsevier 2020.*

can considerably reduce the conductivity of the fracture. The case with 100% rcent ceramic proppants has the maximum conductivity, whereas the case with 100% sands has the lowest conductivity. The fracture conductivity decreases as the number of sand increases at all observed effective closure stresses. The variation in hardness and sphericity of the proppants has a significant impact on the behavior. Sands are crushed into tiny particles, whereas ceramic proppants are crushed into large particles. As a result, debris is more likely to obstruct the sand-propped fracture, resulting in a dramatic drop in fracture conductivity under the functional closure stress (Liang et al., 2020).

Artificially generated fracture networks with appropriate fracture conductivities are required for long-term reservoir production. Due to a strong proppant embedment, fracture conductivity in shale deposits can be severely decreased (Fig. 5.6A and B). Shale flakes move and clog fracture channels as a result of proppant embedment. There is more free area for fluid passage around the shale fracture than in the center (Fig. 5.6B). Because of the spherical form of the proppant, early embedment depth is larger than later embedment depth for a given grain volume loss owing to embedment.

As reservoir pressure drops, effective tension in the rock matrix rises and the magnitude of the load on proppant particle increases. This forces the

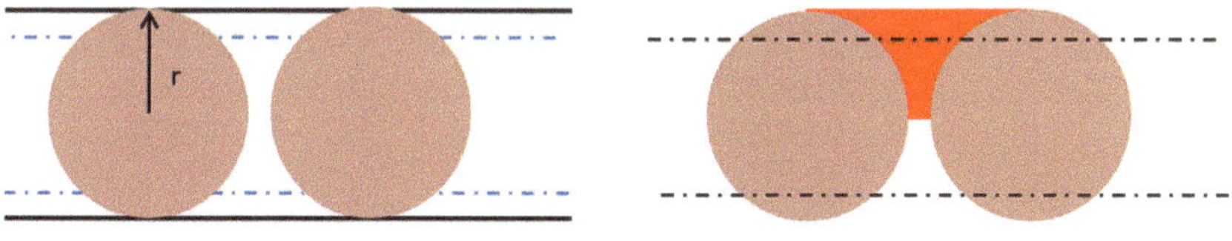

FIG. 5.6 Proppant embedment and crush during placement along the fracture walls.

proppant particles to enter the rock formation and proppant embedment occurs. For many years, proppant embedment has been identified as a source of fracture conductivity degradation in unconventional hydrocarbon reservoirs (Alramahi and Sundberg, 2012). In recent times, many studies on proppant embedment in broad fractures filled with proppant packs are being investigated. The fracture conductivity loss caused by proppant embedment, particularly in narrow fractures, may differ dramatically from fractures filled with multilayer proppant packs. When proppant concentration is low, certain proppant particles can become lodged in the rock formation under high closure pressures. As a result, proppant particles lose their capacity to sustain the fracture, owing to which the fracture aperture and conductivity may reduce significantly. Therefore, identifying proppant embedment processes in small cracks is crucial. Shale with a low Young's modulus might result in considerable proppant embedment and a decrease in fracture conductivity. The conductivity of a monolayer propped fracture is controlled by the Young's modulus and the rock formation's fracture surface roughness. The relationship between proppant embedment volume ratio and fracture conductivity shows that as the embedment volume ratio grows, the fracture conductivity of a monolayer proppant pack falls drastically. The reduction of conductivity about the loss of flow area caused by proppant embedment shows a significant fall as the grain embedment ratio rises.

5.6 Fracturing fluid leak-off and filtrate interaction with the reservoir rocks and fluids

Leak-off refers to the outflow of the fracturing fluid through permeable channels into the deeper sections of the formation. This phenomenon affects proppant transport on multiple scales including the bulk mass transport to particle migration alongside the width of the fracture (Osiptsov, 2017). Furthermore, the rate of fluid leak-off into the reservoir is one of the most important factors affecting fracture geometry and treatment efficiency. Low leak-off may result in inadequate proppant distribution within the fracture as a result of lengthy fracture-closure timings. Furthermore, excessive fluid loss may result in insufficient fracture geometry (McGowen and Vitthal, 1996). Knowledge of a formation's leak-off characteristics is crucial for identifying suitable fluid loss control techniques for treatment and predicting fracture shape. The estimate

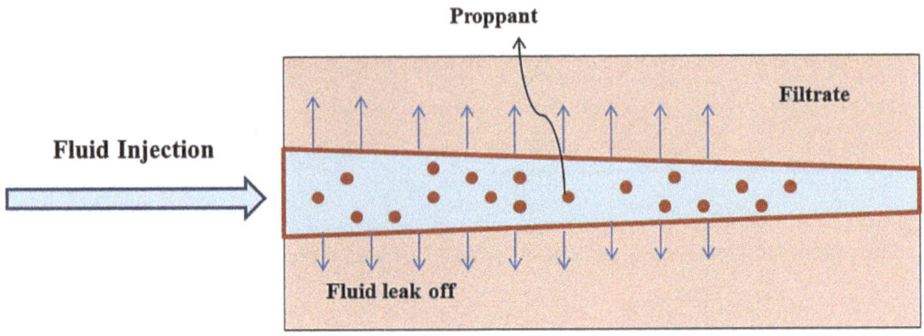

FIG. 5.7 Schematic representation of fracture geometry, fluid leak-off, and filtrate during hydrofracturing.

of formation fluid leak-off rates from pressure decline following injection has been made possible with the recent advancements in pressure analysis. Before each treatment, field measurements (often referred to as "mini fracas") are required to calculate average filtration rates.

Fig. 5.7 depicts a schematic of fracture geometry and fluid leak-off during hydraulic fracturing. During fracturing operations, fracture fluid leak-off and invasion are known to induce several types of damage to the nearby formations that include clay swelling, fines migration, and aqueous trapping. The interaction between the injected fluid and the shale matrix can affect rock strength in addition to modifying or even altering permeability (Corapcioglu et al., 2014; Iriarte and Tutuncu, 2018). Furthermore, it might result in a major reduction in elastic modulus and promote proppant embedment. Owing to this, the conductivity of fractures and the productivity of the formation are significantly affected.

Hydraulic fracturing of petroleum-bearing rock normally results in cross-flow filtration, leading to filter-cake deposition on the porous rock's face and filtrate invasion into the reservoir (Civan, 2016). This filter cake is advantageous because it reduces fluid leakage into the formation, allowing hydraulic pressure in the fracture to be maintained and fracture propagation to proceed. Generally, the fractured formation can be divided into three formations: a filter cake composed of solid additives from the fracturing fluid on the fracture face; a filtrate zone altered by fracturing fluid invasion; and a reservoir zone with the original reservoir fluids (Yarushina et al., 2013). Reservoir fluids are squeezed by the filtrate during leak-off, allowing them to penetrate deep into the reservoir, and thus known as the invaded zone.

5.7 Fracturing fluid residue clean- up

Hydraulic fractures are formed when viscous-laden fluids are injected into a hydrocarbon reservoir. After the treatment, much of the fracture fluid remains in and around the fracture, impeding hydrocarbon flow. For many years, the oil and gas sector has relied on polymers such as guar and its derivatives as a water viscosifying agent. Once the fracturing treatment is complete, decreasing the viscosity of the fluid is crucial for facilitating the flowback of the fracturing fluid so that the hydrocarbons flow from the reservoir to the wellbore with minimal obstruction. Polymer and its derivatives provide many of the requisite properties that are essential for an effective and economic fracturing fluid. However, despite all these benefits, there is one significant drawback: the insoluble residue leftover created during fluid breaking. Because of intermolecular hydrogen bonding, gaur-based polymer molecules tend to coagulate in aqueous solutions with salts, posing a two-fold problem: (1) the collates could clog formation pore spaces and restrict oil and gas flow, and (2) the polymer saturation content required to attain the ideal rheological properties would be greater. Residues from the polymers that are generally of broken, linear, or cross-linked structure contribute greatly to conductivity reduction. According to Bose et al., the adsorption and retention of polymers, as well as the bound water surrounding them, cause nearly 38% permeability impairment on water interaction-sensitive rocks in oil reservoirs (Bose et al., 2015).

Hydraulic fracture clean-up diminishes the flow resistance in the fractures and promotes production efficiency. The volume of the flowback is determined by the formation's characteristics, fracture physical properties, and the fracking operation's design. The proportion of the flowback retrieved from the well at the top might ranges from 10% to 70% of the overall volume of the fluid that was originally injected. The presence of natural microfractures in the formation, as well as elevated matrix capillary pressures, can lead to the formation holding more fluid and, owing to which, less flowback recovery at the surface. The use of clean or residue-free fracturing fluids that are compatible with the formation of rock and reservoir fluid can assist in reducing such damage. However, in certain reservoirs, the injected fracturing fluid does not quickly clean up, and it may take many months of production to reach the maximum gas flow rate. Understanding the underlying reservoir dynamics that influence the flow between the fracture and the formation is critical to designing an effective fracture treatment for a chemical interaction-sensitive reservoir.

5.8 Wettability alteration during hydraulic fracturing operations

Because the permeability of a rock is not constant, using Darcy's equation greatly underestimates the flow rate when compared to an on-field scenario. Permeability determination in shale rocks is a significantly more difficult task because the extra transport mechanism introduces new variables to account for the change owing to organic material dissolution, diffusion, and desorption. Convective flux from natural and hydraulic fractures (by Darcy flow), desorbed gas from clays and organic materials (by surface diffusion), and diffusive flux from the inorganic matrix (by Knudsen diffusion) all contribute to gas production from shale reservoirs (Habana, 2002). During the early flow-back phase, the depletion of compressed fluids deposited in fractures by convective flow raised the initial production rate exponentially (Singh, 2016; Singh and Cai, 2018). Following the initial flow, additional gas generation is caused by surface diffusion, which is a rate-limiting transport (Wang et al., 2016a,b). Furthermore, Knudsen diffusion from extremely small holes in inorganic minerals of shale (function of pore sizes and pore pressure) that do not hold any observable quantity of liquid contributes to the gas created by the matrix, which tends to reduce diffusive flow (Singh et al., 2014). As a result, shale permeability may be characterized essentially in terms of its matrix and crack. Matrix pores are exceedingly tight, finely distributed, and diverse in form. Fracture permeability in a highly porous media, on the other hand, differs from ideal permeability due to the nonideal form of fractures in the formation.

Roshann's wettability alteration studies demonstrate that bivalent ions increase contact angle compared to monovalent ions, implying that ion type and concentration have a major impact on the interaction of the oil/brine/shale system. The findings indicate that divalent cations (Ca^{2+} and Mg^{2+}) electrostatically bind oil and clays and that increasing divalent cations (Ca^{2+} and Mg^{2+}) improves oil–clay adhesion, resulting in a more oil-wet system. The presence of disjoining pressure indicates that increased salinity squeezed the double layer, resulting in a more intense oil-wet system. Also, the surface complexation demonstrates that rising salinity increased the site density of oppositely charged surface species, resulting in a stronger connection between oil and clay. Hence, the increasing salinity improves oil–clay adhesion, resulting in a more oil-wet system (Xie et al., 2017). The surface complexation also demonstrates that pH alters the quantity of charged surface species at oil and clay surfaces, which improves oil–clay adhesion when the surface species on the oil and clay are oppositely charged. Therefore, the additives in the fracturing fluid have a major impact on the wettability alteration phenomenon of the reservoir rocks.

It was discovered that for supercritical carbon dioxide-based fracture fluid, initial reservoir permeability was one of the main factors that could affect wettability alteration because lower permeability cores with more micrometer to nanometer pores were more likely to form adsorption retention in pores and throats, making treated cores much more oleophilic. That is, the low permeability reservoir matrix was significantly more easily destroyed during the SC-CO_2 fracturing process. Singh and Cai (2019a, 2019b) modeled the two-phase relative permeability of brine and sc-CO_2 in water. The spontaneous imbibition tests used to evaluate the impact of wettability alteration inside core samples revealed that as thickener content increased, oil recovery of cores decreased, confirming that wettability alteration prompted by too much thickening agent in SC-CO_2 fracturing fluid would hinder oil flow (Sun et al., 2017).

5.9 Environmental and social impact

Understanding the industry's methods and procedures, as well as their capacity to prevent or mitigate the possible consequences of stimulation, fracturing, and other oil and gas operations on societal health, the environment, and the effect of nearby flora and fauna, is critical. In the 1930s, the idea of (hydraulic) fracturing by pressure rather than explosives gained traction. Water-based fluids have been manufactured since 1953, utilizing various types of gelling agents. A hydrogel viscous fluid-containing filler material (proppants) is pumped into the well with chemical and bioadditives such as breaking agents, scale inhibitors, and other chemicals (as required in small proportions) to keep cracks open. The polymer gel is mostly cross-linked using zirconium or boron complexes in this process (Kreipl and Kreipl, 2017). Major US companies are now driving the market growth for fluid-based fracturing technology. Despite this, there are questions over microseismicity, emissions, water usage, and contaminated groundwater owing to the potential of perforating protective layers and chemicals oozing through the surface. There is no comprehensive national or international regulatory standard framework to oversee, monitor, and analyze the particular environmental and health effects of fracturing (Aczel and Makuch, 2018). Multistage hydraulic fracturing methods are widely employed to produce shale gas, and the fracturing stimulation stage often consumes more than 80% of the total water consumed during the well life cycle. To determine the effect of shale gas production on fluid/water consumption, considerations about the amount of water consumed per unit of energy generated and whether the water usage produces additional water supply stress should be analyzed. Shale gas uses higher water per unit of energy generated than traditional gas but less than crude oil and far less than enhanced/improved oil recovery (EOR/IOR) of crude oil and gas. When used to generate power, shale gas consumes less water than geothermal, nuclear, and solar energy. As a result, shale gas production is not an "elevated water consumption resource" in comparison to conventional oil production and other various kinds of energy production (Zhang and Yang, 2015). Among the possible environmental risks of hydraulic fracturing, water pollution is the most controversial and contentious issue. Several studies have examined water samples from drinking wells within a few kilometers periphery of shale gas fields and no spatial or statistical correlations can be established for methane concentrations, fracturing fluid, or deep saline brine elements presence. However, transmissibility and contamination in shallow aquifers are still plausible issues. The point of contention with hydraulic fracturing is whether the injected fluid and saline formation brine will migrate upward from the deeper sections of the formation to the shallow depth aquifers. The primary elements determining the movement are the connectedness of the formation and aquifers as well as the driving power of transmission. In terms of connectivity, hydraulic induced fracturing is very unlikely to cause fractures that permeate the reservoir to the top surface; microseismic data imaging indicates that the top of fractures is typically thousands of meters under shallow aquifers. Shale formations have fracture depths of more than 1000 m, whereas shallow aquifers have fracturing depths of less than 300 m (Fisher and Warpinski, 2012). Additionally, there are very few such incidents on record. However, the potential of fractures operating as movement channels cannot be fully ruled out. Hydraulic fracturing-induced earthquakes are said to be nondestructive mini quakes with magnitudes of less than 2 M_L and a majority of magnitudes of less than 1 M_L. However, long-term effects are yet to be analyzed and studied in detail.

5.10 Effect of mineralogy on the interaction of acid fracture fluids with carbonates

The porosity and permeability of shale are influenced largely by mineral dissolution and precipitation processes. Delayed or diminished late-stage hydrocarbon production in hydraulically fractured formations has been linked to changes in flow dynamics caused by precipitation processes. One of the most important factors affecting reservoir geochemistry is carbonate mineral reaction. Carbonate minerals play a key role in shale–hydraulic fracture interactions by neutralizing the pH, which has a considerable impact on most reactions in the reservoir and the equilibrium stability of all the other chemically or physically reactive components (Ferguson et al., 2021; Jin et al., 2020). Carbonate minerals in the Marcellus Shale vary in proportion, with calcite ($CaCO_3$) anywhere from 3% to 48% and dolomite ($CaMg(CO_3)_2$) varying between 0% and 10%. Carbonate minerals react with acid via either the dissolution or ion exchange method. Through the release of HCO_3, the reaction neutralizes acid, further buffering the chemistry and regulating the pH. The degree of the reaction between fracture fluid and carbonate grains is determined by the shale formation's carbonate composition and the acid supplied to the formation. Mineral dissolution can either increase pore size, which is beneficial for gas extraction, or cause mechanical instability or failure in microfractures, depending on the borehole's physical dynamics. The acidity, pH, and ionic strength of downhole parameters are controlled by the proportion and kind of carbonate minerals in the matrix as well as the acid injected. Carbonate dissolution by HCl is the primary dissolving process in rich carbonate shale formations, and it can greatly improve porosity/permeability. Herein, it has little or no effect on clay content. However, in poor carbonate, shale HCl can dissolve a significant amount of clay as they are ductile and difficult to fracture. This in turn could enhance permeability and promote fracture propagation. Carbonate content controls the solubility of mineral complexes such as pyrite, calcium, oxalate, and barite either directly or indirectly. These minerals' dissolution could have a major effect on hydrocarbons production. In the absence of carbonates, pyrite dissolution increases due to "chelation" by citrate ions.

5.11 Interactions of spent acid with formation rocks and fluids

An acid fracture job's injection schedule includes a variety of fluids with different features and purposes. To break and fail the mechanical integrity of the formation and create a fracture, an inert viscous fluid (i.e., pad) is usually pumped under high pressure. The acid solution is pumped once pad fluid (initially pumped to initiate fracturing) injection is completed or as an initial stage to creating nonuniform patterns onto the fracture walls. In many circumstances, the acid's reactivity leads the reservoir fracture pressure to sink. This is often caused by the high rate of fluid loss as a result of natural fractures being activated and wormholes being created. During pumping, switching between reactive and nonreactive aqueous solutions decreases the loss of fluid into the formation and retains the fracture open. This increases the acid's infiltration distance, which is especially beneficial in low-permeability reservoirs. It is also thought to help the acid "finger" its way through the viscous pad, enhancing propagation distance. However, determining the diffusion behavior of acid and acid/rock reaction kinetics is critical before injection design. As acid runs through the fracture, it reacts with the fracture face, resulting in a variety of acid concentrations, which causes rock dissolution and fracture conductivity to differ over the length of the fracture (Fig. 5.8).

During an acid-rock reaction, spent acid is described as a partially reacted acid including reaction products such as calcium and magnesium ions. The rock front further reacts with this spent acid as the wormhole and fractures stretch further into the formation. It is preferable to prolong acidification in the presence of carbonates so that the acid fluid can penetrate some distance from the well before interacting with the rock. Emulsifying the aqueous acid phase in kerosene in the form of W/O is one technique of retardation. Petroleum sulfonates or carboxylic acid salts are employed in these emulsions. Because kerosene is a continuous phase, acid–rock interaction is delayed, and the emulsion might travel many meters from the well before all of the acid is used. Alteration in wettability, fluid saturation distribution, and precipitate deposition or emulsification can occur as a result of spent acid and rock/fluid interactions (Portier and Vuataz, 2010; Pournik et al., 2013). When

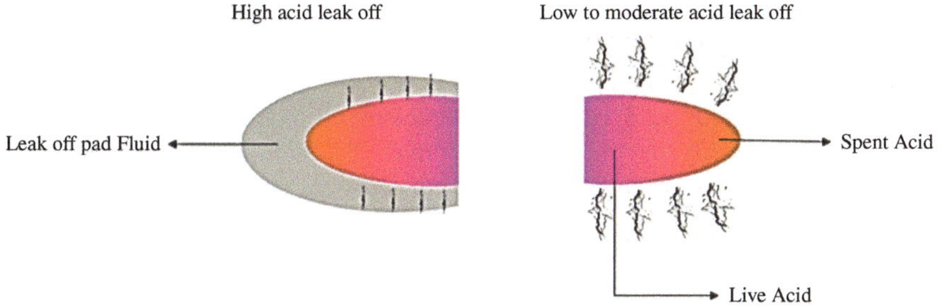

FIG. 5.8 Acid concentration profile along the fracture length and leak-off characteristics.

compared, the spent acid has a higher kinematic viscosity and a lower pH than fresh acids. The field-verified hypothesis that acid can quickly tunnel through the viscous pad fluid next to the fracture, causing the rate of acid fluid loss to be controlled by the spent acid viscosity, is supported by many experimental studies. Spent acid mixed with produced flowing oil can form emulsions resulting in higher resistance to fluid flow. The most stable emulsions are generated with the aid of formation fines. The generation of fines by acidifying sandstone formations with a mixture of hydrofluoric and hydrochloric acid is widely documented. Solid materials' capacity to stabilize emulsions is also well understood. The formation fines appear to be created by a partial breakdown of the formation matrix during the acidizing process. The acidizing process makes the fines somewhat oil-wet, and because of this mixed wettability, they are capable of stabilizing emulsions of a variety of crude oils and generated water. A demulsifier is used to treat stable oil-in-water emulsions or water-in-oil emulsions made with spent mud acids from acidized wells (Royle, 1979). Furthermore, spent acid has a higher dissolving rate and coefficient of diffusion than fresh acid. Because of the prolonged contact duration and high acidity of spent acid near the surface of fractures and wormholes, the chemical reaction becomes more powerful. Acid consumption (spending from live acid) causes a significant decrease in the effective coefficient of the mass transfer rate of hydrogen ions, which slows the acid/rock reaction kinetics and extends the fracture length. Spend acid can shift the wettability of carbonate formations to oil-wet and can cause formation damage as well. Because of the changes in the chemical and physical features of acid systems, acid spending has an impact on the extent of etching and the etching pattern. While freshly prepared acids provide more etching, partially spent acid units generate more conductive patterns. Thus, when designing injection conditions for carbonate-acidizing fracture treatment, the influence of spent acid on acid-rock reaction kinetics should be considered.

5.12 Wettability alteration during acid fracturing operations

Beyond the stimulated zone, spent acid solution may react physically and/or chemically with the formation of fluids and minerals (Aljawad et al., 2019). Alterations in wettability conditions can result from these processes. After treatment, spent acids are generally trapped around the fractures. This build-up causes a blockage in the fluid flow, causing severe damage to the formation. The spent acid trapping mechanism in fractures and wormholes is governed primarily by the capillary forces. To facilitate the flowback of spent acid, capillary pressure must be kept to a bare minimum to mobilize the acid, and the contact angle must be large enough. This means that the rocks should be in a nonwetting phase relative to the fluid flowing through them. However, generally, reservoirs are essentially water wet. Thus, a wettability alteration additive is mixed in the fracture fluid composition to increase productivity. Surfactants are generally employed as additives to interact with the surface of the rocks and alter wettability which is a common technique used in enhanced oil recovery processes (Chowdhury et al., 2021, 2022).

Breakers are chemicals that are added to the fracturing fluid to lower the molecular weight of the polymers employed. These breakers enable proppant pack clean-up by lowering the viscosity of the fracture fluid and enabling residual polymer to blow back. Breakers should ideally be completely dormant throughout treatment and then immediately activate when the pumping stops. However, because breaker activity is strongly reliant on fluid temperature, which fluctuates over time, optimal performance cannot be reached in on-field operation. Ineffective breakers, like all other components of the fracturing fluid, inflict severe damage on the proppant pack, lowering the well's production. Acid, enzymes, and oxidizers are the three most common forms of breakers utilized in modern industry.

Acids such as acetic acid (CH_3COOH) or hydrochloric acid (HCl) bind to the polymer and dissolve it. Because acids are significantly less selective and can induce significant amounts of insoluble material deposition, they are ideal for cleaning fractures damaged by a frac-job when gel did not rupture. Acids are frequently employed after a fracture surgery has been finished, and placement becomes the primary concern.

Enzymes are proteins that aid in the speeding up of metabolism or chemical processes. They function as organic catalysts, attaching to and digesting the leftover polymer. These enzymes were not depleted and will continue to function until no polymer is available for digestion. However, in very acidic or basic conditions, these enzymes undergo denaturation. Because enzymes are protein molecules, they are also particularly sensitive to thermal degradation. Among the enzymes often utilized as breakers in the industry are cellulose, pectinase, amylase, and hemicellulose. As seen in Fig. 5.9, oxidizers, like acids, disrupt the acetyl bond of the polymer backbone (Sarwar et al., 2011).

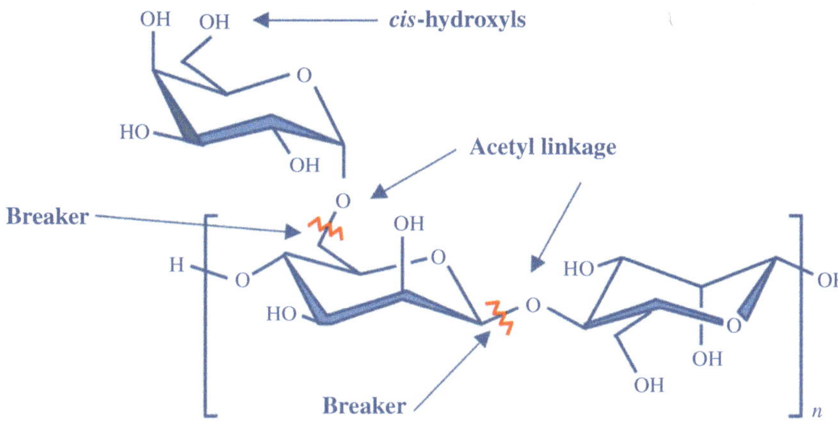

FIG. 5.9 The action of oxidizers on polymers.

At temperatures over 52 °C, powerful oxidizers such as ammonium persulfate [$(NH_4)_2S_2O_8$] produce oxygen-free radicals. These free radicals connect to the polymer strands' acetyl bonds and break them down into their constituent sugars. However, if these sugars are not handled during the formation, they produce insoluble precipitates that impair fracture conductivity. Flowback of the fractured well is thus recommended as soon as the fracture is confirmed to be closed. The oxidation of free radicals is not limited to polymer backbones, and the materials will spend on any accessible free radical acceptor, such as a gel stabilizer. Peroxide oxidizers, unlike persulfate oxidizers, are resistant to temperature fluctuations and are employed in low-temperature applications. In the case of peroxide oxidizers, the breaking action is regulated by the peroxide solution rate in water. When peroxide comes into touch with water, it releases oxygen. Some other acid additives such as corrosion inhibitors and iron control chemicals like chelating agents after spending can also affect the contact angle and wettability conditions of the rocks. Table 5.2 depicts the different additive types, common chemicals, and their purpose in fracturing fluids formulation.

5.13 Combined use of different acid fracturing fluids

The reaction rate of HCl with carbonates is high, limiting acid penetration into the reservoir, generally at lower injection rates. The corrosive nature of HCl is another disadvantage, making the use of organic acids in fracturing processes more enticing (Chang et al., 2008). However, the limitation of organic acids such as formic acid and acetic acid at high acid concentrations and low dissociation constant is a major drawback. Mixing HCl with acid can be more beneficial than using individual acids. Because of its low constant of dissociation, an organic acid does not break down into hydrogen ions when combined with HCl. As a result, the organic acid is preserved until the HCl is nearly exhausted. The carboxylic groups start to disintegrate when the hydrogen ions from HCl generate, which causes further carbonate rock dissolution from the acid front tip and thereby increasing penetration distance.

5.14 Experimental evidence on fracture damage

Experimental studies by You et al. (2016) on shale gas reservoirs discuss the formation damage and its effect on production during hydraulic fracturing operations. The study concluded that residuals of fracturing fluids can readily get lodged in shale microfractures, obstructing gas seepage channels, and lowering gas permeability. Additionally, the shale gas reservoir is damaged by phase entrapment when fracturing fluids are retained in the reservoir. Some permeability damage can be repaired by flowback, but a significant pressure

TABLE 5.2 Different chemical additives are used in fracturing fluid formulation.

Additive type	Main compound(s)	Purpose
Diluted acid (15%)	Hydrochloric acid or muriatic acid	Helps dissolve minerals and initiate cracks in the rock
Biocide	Glutaraldehyde	Eliminates bacteria in the water that produce corrosive byproducts
Breaker	Ammonium persulfate	Allows a delayed breakdown of the gel polymer chains
Corrosion inhibitor	*N,N*-dimethyl formamide	Prevents the corrosion of the pipe
Cross-linker	Borate salts	Maintains fluid viscosity as temperature increases
Friction reducer	Polyacrylamide	Minimizes friction between the fluid and the pipe
Gel	Guar gum or hydroxyethylcellulose	Thickens the water to suspend the sand
Iron control	Citric acid	Prevents the precipitation of metal oxides
KCl	Potassium chloride	Creates a brine carrier fluid
Oxygen scavenger	Ammonium bisulfite	Removes oxygen from the water to protect the pipe from corrosion
pH-Adjusting agent	Sodium or potassium carbonate	Maintains the effectiveness of other components, such as cross-linkers
Proppant	Silica, quartz sand	Allows the fractures to remain open so the gas can escape
Scale inhibitor	Ethylene glycol	Prevents scale deposits in the pipe
Surfactant	Isopropanol	Used to increase the viscosity of the fracture fluid

differential and prolonged immersion in fracturing fluids cause small particles on the fracture surface to break off and migrate, obstructing the cracks. Liang et al. quantified the formation damage induced by drilling and/or fracturing fluids as well as the efficiency of a well-screened acidic treatment fluid in reducing that damage. The findings showed that thinner fractures are more susceptible to fluid damage.

References

Abass, H.H., Al-Mulhem, A.A., Alqam, M.S., Mirajuddin, K.R., 2006. Acid fracturing or proppant fracturing in carbonate formation? A rock mechanic's view. Proc. SPE Annu. Tech. Conf. Exhib. 4, 2520–2528. https://doi.org/10.2523/102590-ms.

Aczel, M.R., Makuch, K.E., 2018. Environmental impact assessments and hydraulic fracturing: lessons from two U.S. States. Case Stud. Environ. 2, 1–11. https://doi.org/10.1525/cse.2017.000638.

Adegbola, K., Boney, C., 2002. Effect of fracture face damage on well productivity. SPE Int. Form. Damage Control Symp. Proc. 1–11. https://doi.org/10.2118/73759-ms.

Al-Ameri, A., Gamadi, T., 2020. Optimization of acid fracturing for a tight carbonate reservoir. Petroleum 6, 70–79. https://doi.org/10.1016/j.petlm.2019.01.003.

Aljawad, M.S., Aljulaih, H., Mahmoud, M., Desouky, M., 2019. Integration of field, laboratory, and modeling aspects of acid fracturing: a comprehensive review. J. Pet. Sci. Eng. 181, 106158. https://doi.org/10.1016/j.petrol.2019.06.022.

Alramahi, B., Sundberg, M.I., 2012. In: Proppant Embedment and Conductivity of Hydraulic Fractures in Shales. 46th US Rock Mech. Geomech. Symp. 2012.

Barati, R., Liang, J.-T., 2014. A review of fracturing fluid systems used for hydraulic fracturing of oil and gas wells. J. Appl. Polym. Sci. 131. https://doi.org/10.1002/app.40735.

Bose, C.C., Gul, A., Fairchild, B., Jones, T., Barati, R., 2015. Nano-proppants for fracture conductivity improvement and fluid loss reduction. In: All Days. SPE, https://doi.org/10.2118/174037-MS.

Chang, F.F., Nasr-El-Din, H.A., Lindvig, T., Qui, X.W., 2008. Matrix acidizing of carbonate reservoirs using organic acids and mixture of HCl and organic acids. In: All Days. SPE. https://doi.org/10.2118/116601-MS.

Chowdhury, S., Srivastava, S., Medhi, S., Sangwai, J.S., 2021. Novel Surfactants for Enhanced Oil Recovery. pp. 207–223. https://doi.org/10.1007/978-3-030-70026-3_7.

Chowdhury, S., Shrivastava, S., Kakati, A., Sangwai, J.S., 2022. Comprehensive review on the role of surfactants in the chemical enhanced oil recovery process. Ind. Eng. Chem. Res. acs.iecr.1c03301. https://doi.org/10.1021/acs.iecr.1c03301.

Civan, F., 2016. Cake filtration. In: Reservoir Formation Damage. Elsevier, pp. 295–341. https://doi.org/10.1016/B978-0-12-801898-9.00012-6.

Corapcioglu, H., Miskimins, J.L., Prasad, M., 2014. Fracturing fluid effects on Young's modulus and embedment in the niobrara formation. In: All Days. SPE, https://doi.org/10.2118/170835-MS.

Elsarawy, A.M., Nasr-El-Din, H.A., Cawiezel, K.E., 2016. The effect of chelating agents on the use of produced water in crosslinked-gel-based hydraulic fracturing. In: All Days. SPE. https://doi.org/10.2118/180215-MS.

Ferguson, B., Agrawal, V., Sharma, S., Hakala, J.A., Xiong, W., 2021. Effects of carbonate minerals on shale-hydraulic fracturing fluid interactions in the Marcellus Shale. Front. Earth Sci. 9, 1–12. https://doi.org/10.3389/feart.2021.695978.

Fisher, K., Warpinski, N., 2012. Hydraulic-fracture-height growth: real data. SPE Prod. Oper. 27, 8–19. https://doi.org/10.2118/145949-PA.

Habana, M.D., 2002. Relative Permeability of Fractured Rock.

Harris, P.C., 1988. Fracturing-fluid additives. J. Pet. Technol. 40, 1277–1279. https://doi.org/10.2118/17112-PA.

Hoefner, M.L., Fogler, H.S., 1989. Fluid-velocity and reaction-rate effects during carbonate acidizing: application of network model. SPE Prod. Eng. 4, 56–62. https://doi.org/10.2118/15573-PA.

Iriarte, J., Tutuncu, A.N., 2018. Geochemical and Geomechanical Alterations Related to Rock-Fluid-Proppant Interactions in the Niobrara Formation. In: Day 2 Thu, February 08, 2018. SPE. https://doi.org/10.2118/189536-MS.

Jew, A.D., Dustin, M.K., Harrison, A.L., Joe-Wong, C.M., Thomas, D.L., Maher, K., Brown, G.E., Bargar, J.R., 2017. Impact of organics and carbonates on the oxidation and precipitation of iron during hydraulic fracturing of shale. Energy Fuels 31, 3643–3658. https://doi.org/10.1021/acs.energyfuels.6b03220.

Jin, X., Zhu, D., Hill, A.D., McDuff, D.R., 2020. Effects of Heterogeneity in Mineralogy Distribution on Acid-Fracturing Efficiency. SPE Prod. Oper. 35, 147–160. https://doi.org/10.2118/194377-PA.

Kreipl, M.P., Kreipl, A.T., 2017. Hydraulic fracturing fluids and their environmental impact: then, today, and tomorrow. Environ. Earth Sci. 76, 160. https://doi.org/10.1007/s12665-017-6480-5.

Li, S., Cai, B., Ding, Y., Lu, Y., He, C., 2016. Damage of fracture face skin in massive hydraulic fracture stimulation. Soc. Pet. Eng. IADC/SPE Asia Pacific Drill. Technol. Conf. 3–12. https://doi.org/10.2118/180561-MS.

Liang, X., Zhou, F., Liang, T., Huang, Y., Wei, D., Ma, S., 2020. The effect of combined proppants upon the fracture conductivity in tight gas reservoirs. Energy Rep. 6, 879–884. https://doi.org/10.1016/j.egyr.2020.04.007.

McGowen, J.M., Vitthal, S., 1996. Fracturing-Fluid Leakoff Under Dynamic Conditions Part 1: Development of a Realistic Laboratory Testing Procedure. vol. 2, pp. 1–16, https://doi.org/10.2118/36492-ms.

Medhi, S., Chowdhury, S., Kumar, A., Gupta, D.K., Aswal, Z., Sangwai, J.S., 2020. Zirconium oxide nanoparticle as an effective additive for non-damaging drilling fluid: a study through rheology and computational fluid dynamics investigation. J. Pet. Sci. Eng. 187, 106826. https://doi.org/10.1016/j.petrol.2019.106826.

Medhi, S., Chowdhury, S., Bhatt, N., Gupta, D.K., Rana, S., Sangwai, J.S., 2021. Analysis of high performing graphene oxide nanosheets based non-damaging drilling fluids through rheological measurements and CFD studies. Powder Technol. 377, 379–395. https://doi.org/10.1016/j.powtec.2020.08.053.

Osiptsov, A.A., 2017. Fluid mechanics of hydraulic fracturing: a review. J. Pet. Sci. Eng. 156, 513–535. https://doi.org/10.1016/j.petrol.2017.05.019.

Portier, S., Vuataz, F.D., 2010. Developing the ability to model acid-rock interactions and mineral dissolution during the RMA stimulation test performed at the Soultz-sous-Forêts EGS site, France. Comp. Rend. Geosci. 342, 668–675. https://doi.org/10.1016/j.crte.2010.04.002.

Pournik, M., Li, L., Smith, B., Nasr-El-Din, H.A.A., 2013. Effect of acid spending on etching and acid-fracture conductivity. SPE Prod. Oper. 28, 46–54. https://doi.org/10.2118/136217-PA.

Royle, R.A., 1979. Demulsifier for Produced Oil-in-Water Emulsions Containing Spent Mud Acids. Patent No. US4402857A.

Sarwar, M.U., Cawiezel, K.E., Nasr-El-Din, H.A., 2011. In: Gel Degradation Studies of Oxidative and Enzyme Breakers to Optimize Breaker Type and Concentration for Effective Break Profiles at Low and Medium Temperature Ranges. Paper presented at the SPE Hydraulic Fracturing Technology Conference, The Woodlands, Texas, USA, January., https://doi.org/10.2118/140520-MS.

Singh, H., 2016. A critical review of water uptake by shales. J. Nat. Gas Sci. Eng. 34, 751–766. https://doi.org/10.1016/j.jngse.2016.07.003.

Singh, H., Cai, J., 2018. Screening improved recovery methods in tight-oil formations by injecting and producing through fractures. Int. J. Heat Mass Transf. 116, 977–993. https://doi.org/10.1016/j.ijheatmasstransfer.2017.09.071.

Singh, H., Cai, J., 2019a. Chapter 6—Permeability of fractured shale and two-phase relative permeability in fractures. In: Cai, J., Hu, X. (Eds.), Petrophysical Characterization and Fluids Transport in Unconventional Reservoirs. Elsevier.

Singh, H., Cai, J., 2019b. Permeability of fractured shale and two-phase relative permeability in fractures. In: Petrophysical Characterization and Fluids Transport in Unconventional Reservoirs. Elsevier, pp. 105–132. https://doi.org/10.1016/B978-0-12-816698-7.00006-1.

Singh, H., Javadpour, F., Ettehadtavakkol, A., Darabi, H., 2014. Nonempirical apparent permeability of shale. SPE Reserv. Eval. Eng. 17, 414–424. https://doi.org/10.2118/170243-PA.

Sun, X., Dai, C., Sun, Y., Du, M., Wang, T., Zou, C., He, J., 2017. Wettability alteration study of supercritical CO_2 fracturing fluid on low permeability oil reservoir. Energy Fuels 31, 13364–13373. https://doi.org/10.1021/acs.energyfuels.7b02534.

Wang, J., Yang, Z., Dong, M., Gong, H., Sang, Q., Li, Y., 2016a. Experimental and numerical investigation of dynamic gas adsorption/desorption–diffusion process in shale. Energy Fuels 30, 10080–10091. https://doi.org/10.1021/acs.energyfuels.6b01447.

Wang, Y., Dasani, D., Tsotsis, T.T., Jessen, K., 2016b. An experimental investigation of desorption kinetics and mass transfer in shale. In: All Days. SPE. https://doi.org/10.2118/180229-MS.

Xie, Q., Brady, P.V., Pooryousefy, E., Zhou, D., Liu, Y., Saeedi, A., 2017. The low salinity effect at high temperatures. Fuel 200, 419–426. https://doi.org/10.1016/j.fuel.2017.03.088.

Xiong, H., Davidson, B., Saunders, B., Holditch, S.A., 1996. A comprehensive approach to select fracturing fluids and additives for fracture treatments. Soc. Pet. Eng. SPE Annu. Tech. Conf. Exhib. 293–301. https://doi.org/10.2523/36603-ms.

Yarushina, V.M., Bercovici, D., Oristaglio, M.L., 2013. Rock deformation models and fluid leak-off in hydraulic fracturing. Geophys. J. Int. 194, 1514–1526. https://doi.org/10.1093/gji/ggt199.

You, L., Xie, B., Yang, J., Kang, Y., Han, H., Wang, L., Yang, B., 2016. Mechanism of fracture damage induced by fracturing fluid flowback in shale gas reservoirs. Nat. Gas. Ind. B 6 (4).

Zhang, D., Yang, T., 2015. Environmental impacts of hydraulic fracturing in shale gas development in the United States. Pet. Explor. Dev. 42, 876–883. https://doi.org/10.1016/S1876-3804(15)30085-9.

Zheng, H., Zhang, J., Qi, Y., 2020. Geology and geomechanics of hydraulic fracturing in the Marcellus shale gas play and their potential applications to the Fuling shale gas development. Energy Geosci. 1, 36–46. https://doi.org/10.1016/j.engeos.2020.05.002.

Chapter 6

Fluid-rock interactions in tight gas reservoirs: Wettability, pore structural alteration, and associated multiphysics transport

Mohammed Abdul Qadeer Siddiqui[a], Syed Oubee Khadri[b], Yueliang Liu[c], Hamid Roshan[a], Klaus Regenauer-Lieb[d], Fadhil Sadooni[e], Ahmad Sakhaee-Pour[f], and Ibnelwaleed A. Hussein[b,g]

[a]*School of Minerals and Energy Resources Engineering, UNSW Australia, Kensington, Sydney, NSW, Australia*

[b]*Gas Processing Center, College of Engineering, Qatar University, Doha, Qatar*

[c]*School of Petroleum Engineering, China University of Petroleum (Beijing), Beijing, PR China*

[d]*WASM: Minerals, Energy and Chemical Engineering, Curtin University, Perth, WA, Australia*

[e]*Environmental Science Center, College of Engineering, Qatar University, Doha, Qatar*

[f]*Petroleum Engineering Department, Houston University, Houston, TX, United States*

[g]*Department of Chemical Engineering, College of Engineering, Qatar University, Doha, Qatar*

6.1 Background of tight gas reservoirs

The term "unconventional" has a different explanation from the perspective of the petroleum industry. In contrast to conventional oil or gas reservoirs, these reserves are generally found in impermeable formations or formations with very low permeability (e.g. <1 mD). Unconventional hydrocarbon reservoirs generally do not conform to the conventional means of hydrocarbon production (Fig. 6.1). These reserves include resources such as oil sands, shale gas, coal-bed methane, gas hydrate, and tight oil. There is a substantial amount of reserves that can address global energy demands but lack the required technology for their efficient production. These types of reservoirs typically consist of complex and intricate rock and fluid properties including very low porosity and permeability (Bera et al., 2022; Siddiqui et al., 2014). For a deeper understanding of such formations, it is essential to understand their flow mechanisms.

Developments in Petroleum Science, Vol. 78. https://doi.org/10.1016/B978-0-323-99285-5.00007-7

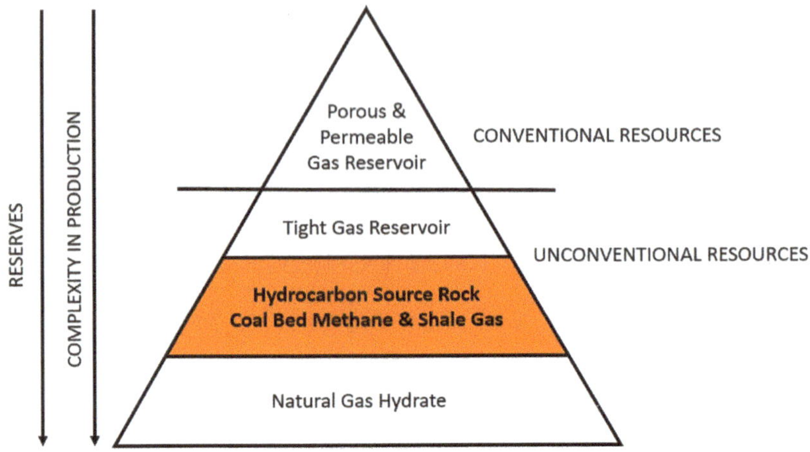

FIG. 6.1 The contrast between conventional and unconventional reservoirs. *Adapted from Norbeck, J.H., States, U., Survey, G., 2015. Identification and characterization of natural fractures while drilling underbalanced. SPE International.*

Owing to the low permeable (tight) nature of the unconventional reservoirs, hydrocarbon extraction often requires a technique known as hydraulic fracturing and sometimes acid stimulation. There are typically three types of unconventional reservoirs—coal seam gases, tight gases, and shale gas. In some cases, unconventional reservoirs contain crude oil typically found in shale formations commonly known as shale oil. Table 6.1 compares the most common types of natural gas-containing unconventional reservoirs.

Unconventional oil reservoirs are typically divided between tight oil reservoirs, heavy oil, oil sands, and shale oil (Fig. 6.2). Although the complexity of these reservoirs is not as much as unconventional gas reservoirs, the production model for these reserves is still complex. Bakken oil reservoir is one of the largest unconventional oil reservoirs containing 300–900 billion barrels trapped in tight formations. For the production of unconventional oil, there are generally three techniques which include hydraulic fracturing (similar to unconventional gas reservoirs), enhanced oil recovery (e.g., acidizing, steam flooding for very highly viscous oil), and horizontal wells.

Tight (unconventional) reservoirs are often characterized by intricate physics. Tight gas reservoirs generally comprise nanopores and micropores. Unique physics principles determine the flow at the nanolevel. The physics at the nanopore level is generally more complex in contrast to the typical macroscale Darcy flow. In shales and coals, sorption is an additional flow complexity. Adsorption/desorption is a phenomenon that affects the mechanical response (and thus the gas flow). Much research is needed to understand how sorption affects the coupled hydromechanical behavior of sorptive rocks. Gas desorption is known to significantly influence the poromechanical

TABLE 6.1 The contrast between unconventional gas reservoirs (Wall et al., 2013).

Feature	Shale gas	Coal seam gas	Tight gas
Source rock	Very low permeability with fine-grained sediment rock as the formation	Formation rock as coal	Low-permeable sandstones and limestones
Typical depth	1000–5000 m	300–1000 m	Greater than 1000 m
Production technology required	Hydraulic fracturing Horizontal drilling	Hydraulic fracturing	Large-scale hydraulic fracturing or horizontal drilling
Water usage	Large volumes of water required	Relatively lower volumes of water	Large volumes of water
Reserves	5760 tcf	30,000 tcf	185 tcf

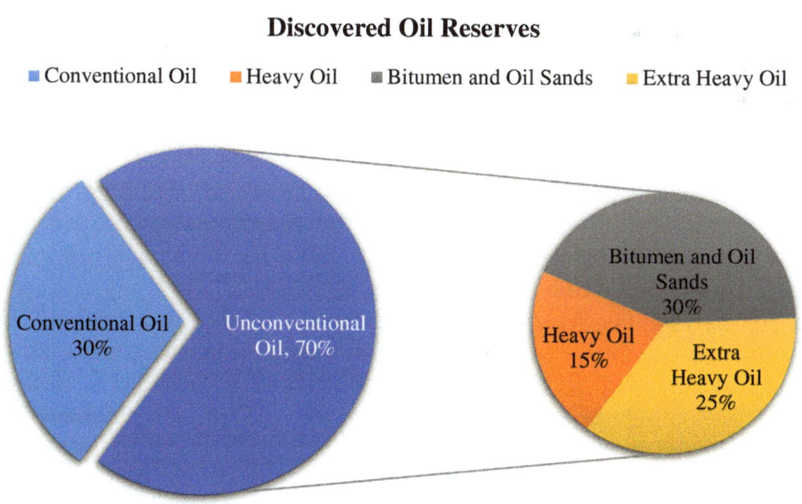

FIG. 6.2 Unconventional oil reservoirs. *Adapted from Golpour, H., Smith, J.D., 2017. Oil shale ex-situ process-leaching study of spent shale. Int. J. Eng. Sci. Invent. 6, 45–53.*

response relative to the magnitude of the sorption-swelling coefficient (Aghighi et al., 2021). The amount of adsorbed gas content in shales is significant (Algazlan et al., 2019; Liu and Rui, 2022) and thus its effect on the constitutive poromechanical behavior of shale cannot be ignored. It, therefore, requires a consistent theoretical approach to characterize such effects.

The complex pore structures and unique storage characteristics of tight rocks have made the flow characterization a complex task. Many challenges persist in developing fundamental theories and models to explain and predict fluid flow in tight reservoirs. Overcoming this challenge requires the expansion of existing theories to a broad multiscale and multiphysics context, and considerable progress is being made in this regard. This chapter will discuss some of the advances in multiscale and multiphysics methods to characterize flow and transport dynamics in tight rocks (especially shales).

6.2 Wettability and pore structural alteration

As mentioned earlier, tight rocks generally have very low permeability (and porosity), which defines the term "tight" in the context. This characteristic is due to the presence of microscale and nanoscale pore throat sizes, which contributes to lower permeability in contrast to typical permeable and porous formations (Mao et al., 2013). Pore throat structure is described as the connectivity between the pores and pore throats, including the geometric shape and distribution of the pores. The pore throat is crucial for studying fluid interactions in confined spaces or tight reservoirs. It is defined by the connectivity between pores and their resulting geometric shape (Fu et al., 2015). The pore systems in tight rocks are highly tortuous with poor connectivity and significant heterogeneities, making the pore morphology very complex (Siddiqui et al., 2020a, b). The reservoir properties such as porosity and permeability are highly correlated to the pore morphology (Zou et al., 2012). To quantify such correlations between reservoir properties and pore morphology, the pore structure must be analyzed quantitatively by characterizing the geometric shape and type of pores connected (Anovitz and Cole, 2015). Various techniques exist for quantifying the pore structure, which include X-ray diffraction (XRD) analysis, interferometry (Siddiqui et al., 2020a), thin sections (Bai et al., 2013), scanning electron microscopy (SEM) (Lin and Cerato, 2014), small-angle neutron scattering (SANS) (Clarkson et al., 2013; Siddiqui et al., 2021), high-pressure mercury intrusion (HPMI) (Li et al., 2017), X-ray computer tomography (Siddiqui et al., 2019), gas adsorption/desorption measurements (Algazlan et al., 2019), (NMR), nuclear magnetic resonance (NMR) (Roshan et al., 2016a), and routine and special core analyses (Honarpour and Mahmood, 1988). Quantification of the microscopic pore structure is essential in understanding many subsurface processes, such as the production of gas and oil, the evolution of steam in geothermal systems, carbon capture processes such as CO_2 geo-storage, low salinity water injection (Xie et al., 2016), and enhanced oil recovery techniques (Siddiqui et al., 2015; Siddiqui and Gajbhiye, 2017). Accurate quantitative predictions of such processes are possible only by an in-depth understanding of the microscopic structure of pore networks.

Over the last decade, research has been growing on the adverse effects of techniques used to produce unconventional resources such as horizontal drilling and multistage fracking (Chen et al., 2021; Pokalai et al., 2015). High-pressure fracturing fluids are used to crack the tight rock formation during hydraulic fracturing. It has been widely reported that most injected fracturing fluid is not recovered during flow back operations, e.g., in shales (Makhanov et al., 2014; Zolfaghari et al., 2016). These trapped fracturing fluids give rise to fluid–rock interactions and have varied effects on the reservoir production efficiency (Roshan et al., 2015).

Various mechanisms have been proposed in different studies that can address and provide background to the cause of fluid loss in shales after hydraulic fracturing, which also includes physio-electro-chemical forces, e.g., clay hydration and osmosis for shale formations (Binazadeh et al., 2016; Engelder et al., 2014; Fakcharoenphol et al., 2014; Makhanov et al., 2014; Roshan et al., 2016c; Xu and Dehghanpour, 2014; Zolfaghari et al., 2016), capillary imbibition (Cheng, 2012; Settari et al., 2002), microfractures initiation (Dehghanpour et al., 2013), and fluid trapped in closed fractures (Ezulike et al., 2016). A comprehensive understanding of shale formation wettability is crucial for production optimization, detailed analysis of fracturing fluid loss, and selection criteria for fracturing fluids and additives based on the rock properties (Lan et al., 2015). There is a general agreement in the scientific community that wettability is the most influencing factor affecting spontaneous imbibition (SI), a process considered the primary mechanism responsible for excess water uptake resulting in tight rocks such as shales (Dutta et al., 2014; Holditch, 1979; Odusina et al., 2011). Despite its significance, the wettability of these unconventional reservoir rock formations (especially shales) is not fully understood (Siddiqui et al., 2018). Different fracturing fluid properties can induce varied fluid–rock interactions affecting the wettability in several ways. Some critical parameters affecting shale wettability are discussed below.

6.2.1 Influence of salinity (ionic concentration) on the wettability of shale

The ionic concentration of fluids interacting with shale is an important factor that must be studied. Engelder et al. (2014) reported the contact angles of air and different brines on Haynesville shale. It was observed that 20% KCl+NaCl which had relatively higher salinity exhibited a lower contact angle (39.92°). The lower saline solution of 2% NaCl brine showed an increased contact angle (40.2°) (Engelder et al., 2014). It was also determined that the lower salinity brine solution was more effectively imbibed than the 2% NaCl brine solution attributed to mechanisms such as diffusion and osmotic effects of the solution. The study concluded that salinity only contributes a slight change in the contact angles of the fluid in question while studying Haynesville Shale. A similar effect on oil recovery was observed by He et al. (2015) and Morsy and Sheng (2014) in Muskwa and Mancos shales, respectively, where low salinity brines exhibited lower imbibition-driven oil recoveries (He et al., 2015; Morsy and Sheng, 2014). Dehghanpour et al. (2013) and Xu and Dehghanpour (2014) also studied the imbibition dependency on the salinity of the brine solution. Their study provided similar conclusions as to some previous studies that there is no consistent dependency on brine salinity for the rate of imbibition in shales (Dehghanpour et al., 2013; Xu and Dehghanpour, 2014).

Liujuan et al. (2016) studied the contact angle change with different salinities of oil-brine-rock on surfaces of varying rock minerals found in shale (albite, pyrite, K-feldspar, calcite, and dolomite). In contrast to the previous studies mentioned, the result provided a more significant observation which concluded that contact angle rises with an increase of salinity up to a limit and then reduces (Liujuan et al., 2016). However, the drawback of using pure minerals for wettability judgment is that a single mineral cannot be considered a representative of the whole shale which consists of different minerals randomly distributed. Mirchi et al. (2014) evaluated wettability at two salinities (0.1 M and 5 M) in both atmospheric and reservoir settings. The contact angle rose with salinity in reservoir conditions, reaching 27.04° at 0.1 M and 69.94° at 5 M (Mirchi et al., 2014). Pan et al. (2018) also determined that the contact angle rose marginally with increasing ionic concentration in their CH_4-brine-shale contact angle studies (Pan et al., 2018).

Based on water adsorption mechanisms in shale, Roshan et al. (2015) studied the wettability of Evergreen shale samples by implementing air–DI water and different ionic solutions. Although these samples were not considered a part of the source rock, they exhibited a high level of TOC (1.935 wt%) (Roshan et al., 2015). Based on different analyses, evergreen shale samples were reported to be composed of kaolinite (19.8 wt%) and illite (12.8 wt%). It was observed that there was an increase in contact angle with a rise in salinity. For example, the contact angles for de-ionized water, 5 wt%, and 10 wt% of NaCl solutions, respectively, were 26°, 30°, and 45°. Their study observed that adding ionic solutions could be considered if a wettability decrease is intended for a particular fluid.

Another research from Roshan et al. (2016a,b,c) provided further insights into the property of wettability for shale formations. The study suggested a physical model that describes the interaction and its physical processes. Based on their analysis, it was observed that the contact angle for air/brine (at the reservoir and ambient conditions) increases with the increase in the concentrations of ions (Roshan et al., 2016a) and that bivalent ionic solutions relatively possess lower wettability than monovalent solutions. This was said to be due to the strong adsorption affinity of divalent ions as a consequence of which the surface potential decreases leading to higher wettability. A further study from Roshan et al. (2016a) focused on water adsorption isotherms with various major cations under certain concentrations. The study observed that monovalent solutions have higher adsorption affinity than bivalent solutions at relatively lower concentrations. But on the contrary, it was also observed that with the increment in salinity, the bivalent solution adsorption becomes higher than monovalent solutions (Roshan et al., 2016b). The wettability alteration with salinity for the above-discussed shales is schematically presented in Fig. 6.3.

6.2.2 Influence of pressure on the wettability of shale

The depth and deposition history of the shale formation can dictate the pressure of in-situ fluids in the shale formation. Also, the operating pressures (e.g., the injection pressures) are varied to achieve different goals. Hence, it is important to understand the role of pressure in dictating the alteration of shale wettability. In this context, Mirchi et al. (2014) analyzed the wettability of shale formation due to the influence of pressure. The study provided conclusive evidence of the proportional relationship exhibited when the pressure is

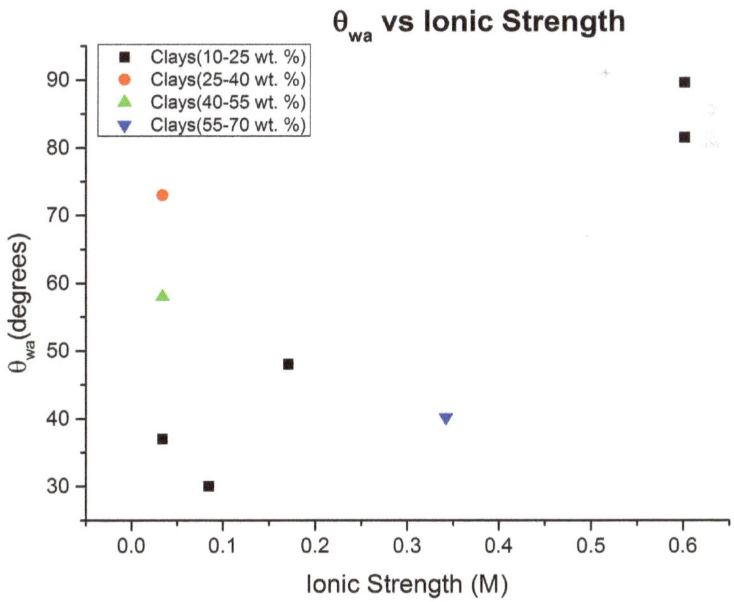

FIG. 6.3 Change of contact angles (degrees) based on ionic strength (M) of solution under ambient conditions with clay contents of 10–25 wt% (*black squares*), 26–40 wt% (*red circle*), 41–55 wt% (*green upper triangle*), and 56–70 wt% (*blue lower triangle*) (Siddiqui et al., 2018).

increased from ambient conditions to reservoir pressure (3000 psi). The general conclusion of the study was that an increase in pressure results in shale losing its affinity to water (Mirchi et al., 2014). Pan et al. (2018) had a similar observation while dealing with the CH_4-brine-shale system under different pressures, i.e., the contact angle increased with pressure (Pan et al., 2018). Roshan et al. (2016a,b,c) also reported a similar behavior while dealing with ionic solutions. The study included ionic solutions such as NaCl, KCl, $CaCl_2$, and $MgCl_2$ (Roshan et al., 2016a). Additionally, the study attributed the increment in contact angle at high pressure to a density difference reduction between the phases present in the system (Al-Yaseri et al., 2016). Fig. 6.4 provides an overview of the analysis done by Roshan et al. (2016a,b,c) and Mirchi et al. (2014).

6.2.3 Influence of temperature on the wettability of shale

The impact of temperature can alter contact angles due to changes in density and, in some cases, viscosity. Roshan et al. (2016a,b,c) and Mirchi et al. (2014) examined the impacts of physical properties such as temperature on shale wettability. The study determined that oil-water (brine) contact angles

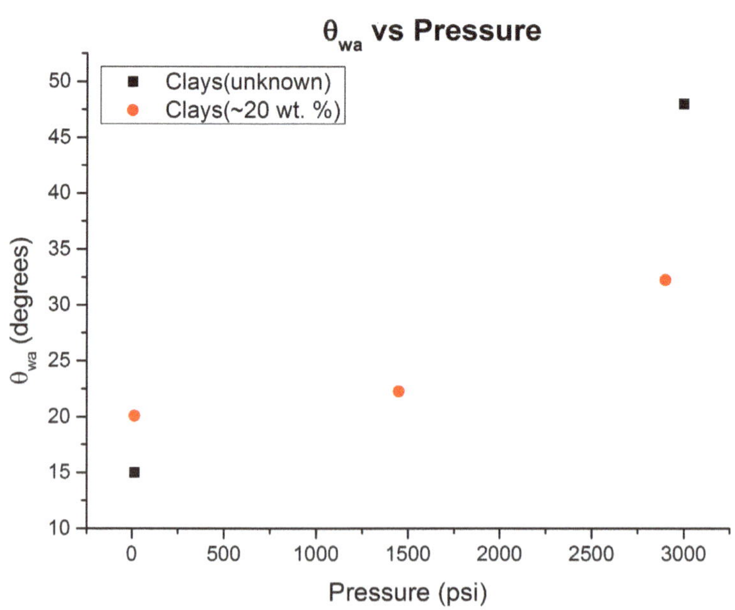

FIG. 6.4 Variation of contact angle (degrees) based on implemented pressure (psi) for shale samples comprising 20% clay content (*red circles*) and unknown clay content (*black squares*) (Siddiqui et al., 2018).

rise at reservoir conditions such as 80°C and improve from 21° to 45°. These observations included the addition of 0.01 wt% surfactant in 5 M brine and under the pressure of 3000 psi (20.7 MPa). Furthermore, Roshan et al. (2016a,b,c) analyzed the observed high contact angles for air–oil and air–brine systems in congruence to four brine solutions which included (NaCl, $CaCl_2$, KCl, and $MgCl_2$) at different concentrations. The study incorporated temperature variations to fully comprehend the effects of physical conditions on the wettability characteristic. It was observed that with temperature increase, the contact angle subsequently increased, which was also ascribed to the density differential between the two phases in the system. However, Pan et al. (2018) and Liang et al. (2016) studied LF and Shengli samples and observed that these samples had an inverse observation with CH_4–brine and air–water. The resulting contact angle decreased with an increase in temperature from ambient conditions to 60°C. Other similar studies were conducted on samples other than shale, which included the inverse between temperature change and contact angles (Arif et al., 2016; De Ruijter et al., 1998; Saraji et al., 2014). Fig. 6.5 depicts the mean contact angle at each temperature for all solutions as provided by Mirchi et al. (2014), Roshan et al. (2016a,b,c), and Liang et al. (2016).

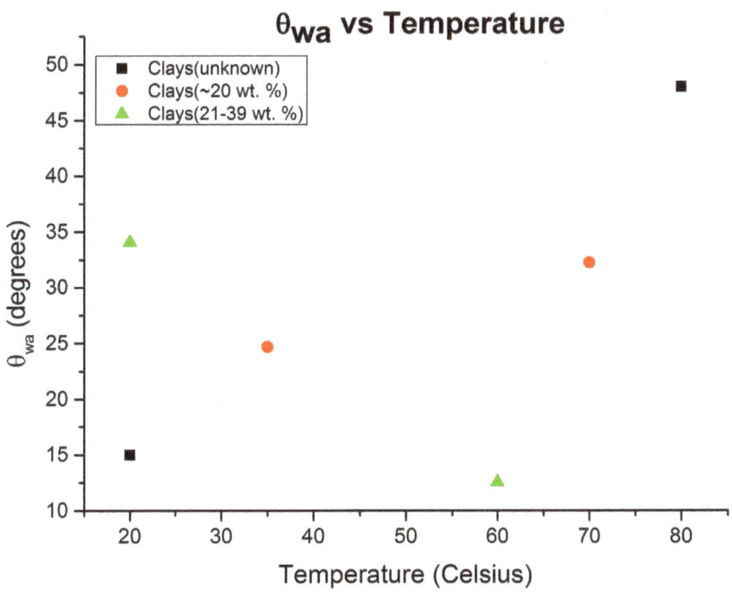

FIG. 6.5 Water–air contact angle (degrees) based on varying temperature (°C) for shales comprising unknown clay content (denoted by *black squares*), *red circles* denote ~20 wt% clay content, and *green upper triangles* denote 21–39 wt% clay content (Siddiqui et al., 2018).

6.2.4 Influence of clay content on the wettability of shale

The clay content of the shale has a unique role to play while in contact with any fluid. Clay swelling is one of the many fluid–clay interactions that could occur. Understanding such interactions to determine the extent of the impact of clay content is crucial for any upstream application for unconventional reservoirs. Borysenko et al. (2009) analyzed a shaly sandstone sample with hydrophobic nature and shale with hydrophilic nature. The samples were obtained from Western Australia. It was observed that the air–water contact angle for the shale was low in contrast to hydrophobic sandstone samples (120°) (Borysenko et al., 2009). Xu and Dehghanpour (2014) utilized M, FS, and OP shale samples based on the order of their clay content. They reported that with a decrease in clay content, there is an increase in the contact angle of air–water on these shale samples (Xu and Dehghanpour, 2014). However, no identifiable trend was observed after a comparative analysis in which the reported contact angles were compiled and plotted against the reported clay content (Figs. 6.6 and 6.7). The oil–water contact angles in Fig. 6.7 were reported elsewhere (Siddiqui et al., 2018).

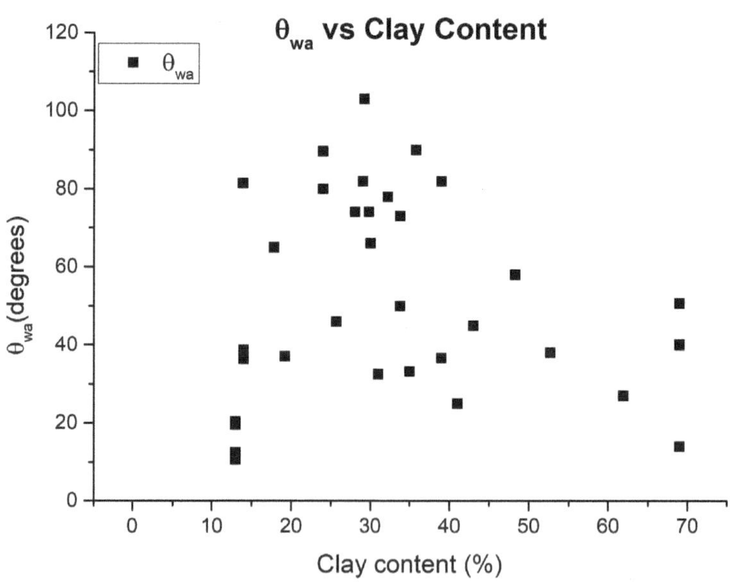

FIG. 6.6 Change in water–air contact angle (degrees) due to varying clay content (wt%) (Siddiqui et al., 2018).

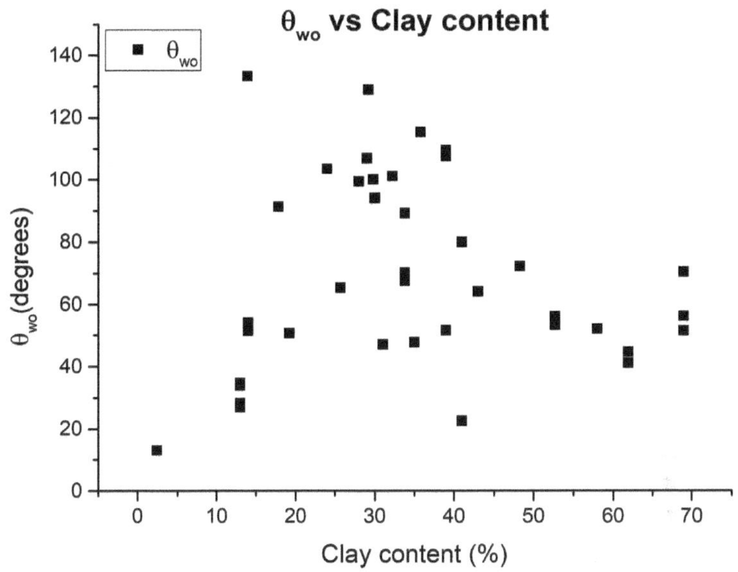

FIG. 6.7 Change in water-oil contact angle (degrees, as calculated in Siddiqui et al. (2018)) due to varying clay content (wt%) (Siddiqui et al., 2018).

6.2.5 Influence of varying organic matter content on the wettability of shale

Organic matter in shale contains the majority of hydrocarbons. Kerogen in organic pores contains adsorbed methane and other hydrocarbons, which get desorbed under pressure drawdown conditions. While dealing with fracturing fluid, it is vital to understand wettability changes due to organic matter as they may directly influence the production of hydrocarbons from nanoscale organic pores. In Green River, Niobrara, Mancos, Bakken, and shale samples, Mokhtari et al. (2013) discovered a favorable association between contact angle (air–water) and total organic content (TOC) (Mokhtari et al., 2013). On Marcellus and Duvernay samples of shale, Yassin et al. (2017) and Roychaudhuri et al. (2013) discovered that air–water contact angles positively correlate with TOC (Roychaudhuri et al., 2013; Yassin et al., 2017). The presence of significant organic matter causes the rock to be more hydrophobic and have a higher affinity for oil, resulting in a surge in contact angle with an increase in TOC. These observations are corroborated by an examination of the water-air (Fig. 6.8) and oil-water contact angles (Fig. 6.9), computed using the equation in Siddiqui et al. (2018), which shows that higher TOC results in larger contact angles. However, as previously noted, there is an observed impact of the pore system on the wettability of shale.

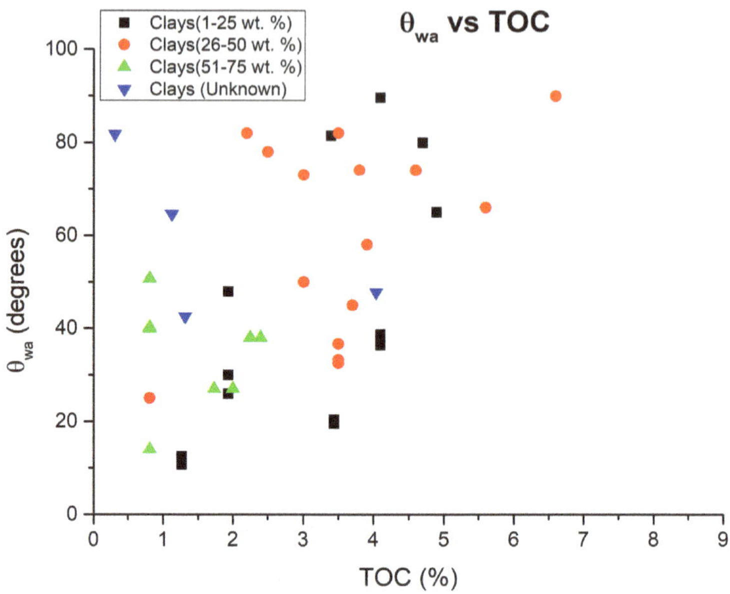

FIG. 6.8 Change in water-air contact angle (degrees) based on varying total organic content (wt%) for shale samples with clay content ranging from 1 to 25 wt% (*black squares*), 26–50 wt% (*red circles*), 51–75 wt% (*green upper triangles*), and unknown clay content (*blue lower triangle*) (Siddiqui et al., 2018).

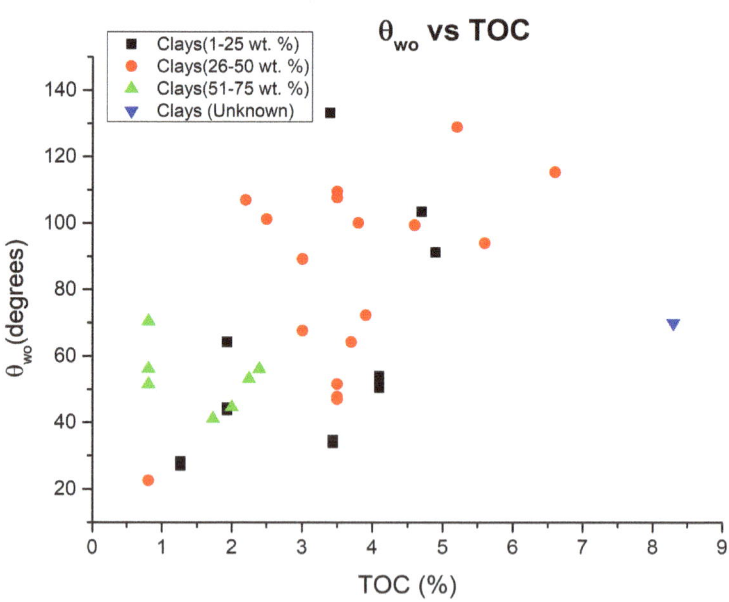

FIG. 6.9 Change in water-oil contact angle (degrees, as determined by Eq. 6.8) based on varying total organic content (wt%) for shales with a clay content of 1–25 wt% (*black squares*), 26–50 wt% (*red circles*), 51–75 wt% (*green upper triangles*), and unknown clay content (denoted by *blue inverted triangle*) (Siddiqui et al., 2018).

6.2.6 Influence of alteration of microstructure due to water imbibition

When dealing with shale, it is vital to understand the role of clay minerals. These sedimentary rocks contain a substantial amount of clay minerals, including montmorillonite, illite, kaolinite, chlorite, palygorskite, and other minerals, combined with minerals such as minerals organic matter, feldspars, quartz, and carbonates (Ougier-Simonin et al., 2016). These clay minerals are generally negatively charged. Montmorillonite is a highly negatively charged clay mineral that exhibits high reactivity while dealing with water. This would lead to microstructural damages in the shale formation caused by swelling of the formation (Gonçalvès et al., 2010; Graham et al., 2001; Mitchell and Soga, 2005; Xie et al., 2018; Zhang et al., 2017). Other shale rock formations containing a lower amount of highly charged clay minerals and a higher amount of lower charge clay minerals have similar issues. It is reported that shale containing a lower density of negatively charged clay minerals such as kaolinite and illite can cause borehole instability while in contact with water-based drilling fluid (Carpacho et al., 2004; Gholami et al., 2018). Based on such observations, it can be deduced that there is a correlation between clay minerals and the microstructure of shale rocks. When exposed to water, cations in the formation connected to clay minerals that are negatively charged, get hydrated, which in turn causes the development of a diffuse double layer (DDL). The type and concentration of cations determine the thickness of this DDL, which can also induce microstructural alterations (Roshan et al., 2016b). Instability issues in illitic or kaolinitic shales suggest that factors apart from clay swelling are at work, including complicated fluid–rock interactions.

Cagnola et al. (2017) demonstrated that several ionic solutions could modify the shale substrate's matrix structure and pore system, resulting in varied imbibition characteristics of the same sample (Cagnola et al., 2017). Furthermore, Zhang et al. (2017) observed that imbibition of water in shale could considerably impact the flow path in shale nanopores due to the system's confinement (Zhang et al., 2017). Siddiqui et al. (2019) performed 3D X-ray microcomputed tomography (XCT) before and after exposure to water in an Australian shale under both unconfined and confined stress conditions (Siddiqui et al., 2019). Fig. 6.10 depicts XCT tomographs for illustration purposes. They studied the imbibition of different ionic solutions and discovered (as observed in Fig. 6.11) that divalent ionic solutions ($CaCl_2$, $MgCl_2$) imbibed faster than monovalent ionic solutions (NaCl, KCl) at all salt concentrations. Furthermore, all ionic solutions except KCl had reduced imbibition at greater salt concentrations (0.1 M vs 0.5 M). Furthermore, it was observed that Soltrol-130 (an iso-paraffin solvent that represents synthetic oil) was imbibed the least amount by the shale.

The XCT analysis significantly altered the shale structure (Fig. 6.12), especially in unconfined conditions (Siddiqui et al., 2019). A confined stress

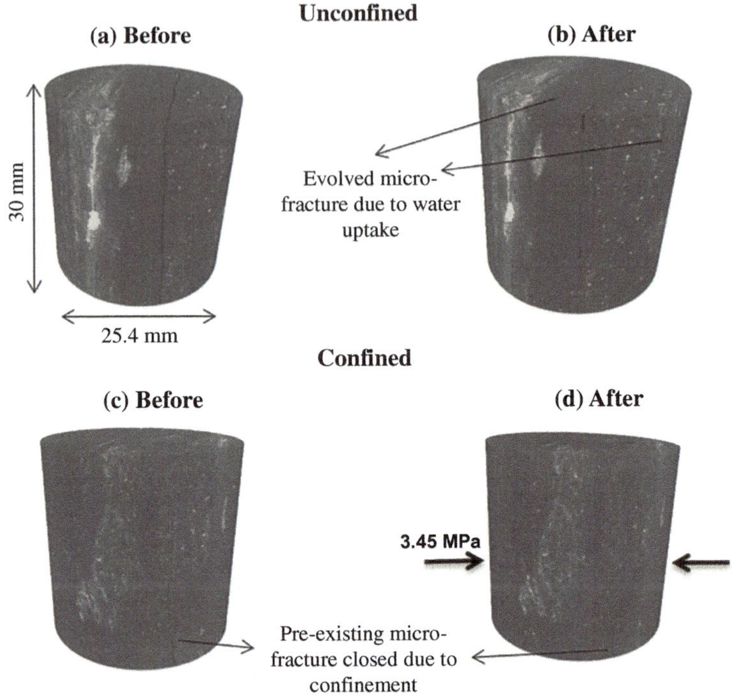

FIG. 6.10 3D visualization of gray scale tomograph of dry and wet samples in unconfined and confined conditions (Siddiqui et al., 2019).

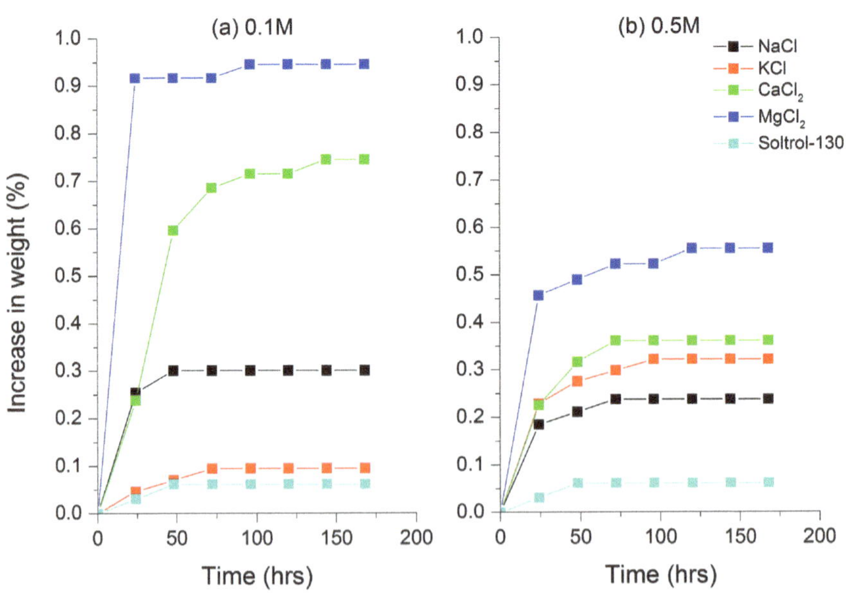

FIG. 6.11 Weight increase of shale sample vs time for (A) 0.1 M solutions and (B) 0.5 M solutions with Soltrol-130 shown on both plots (measured at ambient conditions) (Siddiqui et al., 2019).

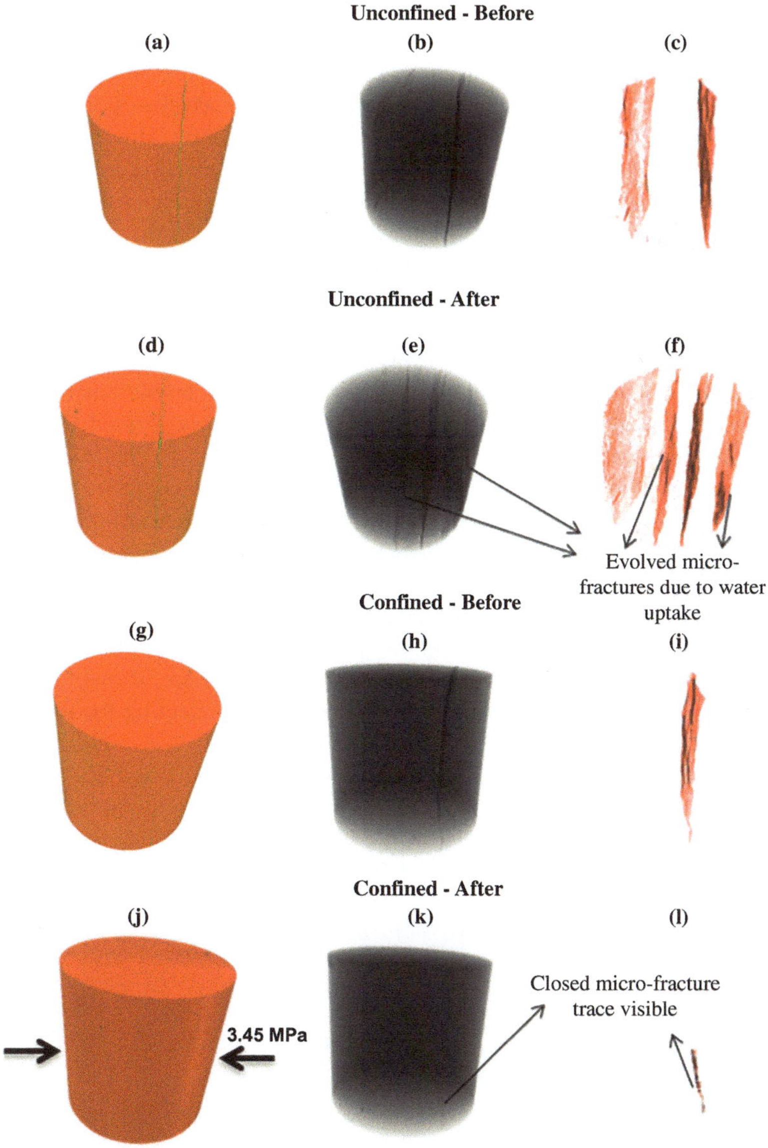

FIG. 6.12 3D segmented image of shale cores depicting microstructural changes caused by the water uptake in unconfined (depicted in (A)–(F)) and confined circumstances (depicted in (G)–(L)). (Note: The first column above in the illustration is the segmented image of the sample, while the second column represents a transparency filter on the core with microfractures inside, and the third column is the obtained micro-fractures.) (Siddiqui et al., 2019).

state was shown to inhibit microstructural alteration in inducing microfractures. At unconfined conditions, the sample's developed microfractures were parallel to the weak bedding planes. When no external pressures are present to counteract the clay minerals' hydration along the weak planes, the resulting pressures are sufficient to form fractures in the rock matrix. However, considering the fact that a 3.45 MPa confinement closed the existing microfractures implies that swelling stresses are much lower in magnitude. Additionally, it was observed that the fracture textures are rough for both newly induced and pre-existing fractures. This roughness may also play a part in determining the degree of fluid–rock interactions (AlRatrout et al., 2018; Khodadadi et al., 2020; Pirzada et al., 2021, 2020).

Digital image correlation (DIC) was further used to quantify the hydration strains. The DIC strain field (Fig. 6.13) revealed considerable positive strain (determined by the red color in Fig. 6.13) around newly formed microfractures. At the same time, the pre-existing microfracture is encircled by the depiction of negative strain. This means that the pre-existing microfracture was closed under compression. On average, the aperture of the existing micro-fracture was decreased by 20 μm. This was measured with the aid of the ImageJ software. In addition to such closure, new microfractures were also created. The observation suggests that examining the local strain is critical for precisely characterizing the permeability change caused by fracture dynamics. Furthermore, no substantial high strain was seen elsewhere, showing that the 3D segmentation of the XCT images was accurate in detecting the developed microfractures.

The linearized Poisson–Boltzmann equation (Hiemenz and Rajagopalan, 1997) was solved to explain the microstructural alteration theoretically:

$$\Pi_{EDL} = 64kTn^{\infty} \tan\mathrm{h}^2\left(\frac{ze\psi_0}{4kT}\right) \exp\left(-\kappa d\right) \tag{6.1}$$

$$\frac{1}{\kappa} = \sqrt{\frac{\varepsilon kT}{2n^{\infty}e^2z^2}} \tag{6.2}$$

where ε is the pore fluid permittivity around charged surfaces separated by a distance d and having surface potential Ψ_0 at temperature T, Z is the charge of the ion, κ is the Debye length, e is the elementary electric charge, n^{∞} is the ionic concentration away from the charged surfaces in bulk solution, and k is the Boltzmann's constant. Eq. 6.1 was then computed for two constant surface charges by utilizing varying ionic concentrations (0.1 M and 0.5 M) of monovalent and divalent solutions.

Table 6.2 displays the value of the various parameters employed, and Fig. 6.14 displays the findings.

According to the study's findings, the formation of microfractures in illite-rich shale was caused due to a large number of nanopores, with 39% of total pores measuring between 2.7 and 7 nm (Roshan et al., 2016c). Hence, in such a tight nanopore system, the overlaying of the generated diffuse double

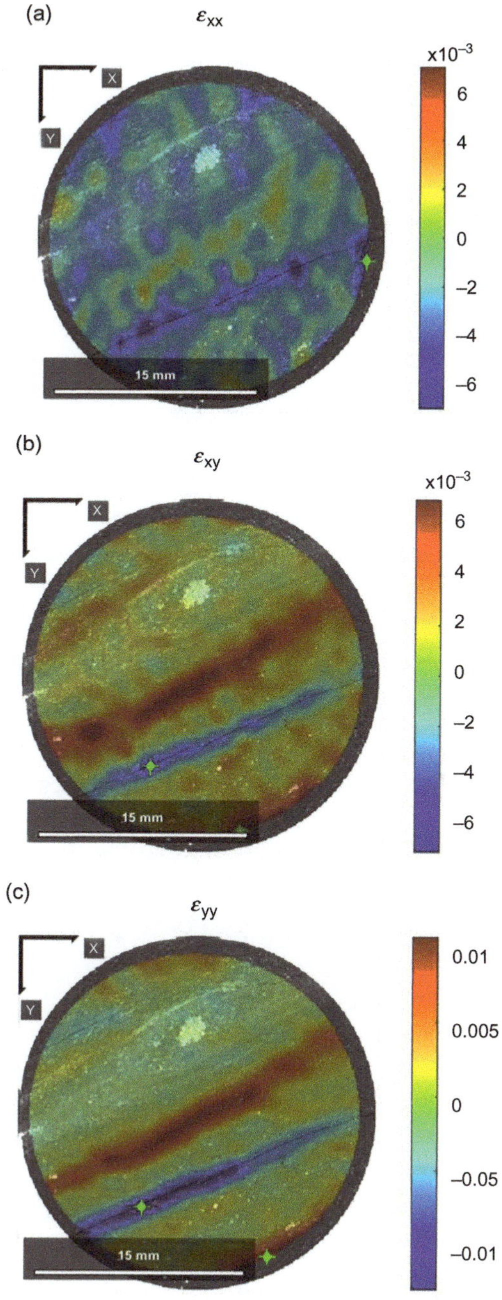

FIG. 6.13 Observed strain-fields (ε) in the directions (A) x-, (B) xy-, and (C) y- of the sample (Siddiqui et al., 2019).

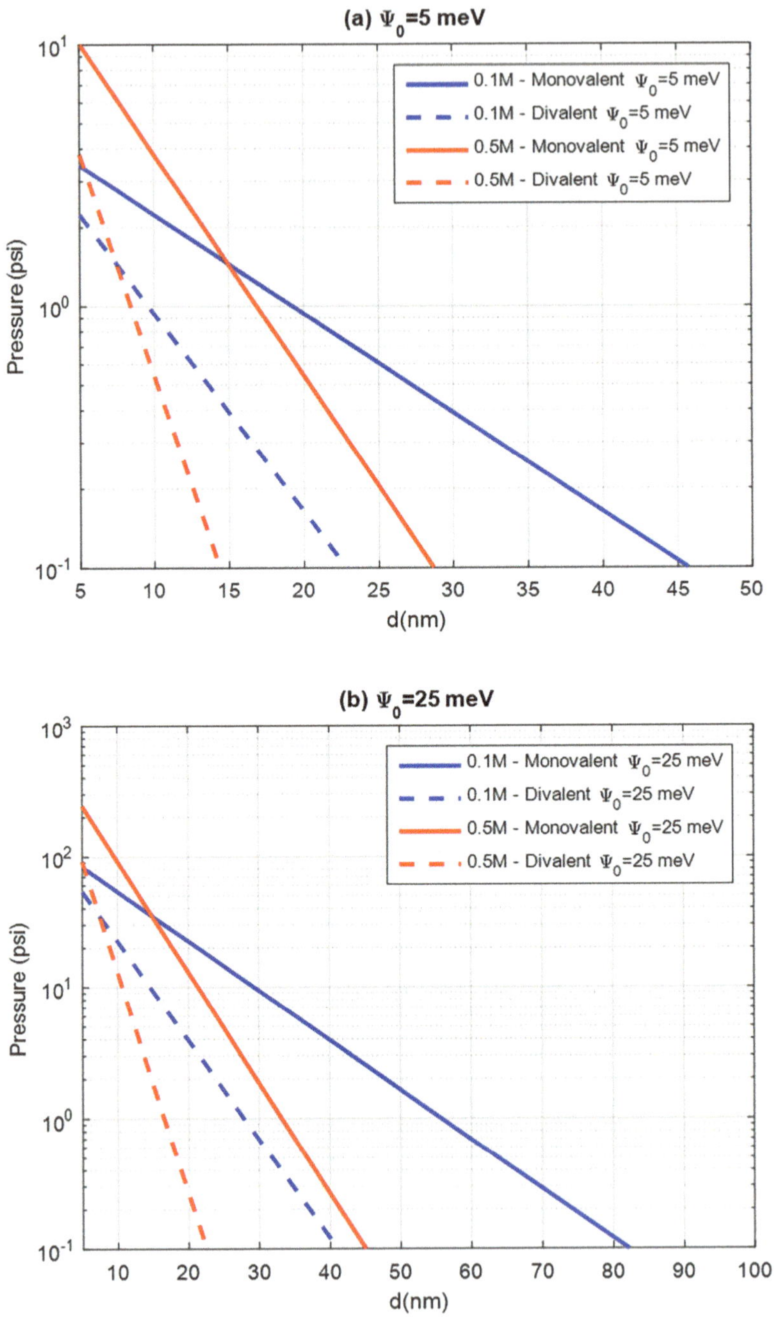

FIG. 6.14 Relationship between two negatively charged surfaces with carrying potential distance relationship and repulsive disjoining pressure (psi) for ionic solutions (monovalent and divalent solutions): (A) $\psi_0 = 5\,\text{meV}$ and (B) $\psi_0 = 25\,\text{meV}$ (Siddiqui et al., 2019).

TABLE 6.2 Parameters utilized in Eqs. 6.1 and 6.2 (Siddiqui et al., 2019).

Parameter	Value	Unit
e	1.60×10^{-19}	Coulomb (C)
T	303	Kelvin (nK)
n^{∞}	100 and 500	mol/m^3
z	+1 and +2	Unitless
ε	6.95×10^{-10}	Farads/m
k	1.38×10^{-23}	Joule/K
Ψ_0	5 and 25	meV

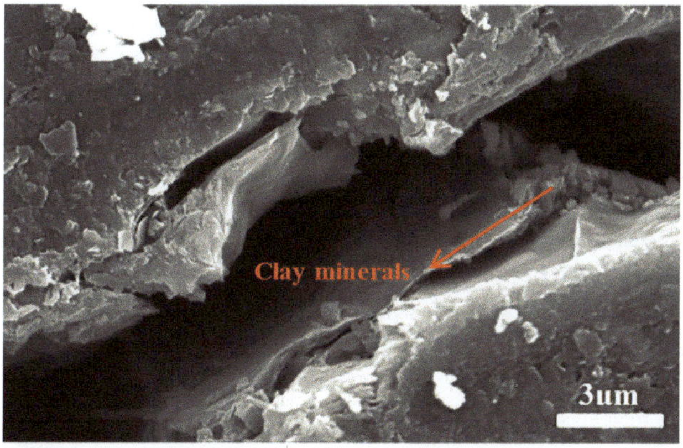

FIG. 6.15 ESEM image of shale sample depicting clay mineral formation in the microfractures (Siddiqui et al., 2019).

layer (DDL) produces additional repulsive pressures to impinge on the pore walls of the formation, fracturing the pore walls under externally unconfined conditions. This is supported by SEM images (example image displayed for 0.5 M NaCl, Fig. 6.15), which show clay minerals present in the shale samples that triggered microfractures in the pore walls (Siddiqui et al., 2019).

6.3 Types of flows in tight/unconventional reservoirs

Shale rocks are an important tight and unconventional energy resource. In shales, huge amounts of gas energy resources exist which have had significant research attention over the years. Depending on the pore sizes, shale gas can be found in either a free or adsorbed state (Wu et al., 2016; Yuan et al., 2015). The multiple pore types and sizes in shales give rise to complicated flow mechanisms that can hinder the progressive exploitation of shale gas (Huang et al., 2015). Therefore, a clear understanding of flow physics at different pore scales is important for the successful development of shale gas resources. To characterize different flow regimes based on pore sizes, a dimensionless factor known as the Knudsen number (K_n) is usually utilized. Knudsnen number is defined as the ratio of mean free path of any moving gas molecule to the pore diameter (Freeman et al., 2011). The different flow regimes that can be characterized using the Knudsen number are shown in Fig. 6.16.

The flow of dominant fluids such as water or oil in shale is relevant to the process of hydraulic fracturing. Such flow assessment would assist in solving any technical and environmental concerns (O'Malley et al., 2016; Singh, 2016). As discussed earlier, there is a persisting discrepancy observed in shale formations around the globe, which is that oil spreads more in contrast to water on shale surfaces, while contrarily capillary uptake of oil is substantially poorer than that of water. Such observation led to increased research into the perception of oil and water diffusion mechanisms in shale rocks. In addition to fluid–rock interactions, e.g., reactivity of water with clays and initiation of subsequent microfracture were found to be not the only cause for the higher water uptake by shales (Siddiqui et al., 2019). Furthermore, it was reported that even when confinement prevented the initiation of new microfractures, shale could be still imbibe more water than oil. In addition to classical Knudsen diffusion regimes, a combination of other diffusion regimes can be active such as shape-changeable diffusion, surface diffusion, and configurational

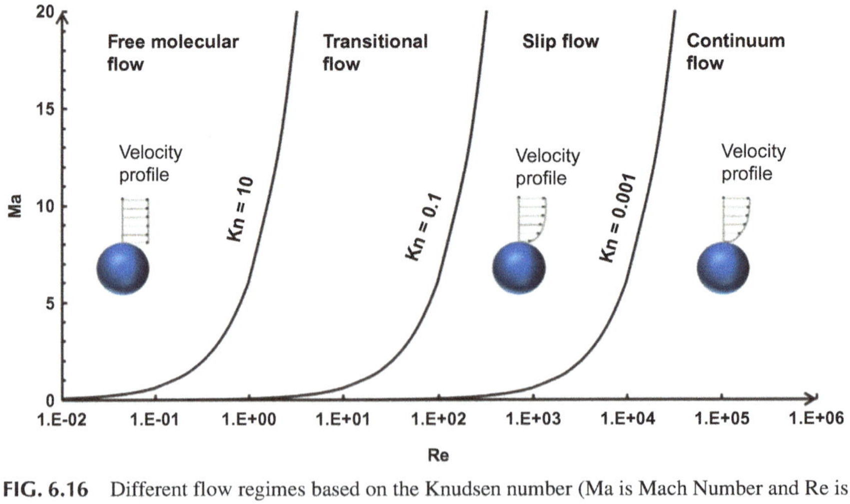

FIG. 6.16 Different flow regimes based on the Knudsen number (Ma is Mach Number and Re is Reynolds Number) (Chadha et al., 2020).

diffusion (Hirschfelder et al., 1955; Kirkaldy et al., 1987; Satterfield and Katzer, 1971). As mentioned earlier, the Knudsen number indicates the type of diffusion mechanism (Kärger et al., 2012).

When studying different diffusion regimes, it is observed that configurational diffusion is the least understood. Compared to configurational diffusion, the physics of other regimes is described accurately while providing numerical solutions to estimate the diffusion in certain cases. The fluid molecule that diffuses through the pore opening is generally smaller than the pore opening size. Configurational diffusion, in contrast, refers to the diffusion type where the molecular size of fluid that is to pass through the pore opening is almost the same size as the pore size (Kärger et al., 2012; Moore and Katzer, 1972; Satterfield et al., 1978). This type of diffusion has been identified to occur frequently in different types of porous media (Furusawa and Smith, 1974; Karger and Ruthven, 1989; Moore and Katzer, 1972; Xiao and Wei, 1992). For example, it was observed that below a limiting pore size of ~4 Å, methane diffusivity decreases significantly (Kärger et al., 2012; Rao and Sircar, 1993). The molecule cannot escape the force field at the pore wall leading to steric effects when the diffusing molecule size and the size of the pore through which diffusion is going to take place are comparable. These steric effects result in a quick and dramatic decrease in the molecule's diffusivity (Kärger et al., 2012). As a result, unexplained experimental discrepancies from published analytical models persist, particularly for complicated fluids (Yang et al., 2020) and tight and heterogeneous porous media like shales (Zhou et al., 2016). However, the experimental exploration of configurational diffusion in shales presents major hurdles. Digital methods appear to be capable of providing crucial data for such investigations. An innovative digital method of combining X-ray and neutron computed tomography (CT) was used by Siddiqui et al. (2021). They combined the experimental spontaneous imbibition of water and oil in shale samples with CT imaging on two shale samples ORS1 and ORS2. Small-angle X-ray scattering (SAXS) and focused ion beam scanning electron microscopy (FIB-SEM) analysis showed that the samples have significant submicron porosity (Fig. 6.17). Fig. 6.18 depicts the gravimetric water and oil uptake curves of two shale samples (ORS1 and ORS2). The benefit of neutron imaging is that it identifies the distribution of hydrogen (water or oil). X-ray images, on the other hand, can accurately describe the fracture spaces. Because the digital resolution cannot trace the flow path of each molecule, the water and oil identified in both continuums are molecular coalescence (Morgano et al., 2018). The 3D neutron and X-ray images of ORS1 and ORS2 were first registered and segmented. Subsequently, the proportion of pre-existing cracks filled with oil and water was calculated (Fig. 6.19). Fig. 6.19 shows that oil absorbed more than water in these pre-existing fractures. The observation from the contact angle measurements corroborates this as the sample had high oil wetness (total spreading) in the ambient setting (presence of air) (Siddiqui et al., 2021).

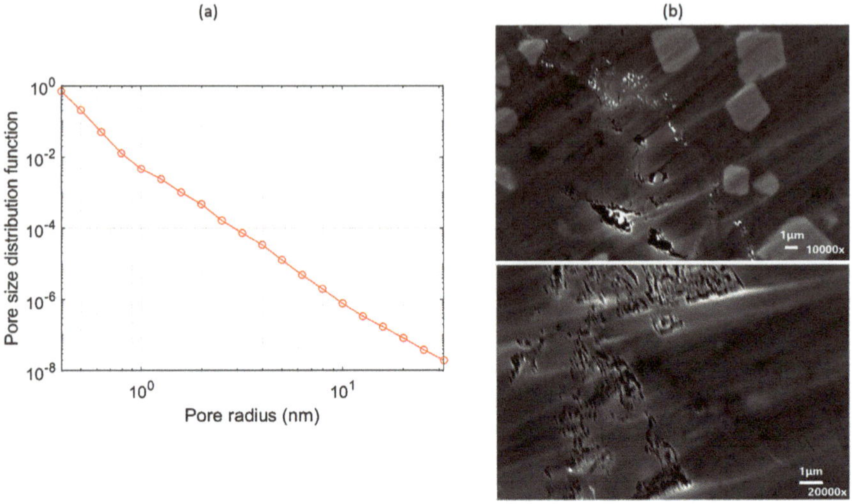

FIG. 6.17 (A) SAXS estimated pore size distribution and (B) FIB-SEM images (Siddiqui et al., 2021).

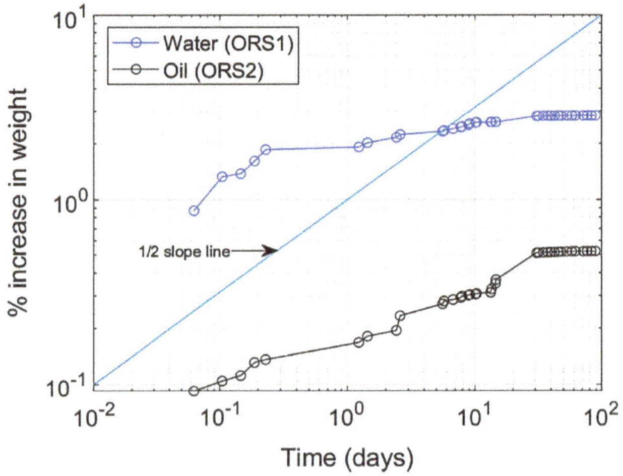

FIG. 6.18 Oil and water uptake of two shape samples (ORS1 and ORS2) (values standardized to initial weight of sample) (Siddiqui et al., 2021).

Despite having very small hydration-induced cracks in ORS1, the gravimetric curves (Fig. 6.18) depicted that the resulting cumulative water uptake was consistently greater than oil. Fig. 6.20 (top) revealed that there is substantial diffusion of water into the shale matrix.

At the initial period of one day, neutron imaging revealed that substantial water had already propagated from the fracture into the matrix. The observation that is presented in the study is supported by the water gravimetric curve

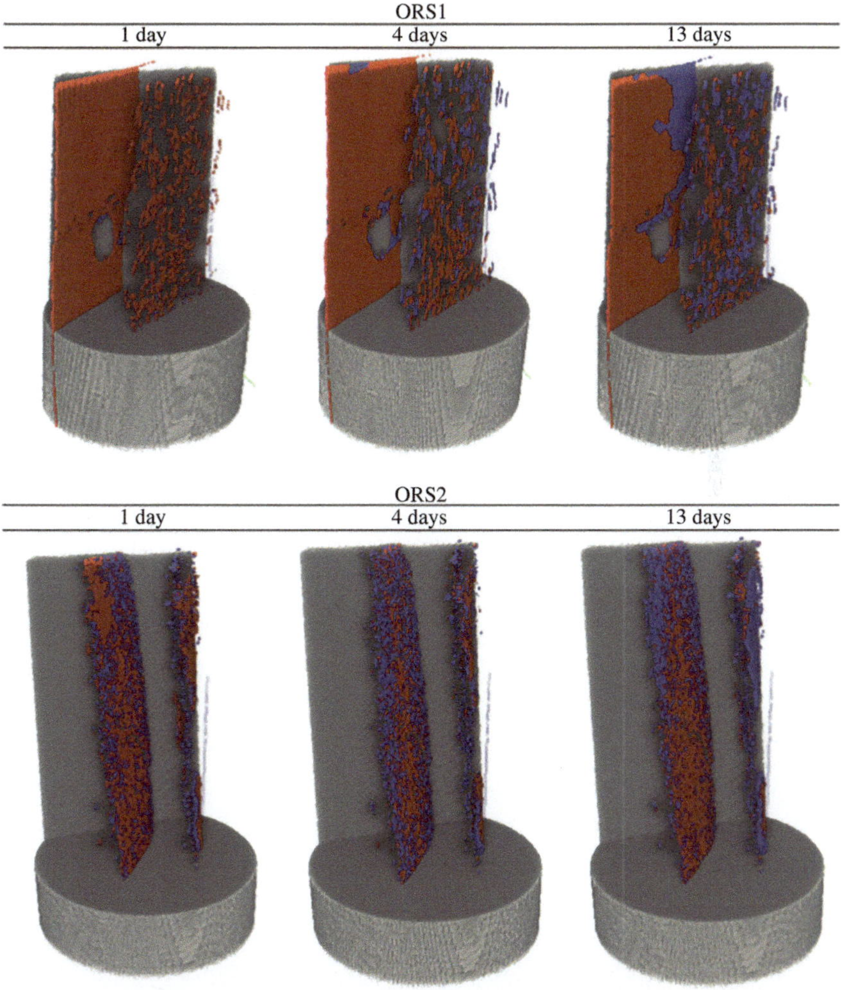

FIG. 6.19 The segmented internal structure of ORS1 (*top*) exposed to water and ORS2 (*bottom*) exposed to oil at 1 day, 4 days, and 13 days, exhibiting empty fracture space (*red*) and the proportion of fracture space filled with water/oil (*blue*). The red color denotes fracture region that is empty, while the blue color reflects fluids that are prevalent in the fractures (Siddiqui et al., 2021).

(Fig. 6.18), which shows a divergence from a half-slope line even before one day of water immersion. Observations for oil portrayed a similar deviation, and as seen, diffusion of oil from the fracture to the matrix occurred, albeit at a much slower rate (Fig. 6.20 (bottom)).

Siddiqui et al. (2021) further used nonequilibrium thermodynamics to explain the impasse of oil in shales and the easy access to water (Siddiqui et al., 2021). They used the following entropy production potential to describe the transport laws:

$$0 \leq T\gamma = -u \cdot \nabla p_{pore} - J^j \cdot \nabla \mu^j \tag{6.3}$$

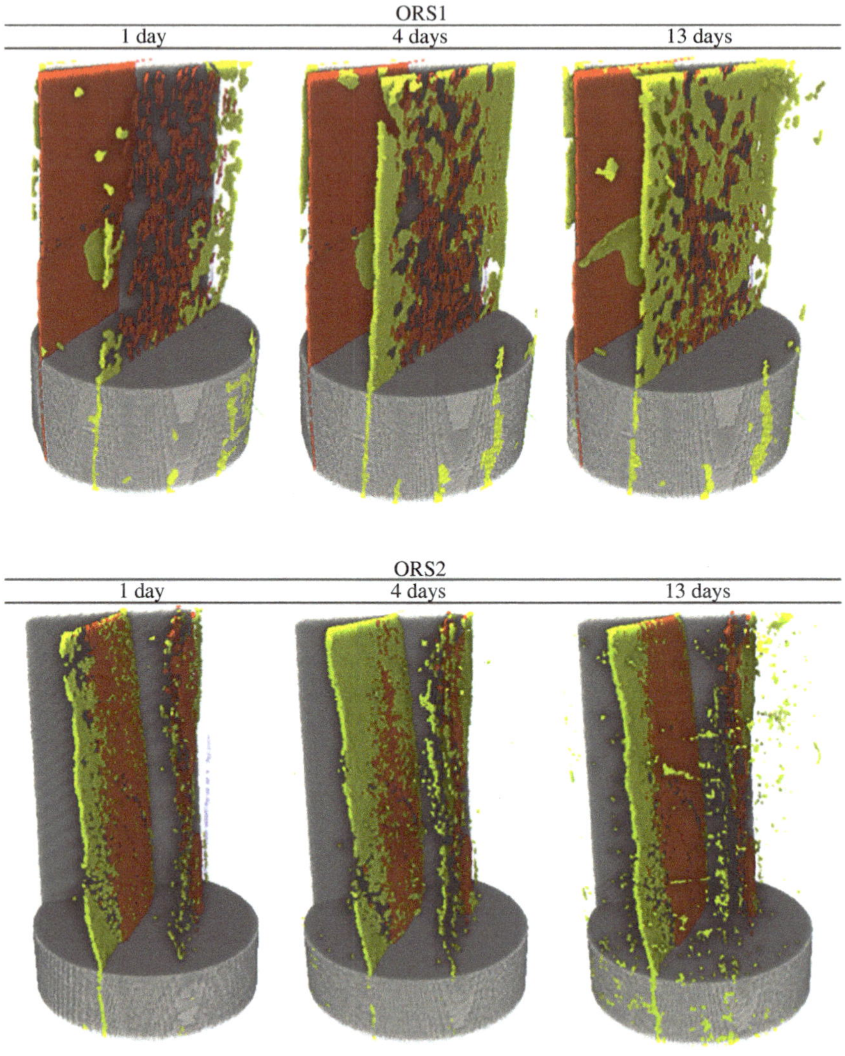

FIG. 6.20 At 1 day, 4 days, and 13 days, the segmented internal structure of ORS1 (*top*) and ORS2 (*bottom*) shows an empty plus water/oil filled fracture area (*red*) and water/oil diffusion into matrix (*yellow*). The red color symbolizes fracture space and the fluids that reside within it, while the yellow color reflects fluids that have dispersed into the matrix near to the fractures (Siddiqui et al., 2021).

where ∇p_{pore} is the fluid pressure in the porous medium and superscript $j=w$, o represents de-ionized water (DIW) or oil (Soltrol-130), respectively. While dealing with spontaneous imbibition (SI), the imbibition is induced due to capillary pressure gradient (∇P_c). The dissipation potential in Eq. 6.3 identified two forces, i.e., capillary pressure ($-\nabla P_c$) and chemical potential ($-\nabla \mu^j$).

Next, based on the theory of irreversible processes, a linear relationship can be assumed to exist between each of the fluxes and the forces, such as

$$u = L_{11} \nabla P_c + L_{12} \nabla \mu^j \tag{6.4}$$

$$J^j = L_{21} \nabla P_c + L_{22} \nabla \mu^j \tag{6.5}$$

where L_{mn} are phenomenological transport coefficients. The above equations reflect the transport fluxes due to capillarity and diffusion, respectively. Some assumptions were made to simplify the above transport equations (Eqs. 6.4 and 6.5). First, it was assumed that capillary and diffusive flow due to chemical potential change is insignificant due to the fluid's homogeneous chemical potential, especially when there are no dissolved species, i.e., $\nabla \mu^j = 0$. It should be noted that the Onsager symmetry was not utilized in this case (Onsager, 1931). Second, capillary flow can be assumed to be laminar allowing the estimation of the transport coefficients (L_{11},L_{21}) through the Poiseuille's law (assuming parallel cylindrical pores):

$$u^j = \frac{A_p r^2}{8\mu_j} . \nabla P_c \tag{6.6}$$

where A_p represents the total cross-sectional pore area, μ_j represents fluid viscosity, r represents the radius of the pore, and ∇P_c represents the gradient of the capillary pressure. Third, when the fluid molecules reach pores having similar dimensions as the size of the molecule, the same capillary pressure gradient changes the laminar flow into a diffusion process. The fluid molecule diffusion flow rate can be written as (Chinard, 1952; Pappenheimer, 1953)

$$J^j = \frac{D_j \bar{V}_j A_p}{RT} . \nabla P_c \tag{6.7}$$

where D_j is the fluid diffusivity, R is the universal gas constant, $\bar{V}_j$ its molar volume, and T is temperature. The phenomenological coefficients (L_{11}, L_{21}) are then the coefficients of ∇P_c in Eqs. (6.6) and (6.7), respectively. The relative ratio can be used to evaluate the contribution of capillary and diffusive water flow to overall flow. Because the fluid characteristics of water are more easily known than those of oil, we have for water ($j = w$)

$$\frac{u^w}{u^w + J^w} = \frac{1}{1 + \dfrac{8\eta_w D_{H_2O} \bar{V}_{H_2O}}{r^2 RT}} \tag{6.8}$$

By utilizing the mean values of viscosity, diffusivity, molar volume, and RT at 25°C: $D_{H_2O} = 2.295 \times 10^{-9}$ m^2/s, $\eta_w = 8.9 \times 10^{-4}$ Pa s, $\bar{V}_{H_2O} =$ 0.000018 m^3/mol, and $RT = 2477.6$ J/mol, the contribution of diffusive flow to the total flow can be determined from Eq. 6.8. The plot of Eq. 6.8 (Fig. 6.21) shows that Darcy's flow is the dominant flow mechanism beyond

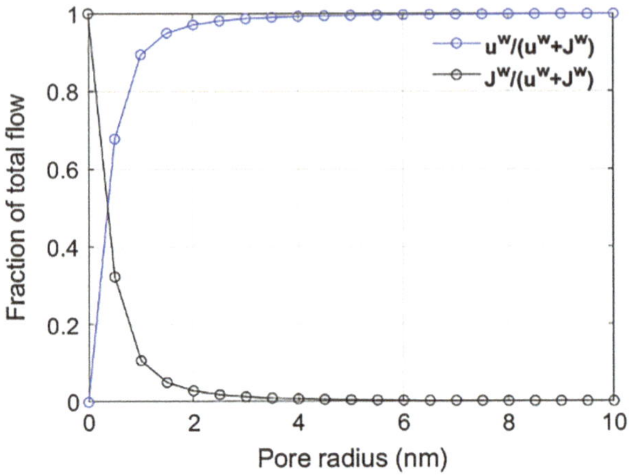

FIG. 6.21 Variation of fraction of total flow based on capillarity and diffusion (Siddiqui et al., 2021).

a pore size of 2nm. Below the threshold of 2nm, it is observed that diffusive flow becomes dominant.

Configurational diffusion is significant at smaller pore sizes that are similar to the size of molecules, and hence such molecules cannot pass easily through these pores. When molecular sizes are significantly greater than the pore size, strong repulsion results in complete denial of entry to the fluid molecule. Similar observations have been made in biological research, where it was discovered that molecular diffusion is significantly impeded at molecular sizes that are 20 times lower than pore sizes (Manegold, 1929; Pappenheimer, 1953). This happens as a result of a combination of viscous drag and steric forces. The fluid diffusion coefficient in a porous medium cannot be depicted as one value but rather varies according to the pore size distribution. Configurational diffusion is important below a threshold pore size and must be accounted for in fluid flow models. Pappenheimer et al. (1951) created a simpler limitation factor for effective diffusion (Pappenheimer et al., 1951):

$$\frac{D_c}{D_e} = \frac{(1-\lambda)^2}{1+2.4\lambda} \tag{6.9}$$

where D_c denotes the configuration diffusion coefficient, D_e is the unconstrained diffusion coefficient in bigger pores, and $\lambda = d_m/d_p$ is the ratio of the fluid molecular diameter (d_m) to the diameter of the pore (d_p). Fig. 6.22 (based on computation from Eq. (6.9)) concludes how pore radius affects effective diffusion D_c/D_e. Because of their larger molecular sizes, hydrocarbons have always lower diffusivity than water, as similarly shown for other hydrocarbon molecules too. This implies that configurational diffusion is more prominent for heavier and bigger molecules like oil (C_{12}–C_{14}) than for

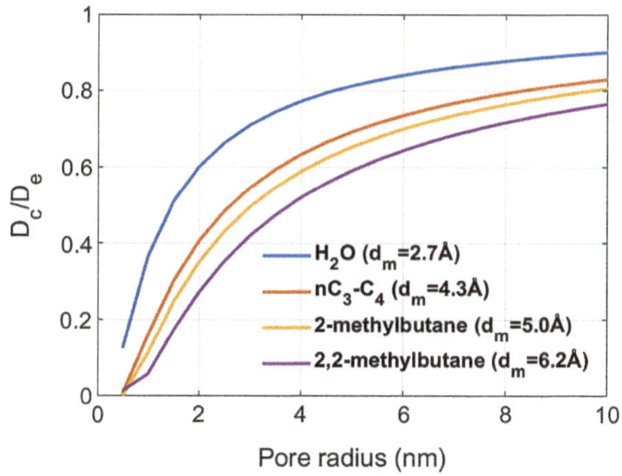

FIG. 6.22 Restricted diffusion (configurational diffusion) based on pore size for different sizes of molecules (Siddiqui et al., 2021).

water. The experimental findings corroborate this theoretical explanation. These findings also provide new insights into why oil always imbibes less than water into the shale matrix although spreads almost completely on the shale surface where water resists spreading by having higher contact angles.

It should be emphasized that configurational diffusion theory is based on certain simplified assumptions, for instance, supposing pores to be cylindrical and molecules to be spherical. As a result, it can only forecast a portion of the molecular impediment (Ferry, 1936; Pappenheimer et al., 1951). This suggests that the configurational diffusion, in reality, could be significantly higher than that predicted by theory. It is emphasized that the need for experiments and modeling is warranted to further the understanding of configurational diffusion.

6.4 Multiphysics flow in tight/unconventional reservoirs: Incorporation of chemical damage

Tight or unconventional reservoirs are characterized by significant amounts of clay minerals (Siddiqui et al., 2020b). These rocks exhibit significant complexity in their behavioral responses to different physical loadings (thermal (T), hydraulic (H), chemical (C), mechanical (M), and electrical (E)) due to clay minerals in their structure. The electrical effects are usually insignificant, but the thermo-hydro-chemo-mechanical effects can be critical (Roshan and Aghighi, 2012; Roshan and Oeser, 2012). These clay-induced complexities make the prediction of tight/unconventional rock behavior through basic phenomenological constitutive models very difficult. The coupled action of various physical phenomena active in such rocks is termed multiphysics (Jacquey and Regenauer-Lieb, 2021). They affect each other and the rock

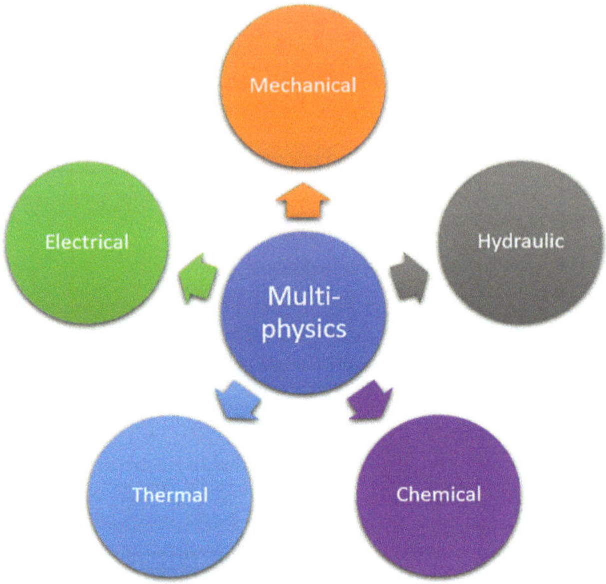

FIG. 6.23 Schematic explaining the meaning of multiphysics: A manifestation of different physics active together in the rock.

in a coupled manner which can only be accurately accounted for using thermodynamically consistent multiphysics constitutive models. The meaning of multiphysics implied in this chapter is schematically shown in Fig. 6.23.

Understanding the specific role that clay minerals play in tight unconventional rocks is critical to incorporating their physics in constitutive models. The swelling characteristics of clays are the main physics that necessitated incorporation into models that describe the behavior of clay-rich rocks. Incorporating clay swelling into constitutive models is necessary because of the issues it causes. For example, borehole instability in clay-rich formations (e.g., shale) is a critical issue the geo-energy sector faces. It is the main cause of the stuck pipe problem and overgauged boreholes. It also causes loss of circulation due to borehole breakdown failure. Such problems are brought about by the rock stress exceeding the rock strength during drilling operations (Aadnøy and Belayneh, 2004; McLean and Addis, 1990; Zhang et al., 2008). Clay-rich rocks are considered to be chemically active due to the water-adsorbing characteristic of their basal and interlayer surfaces. Two mechanisms may cause instability: (i) change in stresses resulting from coupled action of hydraulic, mechanical, thermal, and chemical interaction, which can be formulated using the concept of nonequilibrium thermodynamics (Aghighi et al., 2021; Heidug and Wong, 1996) and (ii) the alteration of rock mechanical properties as a result of the fluid–rock interactions (Roshan and Rahman, 2013).

In a multiscale shale (or any other tight unconventional) porous medium, the dynamic movement of the fluid causes dissipation. An example of such dissipation is the friction between gas and solid phase during transport and enhanced molecule–pore wall interactions in shale nanopores giving rise to energy dissipation. Such coupled (interdependent) processes cause inherent difficulties in studying shale fluid transport. Thus, the entropy maximization principle can be utilized to develop thermo-hydro-chemo-mechaniccal (THCM) flow models. The first proof of concept using energy dissipation to characterize coupled fluid flows was presented by Lars Onsager (Onsager, 1931). The reciprocal relations presented by Onsager combine thermodynamics, transport theories, and statistical mechanics. Hence, the flow theories developed using such an approach will be thermodynamically consistent and never violate the second law of thermodynamics. Many famous transport laws, such as the Soret and Dufour effects, are easily re-derived from the Onsager reciprocity relations (Monroe and Newman, 2007). Onsager's reciprocity relations relate the variable fluxes to their thermodynamic driving forces (Eq. 6.10). Conservation laws can then be used to set partial differential equations that describe the macroscopic evolution of nonequilibrium states, which can simulate macroscopic fluid flow at continuum (REV) scales.

$$\gamma = \sum_{k=1}^{N} \aleph_k D_k \geq 0 \tag{6.10}$$

where γ is the entropy production rate. It is noted that the sum of the product $\aleph_k \dot{D}_k$ (product of thermodynamic forces and corresponding fluxes) yields the dissipation in the form of nonthermal energy transformed into heat, while $\aleph_k$ represents the vector of various thermodynamic forces causing the flow and D_k represents the vector of respective fluxes. The form of entropy production (Eq. 6.10) suggests that the thermodynamic forces and the corresponding fluxes must be constitutively related to each other through appropriate coefficients:

$$D_k = \sum_j L_{kj} \aleph_j \tag{6.11}$$

where L_{kj} is the coefficient matrix. Onsager proved that the coefficient matrix L, relating the thermodynamic fluxes and forces, is reciprocal. The matrix of coefficients L_{kj} is known as the Onsager conductivity matrix and does not depend on the forces. The important practicality of the Onsager reciprocal relations is to offer alternative measurements for the same property. For instance, if it is difficult to measure the coefficient L_{kj}, one may rather measure L_{jk} (Flem et al. 1996, 2001). Thus, the energy dissipation during fluid flow has been utilized in the context of irreversible thermodynamics to develop new theoretical flow models with the inclusion of multiscale and multiphysics effects in shales (Siddiqui et al., 2021).

Although numerous THCM models exist, not much attention has been given to a critical multiphysics phenomenon arising from fluid–rock interactions in clay-rich tight unconventional rocks, i.e., chemical damage. Chemical damage implies that the microstructure of the rock deteriorates as a result of clay-water reactivity. Microstructural deterioration results in the development of microcracks. Despite the microcracks development, a measurable permanent deformation is not exhibited. However, it can result in a substantial loss in the material's strength. This loss of strength can occur even when no mechanical loads are present (Browning et al., 2017; Jia et al., 2020; Khan et al., 2004, 2007; Murakami, 2012). At the continuum scale for an isotropic material, a continuous macroscopic damage variable is used to quantitatively characterize the fluid–rock interactions-induced defects in the rock microstructure (Lemaitre, 1985, 1992). In this context, the following thermodynamic potential was used by Siddiqui and Roshan (Siddiqui and Roshan, 2022) to develop constitutive equations for a clay-rich chemically active shale:

$$W = W\left(\boldsymbol{\varepsilon}, D, p_w, p_g, \mu^\beta\right) \tag{6.12}$$

where $\boldsymbol{\varepsilon}, D, p_w, p_g, \mu^\beta$ are, respectively, strain, damage variable, pore pressure of water, pore pressure of gas, and chemical potential (β = gas, water, solute), and they represent the state variables of the rock skeleton system. The constitutive equations derived from this potential were similar to Coussy (Coussy, 2005)). The transport equations for the three phases (water (u^w), gas (u^g), and solute (J^c)) that were derived using irreversible thermodynamics were (Siddiqui and Roshan, 2022):

$$u^w = -\frac{kk_{rw}}{\tau_w}\left(\nabla p_w + \mathfrak{R}\frac{RT}{M^S\bar{C}^w\bar{C}^c}\nabla C^c\right) \tag{6.13}$$

$$u^g = -\frac{kk_{rg}}{\tau_g}\nabla p_g \tag{6.14}$$

$$J^c = -(1-\mathfrak{R})D_e\nabla C^c \tag{6.15}$$

where k is the absolute shale permeability, k_{rw} and k_{rg} are the relative permeability of water and gas, respectively, $\mathfrak{R}$ is the shale membrane efficiency, D_e is the effective solute diffusion coefficient, and τ_w and τ_g are the dynamic water and gas viscosities, respectively. The final field equations were derived after their substitution in balance laws. Also, using the damage entropy production, the following damage evolution law was derived (Siddiqui and Roshan, 2022):

$$\dot{D} = \frac{2RT(1-D)\bar{C}^c\dot{C}^c}{S_w\omega^0C^c} \tag{6.16}$$

The clay–fluid interactions induced chemical damage (in the form of micro-cracks) causing alteration of the rock permeability. This alteration is given by Choinska et al. (2007) and Zhang et al. (2007):

$$k = k_0 \left(\frac{\phi}{\phi_0}\right)^3 e^{\alpha_D * D} \tag{6.17}$$

where k_0 is the initial permeability, ϕ_0 is the initial porosity, and α_D is a constant. Hence, with the evolution of the chemical damage variable D, the absolute permeability of the shale matrix is updated according to Eq. (6.17).

The finite element simulation of the field equations (details of simulation geometry and input parameters can be found in Siddiqui and Roshan (2022)) showed that the poroelastic behavior significantly differs from when damage is neglected (Fig. 6.24). In the region of the simulated representative elementary volume (REV) saturated with water, the mechanical stresses are altered substantially when the damage-induced permeability is incorporated. On the other hand, the region which is saturated with gas remains unperturbed (Fig. 6.24). The strength of the rock is reduced due to the evolution of chemical damage. This means that with time, as more region of the REV is saturated with water, more of the REV experiences deterioration of the microstructure. Similar experimental observations were made by David et al. (2015). They observed that mechanical instability occurred only in regions saturated with water (David et al., 2015).

The evolution of the damage variable does not depend on the confining conditions. The value of the chemical damage variable ($D = D_{chem}$) is always highest where there is a maximum water saturation level, i.e., at the interface (Fig. 6.25A). The water-saturated region experienced a significant permeability increase with time while the gas-saturated region's permeability remained

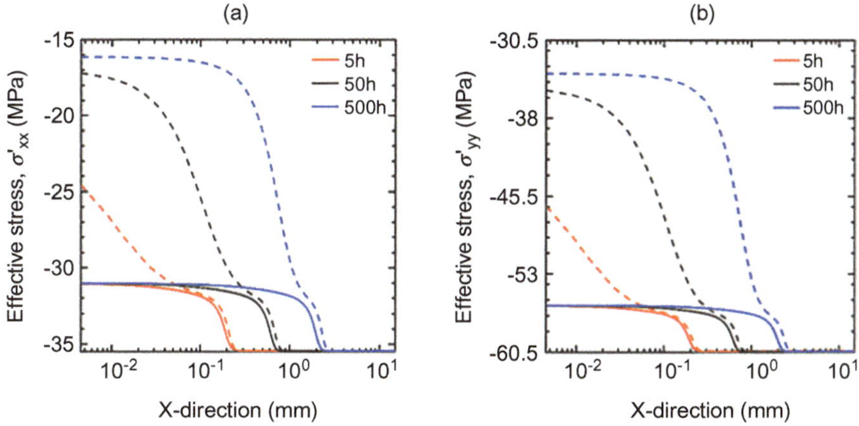

FIG. 6.24 (A) σ'_{xx} and (B) σ'_{yy} effective stress responses without damage (*solid lines*) and with damage (*dashed lines*) and along the REV's X-direction (Siddiqui and Roshan. 2022).

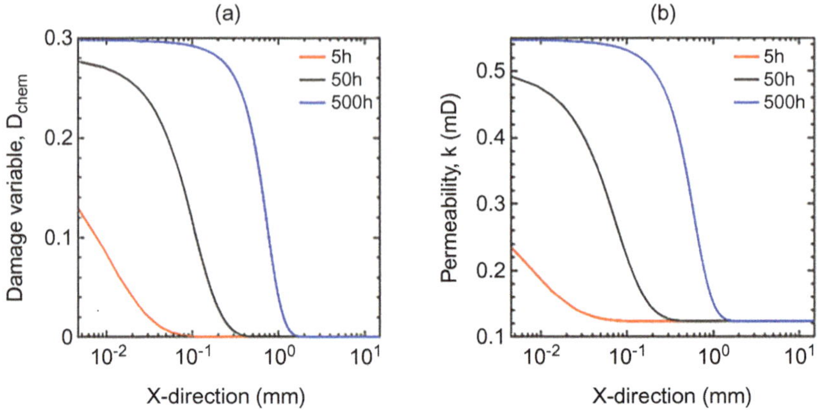

FIG. 6.25 (A) Damage variable profile and (B) permeability evolution profile along the REV's X-direction (Siddiqui and Roshan, 2022).

unchanged (Fig. 6.25B). This is because damage occurs only in the water-saturated region which experienced a ~5.5 times increase of permeability after 500 h.

The fluid–rock chemical interaction-induced chemical damage was found to vary with the chemical solute diffusivity. The effective solute diffusion coefficient characterizes solute diffusivity. The value of the diffusion coefficient is a function of the rock tortuosity (Barone et al., 1990) which can be significant in tight shale rocks. The value of D_e thus reflects the extent of tortuosity of the rock. Much slower evolution of the damage variable was observed when the value of D_e was decreased from $D_e = 3.3 \times 10^{-14}\ m^2/s$ to $D_e = 3.3 \times 10^{-16}\ m^2/s$ (Fig. 6.26A). This implies the slower solute transport due to high tortuosity (and low diffusivity). A point close to the fracture-matrix interface $(x, y) = (0.02, 7.5)$ was chosen to observe the behavior of the damage variable evolution. The point lies in the water-saturated region. The plot (Fig. 6.26B) clearly shows that when the diffusion coefficient is decreased (or tortuosity is increased), the maximum limit of the chemical damage variable is attained with much delay. This suggests that chemical damage due to fluid–rock interactions in tight shale rocks is delayed significantly if they are highly tortuous.

The main chemical force causing the microstructural alteration is the solute concentration gradient between the fracturing fluid and in-situ pore water. Thus, when the solute concentration gradient is affected by the change in solute diffusion coefficient, the evolution of rock's permeability is also significantly altered (Fig. 6.27). This substantial increase in permeability can cause water entrapment during the well shut-in periods. The entrapped water cannot flow back owing to strong wettability (capillary) forces. The results indicate that ignoring critical fluid–rock interactions-induced physical processes in numerical modeling may not reveal the actual causes of shale water loss.

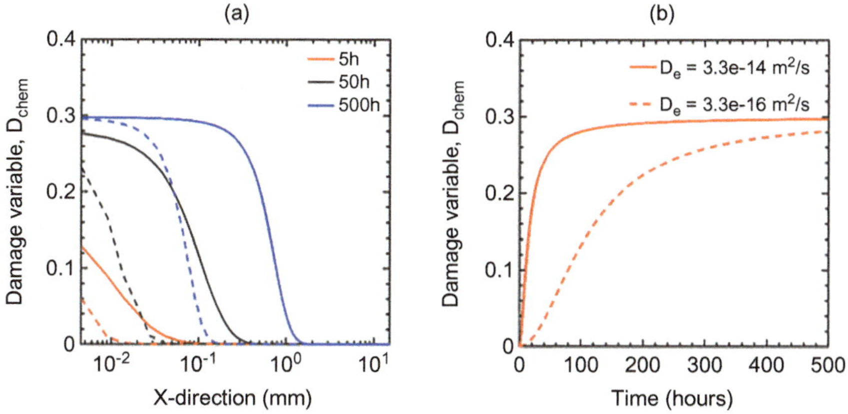

FIG. 6.26 (A) Damage variable profiles (*solid lines*: higher diffusion coefficient, *dashed lines*: lower diffusion coefficient) and (B) damage evolution with time at point (0.02, 7.5) (Siddiqui and Roshan, 2022).

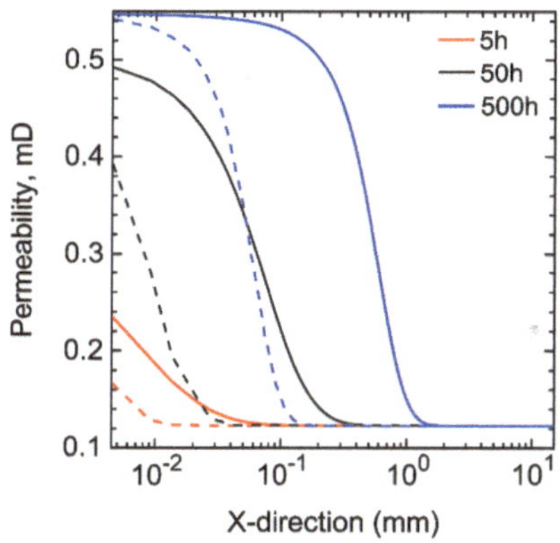

FIG. 6.27 Permeability profiles for $D_e = 3.3 \times 10^{-16}$ m^2/s (*dashed lines*) and $D_e = 3.3 \times 10^{-14}$ m^2/s (*solid lines*) (Siddiqui and Roshan, 2022).

6.5 Multiphysics flow in sorptive dual-porosity tight rocks

When studying unconventional reservoirs such as shales, coals, and tight reservoirs, these formations typically possess dual porosity systems (Roshan et al., 2019; Wei and Zhang, 2010; Yucel Akkutlu and Fathi, 2012). As mentioned earlier, flow mechanisms in shale formations or in formations with micropore and nanopore systems influence the flow adversely by coupled processes, which include matric deformations and the formation of fractures. This implies a multiphysics scenario where geomechanics and fluid dynamics (adsorption and diffusion) are coupled (Jiang and Younis, 2016). When studying shale formations, the gas exists as free gas in the pore system (organic pores) and pre-existing natural fractures. On the other hand, the gas is adsorbed in significant amounts in the kerogen or on the organic pore surface.

The observed distribution of original gas in place (OGIP) as adsorbed gas is up to 35% to 58% (in Barnett Shale formation) and can reach up to 60% to 85% (in Lewis Shale formation) (Jiang and Younis, 2016). Any resulting changes due to gas diffusion through the micro- or nanopore systems in shale are determined by the coupled physics of geomechanics (deformations) and flow processes (adsorption and desorption). It is also crucial to note that these physics are coupled in such a manner that each influences the other (Algazlan et al., 2022).

For sorptive dual-porosity rocks, Algazlan et al. (2022) developed a sorption kinetics poroelastic coupled model that utilized the aspects of non-equilibrium thermomechanics (combination of continuum mechanics and thermodynamics). Their model advanced the characterization of the influence of sorption kinetics and its resulting deformation on the shale rock (Algazlan et al., 2022). The study provided a set of constitutive equations that include time dependence of the coupling processes, thus providing a transient approach while also being thermodynamically consistent (Aghighi et al., 2021, b). The dual porosity REV used by Algazlan et al. (2022) is shown in Fig. 6.28.

The finite element simulation results on a wellbore geometry depict that the gas pressure in the bedding planes depletes slowly due to desorption of gas, prominent at late times (Fig. 6.29B). This effect was masked in the case where sorption kinetics was not considered. The slower pressure depletion is due to feeding of the gas (methane) desorbed from the matrix and into the bedding planes. This is also evident from the higher amount of matrix gas content when considering sorption kinetics (Fig. 6.29A). Furthermore, it was also determined that when no gas desorption was considered, the effective tangential stresses were wrongly estimated (Fig. 6.29C). This is due to the effect of matrix shrinkage as a result of the gas desorption.

Algazlan et al. (2022) further studied the influence of sorption kinetics on the pressure of gas in the bedding planes, free and adsorbed gas concentrations, and tangential stresses (effective) for a one-year period of production. They varied the effective diffusion coefficient (D_e) and desorption rate (K_d) (Fig. 6.30). It was observed in the study that a higher K_d does not cause any

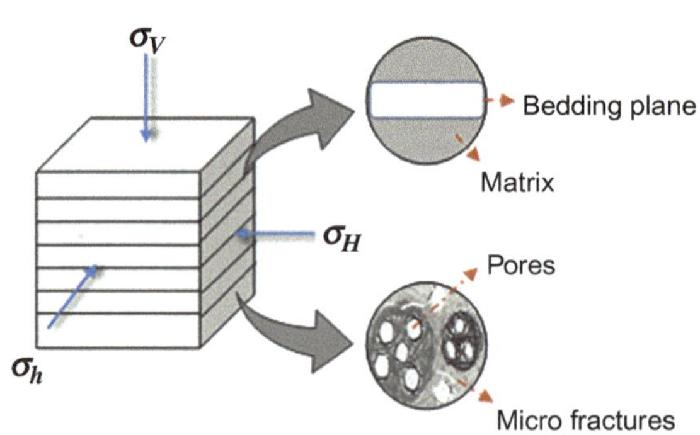

FIG. 6.28 REV for a dual-porosity system at reference configuration (Algazlan et al., 2022).

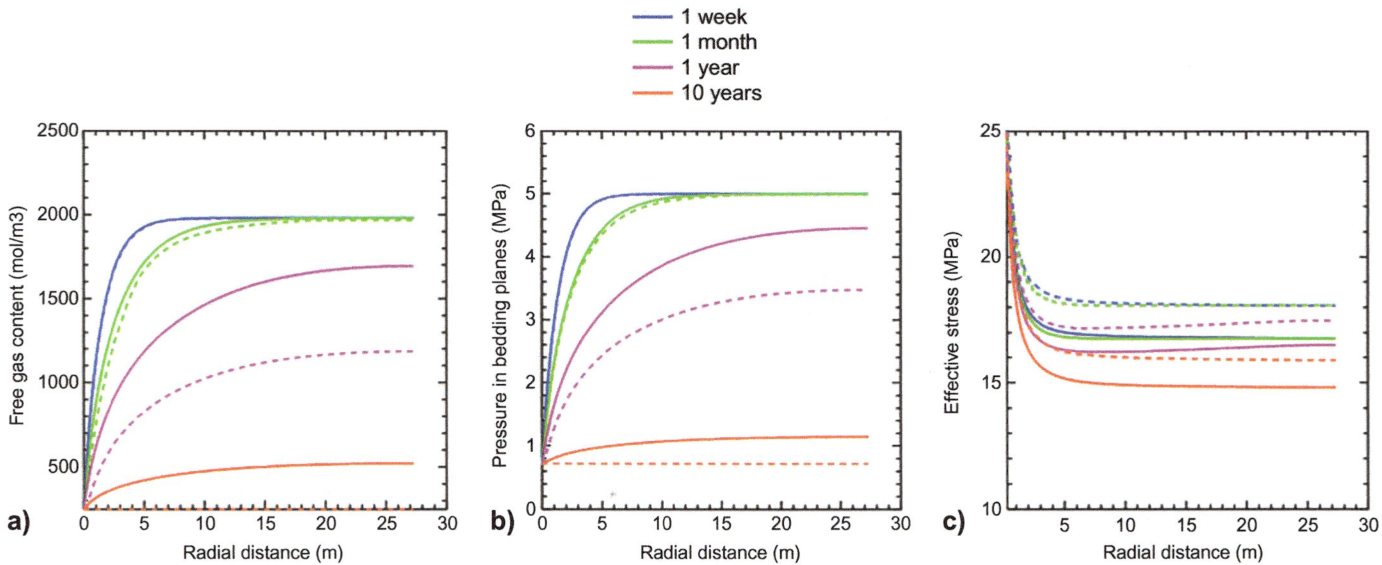

FIG. 6.29 Comparison between sorptive poroelastic (*solid lines*) and pure dual-poroelastic (*dashed lines*) for (A) free gas content, (B) pressure in the bedding planes, and (C) effective tangential stress (Algazlan et al., 2022).

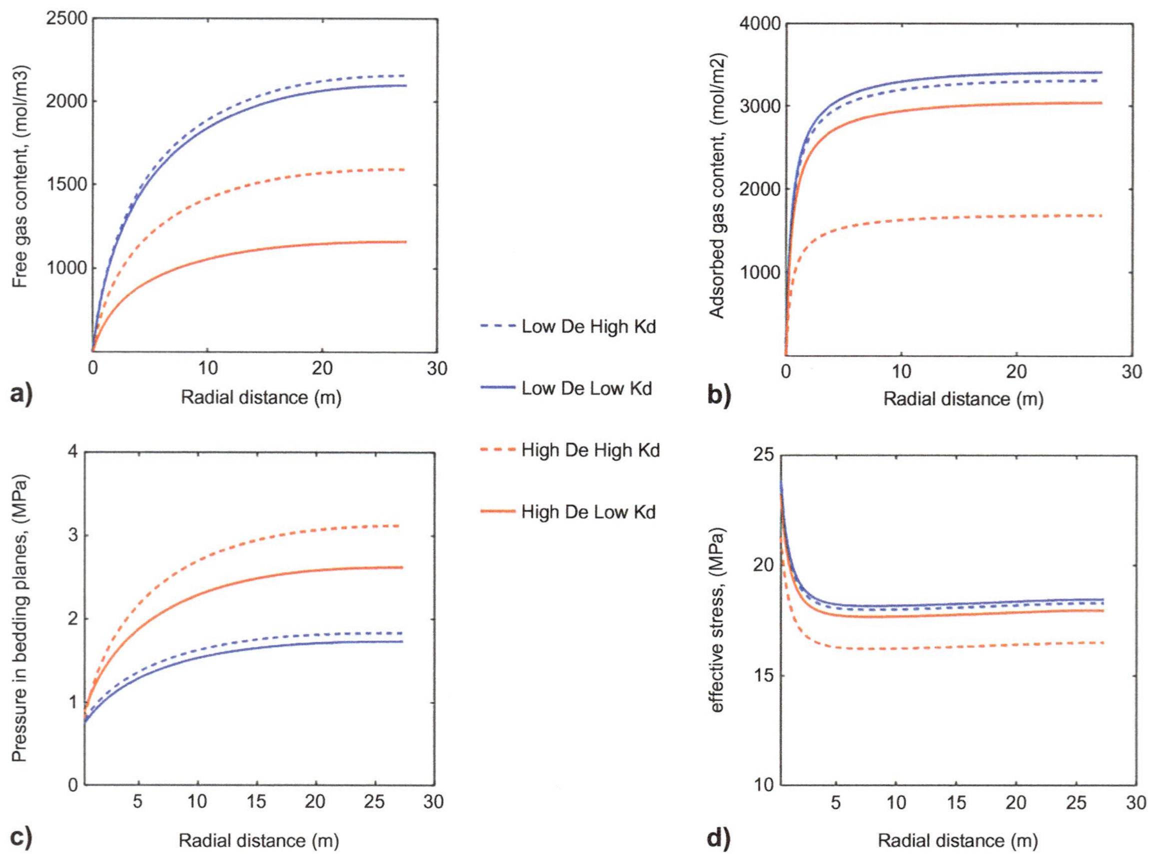

FIG. 6.30 Comparison between sorptive poroelastic (*solid lines*) and pure dual-poroelastic (*dashed lines*) for (A) free gas content, (B) pressure in the bedding planes, and (C) effective tangential stress (Algazlan et al., 2022).

substantial changes compared to a lower K_d. A higher value of K_d only minimally reduces the adsorbed gas concentration staying in place for low diffusivity values (blue curves in Fig. 6.30). Furthermore, increasing D_e (red curves in Fig. 6.30) while maintaining K_d constant (solid red curve) results in the reduction of the adsorbed gas concentration (Algazlan et al., 2022). However, increasing K_d at high D_e (dashed red curve in Fig. 6.30) significantly reduces the quantity of adsorbed gas in the matrix. This finding stresses the importance of including the gas sorption kinetics in the constitutive model for the correct prediction of rock and flow behavior due to sorptive fluid–rock interactions at a significantly greater diffusion rate in the matrix.

The feeding of the desorbed gas enhances the concentration of bedding planes gas (free gas) after one year of production (Fig. 6.30A). This effect is more pronounced in the case of high D_e. When K_d is low, increasing D_e considerably decreases the amount of free gas in the bedding planes. The gas escaping from the matrix flows into the bedding planes, which results in a slower bedding plane pressure depletion (Fig. 6.30C). Increased K_d reduces pressure depletion, mainly when D_e is high. Desorbing more gas results in enhanced shrinking of the matrix, which lowers the effective stresses (Fig. 6.30D). This consequence of stress alleviation is larger when D_e is higher, similar to the adsorbed gas behavior in Fig. 6.30B. This demonstrates the significance of sorption kinetics on effective stresses, particularly when the effective diffusivity values are large.

6.6 Conclusion

Several advances have been made in the field of thermodynamically consistent characterization of clay–fluid interactions in tight rocks (especially shales). The main focus was to reveal how different shale complexities, especially flow therein, are handled using advanced multiphysics modeling and multiscale experiments. Some new developments were discussed in this chapter in the context of shale gas production and shale water loss that are exhaustive yet incomplete. The investigations of wettability and associated flow complexities in shales were discussed as a combination of simple and advanced digital techniques. The wettability has been a complex problem for shale researchers. The chapter also presented the complex clay effects coupled with the fluid flow that give rise to nonequilibrium behavior such as chemical damage. It was discussed how the neglection of chemical damage in flow models deteriorates model accuracy. The advances presented herein, in the context of improved flow models, account for such behavior in a thermodynamically consistent manner. Tight rocks such as shales and coals are highly sorptive and this particular sorptive fluid–rock interaction was also discussed in the realm of nonequilibrium thermodynamics. The simulation of the constitutive model revealed that sorption kinetics significantly impacts the overall poromechanical behavior and must be accounted for in flow model development. Overall, the transport behavior in shales (and other tight rocks) is highly complex occurring at multiscales and affected by multiphysics with many more advancements yet to be made.

Acknowledgments

The authors would like to acknowledge the support of the Qatar National Research Fund (a member of the Qatar Foundation) through Grant # NPRP12S-0305-190235 and NPRP11S-1228-170138. The authors, particularly M.A.Q.S, H.R., and K.R., also acknowledge the support of the Australian Research Council (ARC) through the Discovery Project Grant (DP200102517). The findings achieved herein are solely the responsibility of the authors.

References

Aadnøy, B.S., Belayneh, M., 2004. Elasto-plastic fracturing model for wellbore stability using non-penetrating fluids. J. Pet. Sci. Eng. https://doi.org/10.1016/j.petrol.2004.07.006.

Aghighi, M.A., Lv, A., Roshan, H., 2021. Non-equilibrium thermodynamics approach to mass transport in sorptive dual continuum porous media: a theoretical foundation and numerical simulation. J. Nat. Gas Sci. Eng. 87, 103757. https://doi.org/10.1016/j.jngse.2020.103757.

Algazlan, M., Pinetown, K., Grigore, M., Chen, Z., Sarmadivaleh, M., Roshan, H., 2019. Petrophysical assessment of Australian organic-rich shales: beetaloo, cooper and perth basins. J. Nat. Gas Sci. Eng. https://doi.org/10.1016/j.jngse.2019.102952.

Algazlan, M., Abdul, M., Siddiqui, Q., Roshan, H., 2022. A sorption-kinetics coupled dual-porosity poromechanical model for organic-rich shales. Comput. Geotech. 147, 104755. https://doi.org/10.1016/j.compgeo.2022.104755.

AlRatrout, A., Blunt, M.J., Bijeljic, B., 2018. Wettability in complex porous materials, the mixed-wet state, and its relationship to surface roughness. Proc. Natl. Acad. Sci. U. S. A. 115, 8901–8906. https://doi.org/10.1073/pnas.1803734115.

Al-Yaseri, A.Z., Roshan, H., Lebedev, M., Barifcani, A., Iglauer, S., 2016. Dependence of quartz wettability on fluid density. Geophys. Res. Lett. 43, 3771–3776. https://doi.org/10.1002/2016GL068278.

Anovitz, L.M., Cole, D.R., 2015. Characterization and analysis of porosity and pore structures. Rev. Min. Geochem. 80. https://doi.org/10.2138/rmg.2015.80.04.

Arif, M., Al-Yaseri, A.Z., Barifcani, A., Lebedev, M., Iglauer, S., 2016. Impact of pressure and temperature on CO_2-brine-mica contact angles and CO_2-brine interfacial tension: implications for carbon geo-sequestration. J. Colloid Interface Sci. 462, 208–215. https://doi.org/10.1016/j.jcis.2015.09.076.

Bai, B., Zhu, R., Wu, S., Yang, W., Gelb, J., Gu, A., Zhang, X., Su, L., 2013. Multi-scale method of nano(Micro)-CT study on microscopic pore structure of tight sandstone of Yanchang formation, Ordos Basin. Pet. Explor. Dev. https://doi.org/10.1016/S1876-3804(13)60042-7.

Barone, F.S., Rowe, R.K., Quigley, R.M., 1990. Laboratory determination of chloride diffusion coefficient in an intact shale. Can. Geotech. J. https://doi.org/10.1139/t90-021.

Bera, A., Kumar, S., Foroozesh, J., Gharavi, A., 2022. Journal of Natural Gas Science and Engineering Multiphysics gas transport in nanoporous unconventional reservoirs: challenges of mathematical modelling. J. Nat. Gas Sci. Eng. 103, 104649. https://doi.org/10.1016/j.jngse.2022.104649.

Binazadeh, M., Xu, M., Zolfaghari, A., Dehghanpour, H., 2016. Effect of electrostatic interactions on water uptake of gas shales: the interplay of solution ionic strength and electrostatic double layer. Energy Fuels 30, 992–1001. https://doi.org/10.1021/acs.energyfuels.5b02990.

Borysenko, A., Clennell, B., Sedev, R., Burgar, I., Ralston, J., Raven, M., Dewhurst, D., Liu, K., 2009. Experimental investigations of the wettability of clays and shales. J. Geophys. Res. Solid Earth 114, 1–11. https://doi.org/10.1029/2008JB005928.

Browning, J., Meredith, P.G., Stuart, C.E., Healy, D., Harland, S., Mitchell, T.M., 2017. Acoustic characterization of crack damage evolution in sandstone deformed under conventional and true triaxial loading. J. Geophys. Res. Solid Earth. https://doi.org/10.1002/2016JB013646.

Cagnola, A., Li, Z., Roshan, H., Masoumi, H., 2017. Microstructural evolution of organic matter-rich shales by ionic solutions. In: 51st US Rock Mechanics/Geomechanics Symposium. ARMA. American Rock Mechanics Association, San Francisco, CA, USA.

Carpacho, C., Ramirez, M., Osorio, J., Kenny, P., 2004. Replacing potassium with aluminium complex overcomes wellbore instability problems in kaolinitic shales in South America. In: AADE-04-DF-HO-17, American Association of Drilling Engineers 2004 Drilling Fluids Conference. American Association of Drilling Engineers, Houston, Texas.

Chadha, S., Jefferson-Loveday, R., Hussain, T., 2020. Modelling Knudsen number effects in suspension high velocity oxy fuel thermal spray. Int. J. Heat Mass Transf. 152, 119454. https://doi.org/10.1016/j.ijheatmasstransfer.2020.119454.

Chen, B., Barboza, B.R., Sun, Y., Bai, J., Thomas, H.R., Dutko, M., Cottrell, M., Li, C., 2021. A review of hydraulic fracturing simulation. Arch. Comput. Methods Eng. https://doi.org/10.1007/s11831-021-09653-z.

Cheng, Y., 2012. Impact of water dynamics in fractures on the performance of hydraulically fractured wells in gas-shale reservoirs. J. Can. Pet. Technol. 51, 143–151. https://doi.org/10.2118/127863-PA.

Chinard, F.P., 1952. Derivation of an expression for the rate of formation of glomerular fluid (GFR). Applicability of certain physical and physico-chemical concepts. Am. J. Physiol. Content. https://doi.org/10.1152/ajplegacy.1952.171.3.578.

Choinska, M., Khelidj, A., Chatzigeorgiou, G., Pijaudier-Cabot, G., 2007. Effects and interactions of temperature and stress-level related damage on permeability of concrete. Cem. Concr. Res. https://doi.org/10.1016/j.cemconres.2006.09.015.

Clarkson, C.R., Solano, N., Bustin, R.M., Bustin, A.M.M., Chalmers, G.R.L., He, L., Melnichenko, Y.B., Radliński, A.P., Blach, T.P., 2013. Pore structure characterization of North American shale gas reservoirs using USANS/SANS, gas adsorption, and mercury intrusion. Fuel. https://doi.org/10.1016/j.fuel.2012.06.119.

Coussy, O., 2005. Poromechanics., https://doi.org/10.1002/0470092718.

David, C., Dautriat, J., Sarout, J., Delle Piane, C., Menéndez, B., Macault, R., Bertauld, D., 2015. Mechanical instability induced by water weakening in laboratory fluid injection tests. J. Geophys. Res. Solid Earth. https://doi.org/10.1002/2015JB011894.

De Ruijter, M., Kölsch, P., Voué, M., De Coninck, J., Rabe, J.P., 1998. Effect of temperature on the dynamic contact angle. Colloids Surf. A Physicochem. Eng. Asp. 144, 235–243. https://doi.org/10.1016/S0927-7757(98)00659-1.

Dehghanpour, H., Lan, Q., Saeed, Y., Fei, H., Qi, Z., Tg, A., Simpson, F., Park, L.O., 2013. Spontaneous Imbibition of Brine and Oil in Gas Shales : Effect of Water Adsorption and Resulting Microfractures.

Dutta, R., Lee, C.-H., Odumabo, S., Ye, P., Walker, S.C., Karpyn, Z.T., Ayala, H., Ayala, L.F., 2014. Experimental investigation of fracturing-fluid migration caused by spontaneous imbibition in fractured low-permeability sands. SPE Reserv. Eval. Eng. 17, 74–81. https://doi.org/10.2118/154939-PA.

Engelder, T., Cathles, L.M., Bryndzia, L.T., 2014. The fate of residual treatment water in gas shale. J. Unconv. Oil Gas Resour. 7, 33–48. https://doi.org/10.1016/j.juogr.2014.03.002.

Ezulike, O., Dehghanpour, H., Virues, C., Hawkes, R.V., Jones, R.S., 2016. Flowback fracture closure: a key factor for estimating effective pore volume. SPE Reserv. Eval. Eng. 19, 567–582. https://doi.org/10.2118/175143-PA.

Fakcharoenphol, P., Kazemi, H., Charoenwongsa, S., Wu, Y.-S., Kurtoglu, B., Kazemi, H., Charoenwongsa, S., Wu, Y.-S., 2014. SPE 198998 The Effect of Osmotic Pressure on Improve Oil Recovery from Fractured Shale Formations. SPE Unconv. Resour. Conf. – USA. pp. 1–12, https://doi.org/10.2118/168998-MS.

Ferry, J.D., 1936. Statistical evaluation of sieve constants in ultrafiltration. J. Gen. Physiol. https://doi.org/10.1085/jgp.20.1.95.

Flem, B.E., Ratkje, S.K., Sterten, A., 1996. Peltier heats in cryolite melts with alumina. In: Light Metals: Proceedings of Sessions, TMS Annual Meeting. Warrendale, Pennsylvania.

Flem, B.E., Xu, Q., Kjelstrup, S., Sterten, Å., 2001. Thermoelectric powers of cells with NaF-AlF_3-Al_2O_3 melts. J. Non-Equilibrium Thermodyn. https://doi.org/10.1515/JNETDY.2001.010.

Freeman, C.M., Moridis, G.J., Blasingame, T.A., 2011. A numerical study of microscale flow behavior in tight gas and shale gas reservoir systems. Transp. Porous Media. https://doi.org/10.1007/s11242-011-9761-6.

Fu, H., Wang, X., Zhang, L., Gao, R., Li, Z., Xu, T., Zhu, X., Xu, W., Li, Q., 2015. Investigation of the factors that control the development of pore structure in lacustrine shale: a case study of block X in the Ordos Basin, China. J. Nat. Gas Sci. Eng. https://doi.org/10.1016/j.jngse.2015.07.025.

Furusawa, T., Smith, J.M., 1974. Intraparticle mass transport in slurries by dynamic adsorption studies. AIChE J. https://doi.org/10.1002/aic.690200111.

Gholami, R., Elochukwu, H., Fakhari, N., Sarmadivaleh, M., 2018. A review on borehole instability in active shale formations: interactions, mechanisms and inhibitors. Earth-Sci. Rev. 177, 2–13. https://doi.org/10.1016/j.earscirev.2017.11.002.

Gonçalvès, J., Rousseau-Gueutin, P., De Marsily, G., Cosenza, P., Violette, S., 2010. What is the significance of pore pressure in a saturated shale layer? Water Resour. Res. 46. https://doi.org/10.1029/2009WR008090.

Graham, J., Yuen, K., Goh, T.B., Janzen, P., Sivakumar, V., 2001. Hydraulic conductivity and pore fluid chemistry in artificially weathered plastic clay. Eng. Geol. 60, 69–81. https://doi.org/10.1016/S0013-7952(00)00090-9.

He, K., Nguyen, C., Kothamasu, R., Xu, L., 2015. Insights into whether low salinity brine enhances oil production in liquids-rich shale formations. In: SPE EUROPEC. Society of Petroleum Engineers, Madrid, Spain, pp. 1–4.

Heidug, W.K., Wong, S.-W., 1996. Hydration swelling of water-absorbing rocks: a constitutive model. Int. J. Num. Anal. Methods Geomech. https://doi.org/10.1002/(sici)1096-9853(199606)20:6<403::aid-nag832>3.0.co;2-7.

Hiemenz, P.C., Rajagopalan, R., 1997. Principles of colloid and surface chemistry. Marcel Dekker. https://doi.org/10.1016/0021-9797(79)90045-6.

Hirschfelder, J.O., Curtiss, C.F., Bird, R.B., 1955. Molecular theory of gases and liquids. Phys. Today. https://doi.org/10.1063/1.3061949.

Holditch, S.A., 1979. Factors affecting water blocking and gas flow from hydraulically fractured gas wells. SPE J. Pet. Technol. 31, 1515–1524. https://doi.org/10.2118/7561-pa.

Honarpour, M., Mahmood, S.M., 1988. Relative-permeability measurements: an overview. J. Pet. Technol. 40, 963–966. https://doi.org/10.2118/18565-PA.

Huang, T., Guo, X., Chen, F., 2015. Modeling transient flow behavior of a multiscale triple porosity model for shale gas reservoirs. J. Nat. Gas Sci. Eng. https://doi.org/10.1016/j.jngse.2015.01.022.

Jacquey, A.B., Regenauer-Lieb, K., 2021. Thermomechanics for geological, civil engineering and geodynamic applications: rate-dependent critical state line models. Rock Mech. Rock Eng. 54, 5355–5373. https://doi.org/10.1007/s00603-021-02397-z.

Jia, S., Zhao, Z., Wu, G., Wu, B., Wen, C., 2020. A coupled elastoplastic damage model for clayey rock and its numerical implementation and validation. Geofluids. https://doi.org/10.1155/2020/9853782.

Jiang, J., Younis, R.M., 2016. Compositional modeling of enhanced hydrocarbons recovery for fractured shale gas-condensate reservoirs with the effects of capillary pressure and multicomponent mechanisms. J. Nat. Gas Sci. Eng. https://doi.org/10.1016/j.jngse.2016.08.006.

Karger, J., Ruthven, D.M., 1989. On the comparison between macroscopic and NMR measurements of intracrystalline diffusion in zeolites. Zeolites. https://doi.org/10.1016/0144-2449(89)90071-7.

Kärger, J., Ruthven, D.M., Theodorou, D.N., 2012. Diffusion in Nanoporous Materials, Diffusion in Nanoporous Materials., https://doi.org/10.1002/9783527651276.

Khan, A.-R., Al-Gadhib, A.H., Baluch, M.H., 2004. Elasto-damage model for high strength concrete subjected to uniaxial cyclic compression. In: SEM X International Congress & Exposition on Experimental & Applied Mechanics—Addressing Future Experimental Mechanics Challenges With Special Focus on Extreme Environments.

Khan, A.R., Al-Gadhib, A.H., Baluch, M.H., 2007. Elasto-damage model for high strength concrete subjected to multiaxial loading. Int. J. Damage Mech. https://doi.org/10.1177/1056789506065914.

Khodadadi, B., Qadeer Siddiqui, M.A., Pirzada, M.A., Le-Hussain, F., Iglauer, S., Roshan, H., 2020. Direct observation of two-phase flow in deformable fractures of shales: a Utica shale example. J. Pet. Sci. Eng. 194, 107487. https://doi.org/10.1016/j.petrol.2020.107487.

Kirkaldy, J.S., Young, D.J., Lane, J.E., 1987. Diffusion profiles associated with the onsager matrix in non-equilibrium A-A*-vacancy and A-B-vacancy solutions. Acta Metall. https://doi.org/10.1016/0001-6160(87)90008-3.

Lan, Q., Xu, M., Binazadeh, M., Dehghanpour, H., Wood, J.M., 2015. A comparative investigation of shale wettability: the significance of pore connectivity. J. Nat. Gas Sci. Eng. 27, 1174–1188. https://doi.org/10.1016/j.jngse.2015.09.064.

Lemaitre, J., 1985. A continuous damage mechanics model for ductile fracture. J. Eng. Mater. Technol. Trans. ASME. https://doi.org/10.1115/1.3225775.

Lemaitre, J., 1992. A Course on Damage Mechanics, A Course on Damage Mechanics., https://doi.org/10.1007/978-3-662-02761-5.

Li, P., Zheng, M., Bi, H., Wu, S., Wang, X., 2017. Pore throat structure and fractal characteristics of tight oil sandstone: a case study in the Ordos Basin, China. J. Pet. Sci. Eng. 149, 665–674. https://doi.org/10.1016/j.petrol.2016.11.015.

Liang, L., Luo, D., Liu, X., Xiong, J., 2016. Experimental study on the wettability and adsorption characteristics of Longmaxi Formation shale in the Sichuan Basin, China. J. Nat. Gas Sci. Eng. 33, 1107–1118. https://doi.org/10.1016/j.jngse.2016.05.024.

Lin, B., Cerato, A.B., 2014. Applications of SEM and ESEM in microstructural investigation of shale-weathered expansive soils along swelling-shrinkage cycles. Eng. Geol. 177, 66–74. https://doi.org/10.1016/j.enggeo.2014.05.006.

Liu, Y., Rui, Z., 2022. In: Liu, Y., Rui, Z. (Eds.), Chapter 2—Confined fluid-phase behavior in shale. Gulf Professional Publishing, pp. 9–55, https://doi.org/10.1016/B978-0-323-91660-8.00002-6.

Liujuan, X.I.E., Shuangfang, L.U., Junqian, L.I., Yongjie, H.U., Pengfei, Z., 2016. Experimental investigations of the mineral wettability in shale and its influence factors. Acta Geol. Sin. 89, 2014–2015. https://doi.org/10.1111/1755-6724.12303_9.

Makhanov, K., Habibi, A., Dehghanpour, H., Kuru, E., 2014. Liquid uptake of gas shales: a workflow to estimate water loss during shut-in periods after fracturing operations. J. Unconv. Oil Gas Resour. 7, 22–32. https://doi.org/10.1016/j.juogr.2014.04.001.

Manegold, E., 1929. Ueber Kollodiummembranen. III—Die Dialyse durch Kollodiummembranen und der Zusammenhang zwischen Dialyse, Diffusion und Membranstruktur. Kolloid-Zeitschrift. https://doi.org/10.1007/BF01553300.

Mao, Z.Q., Xiao, L., Wang, Z.N., Jin, Y., Liu, X.G., Xie, B., 2013. Estimation of permeability by integrating nuclear magnetic resonance (NMR) logs with mercury injection capillary pressure (MICP) data in tight gas sands. Appl. Magn. Reson. https://doi.org/10.1007/s00723-012-0384-z.

McLean, M.R., Addis, M.A., 1990. Wellbore Stability: The Effect of Strength Criteria on Mud Weight Recommendations., https://doi.org/10.2118/20405-ms.

Mirchi, V., Saraji, S., Goual, L., Piri, M., 2014. Dynamic interfacial tensions and contact angles of surfactant-in-brine/oil/shale systems: implications to enhanced oil recovery in shale oil reservoirs. SPE Conf. 1–15. https://doi.org/10.2118/169171-MS.

Mitchell, J.K., Soga, K., 2005. Soil composition and engineering properties. In: Fundamentals of Soil Behavior, third ed, pp. 83–108.

Mokhtari, M., Alqahtani, A.A., Tutuncu, A.N., Yin, X., 2013. Stress-dependent permeability anisotropy and wettability of shale resources. In: Unconv. Resour. Technol. Conf. Denver, Color. 12-14 August 2013, pp. 2713–2728, https://doi.org/10.1190/urtec2013-283.

Monroe, C.W., Newman, J., 2007. An introduction to the onsager reciprocal relations. Chem. Eng. Educ. 41, 233–238.

Moore, R.M., Katzer, J.R., 1972. Counterdiffusion of liquid hydrocarbons in type Y zeolite: effect of molecular size, molecular type, and direction of diffusion. AIChE J. https://doi.org/10.1002/aic.690180426.

Morgano, M., Trtik, P., Meyer, M., Lehmann, E.H., Hovind, J., Strobl, M., 2018. Unlocking high spatial resolution in neutron imaging through an add-on fibre optics taper. Opt. Express. https://doi.org/10.1364/oe.26.001809.

Morsy, S., Sheng, J.J., 2014. Effect of water salinity on shale reservoir. Productivity 8, 9–14. https://doi.org/10.3968/5604.

Murakami, S., 2012. Continuum damage mechanics. Solid Mech. Appl. https://doi.org/10.1002/9781118097298.weoc055.

O'Malley, D., Karra, S., Currier, R.P., Makedonska, N., Hyman, J.D., Viswanathan, H.S., 2016. Where does water go during hydraulic fracturing? Groundwater 54, 488–497. https://doi.org/10.1111/gwat.12380.

Odusina, E.O., Sondergeld, C.H., Rai, C.S., 2011. NMR study of shale wettability. Can. Unconven. Res. Conf. https://doi.org/10.2118/147371-MS.

Onsager, L., 1931. Reciprocal relations in irreversible processes. II. Phys. Rev. https://doi.org/10.1103/PhysRev.38.2265.

Ougier-Simonin, A., Renard, F., Boehm, C., Vidal-Gilbert, S., 2016. Microfracturing and microporosity in shales. Earth Sci. Rev. https://doi.org/10.1016/j.earscirev.2016.09.006.

Pan, B., Li, Y., Wang, H., Jones, F., Iglauer, S., 2018. CO_2 and CH_4 wettabilities of organic-rich shale. Energy Fuels. https://doi.org/10.1021/acs.energyfuels.7b01147. acs.energyfuels.7b01147.

Pappenheimer, J.R., 1953. Passage of molecules through capillary walls. Physiol. Rev. https://doi.org/10.1152/physrev.1953.33.3.387.

Pappenheimer, J.R., Renkin, E.M., Borrero, L.M., 1951. Filtration, diffusion and molecular sieving through peripheral capillary membranes; a contribution to the pore theory of capillary permeability. Am. J. Physiol. https://doi.org/10.1152/ajplegacy.1951.167.1.13.

Pirzada, M.A., Roshan, H., Sun, H., Oh, J., Andersen, M.S., Hedayat, A., Bahaaddini, M., 2020. Effect of contact surface area on frictional behaviour of dry and saturated rock joints. J. Struct. Geol. 135. https://doi.org/10.1016/j.jsg.2020.104044.

Pirzada, M.A., Bahaaddini, M., Moradian, O., Roshan, H., 2021. Evolution of contact area and aperture during the shearing process of natural rock fractures. Eng. Geol. 291, 106236. https://doi.org/10.1016/j.enggeo.2021.106236.

Pokalai, K., Fei, Y., Ahmad, M., Haghighi, M., Perdomo, M.E.G., 2015. Design and Optimisation of Multi-Stage Hydraulic Fracturing in a Horizontal Well in a Shale Gas Reservoir in the Cooper Basin, South Australia., p. 1947.

Rao, M.B., Sircar, S., 1993. Nanoporous carbon membranes for separation of gas mixtures by selective surface flow. J. Memb. Sci. https://doi.org/10.1016/0376-7388(93)85279-6.

Roshan, H., Aghighi, M.A., 2012. Analysis of pore pressure distribution in shale formations under hydraulic, chemical, thermal and electrical interactions. Transp. Porous Media. https://doi.org/10.1007/s11242-011-9891-x.

Roshan, H., Oeser, M., 2012. A non-isothermal constitutive model for chemically active elasto-plastic rocks. Rock Mech. Rock Eng. https://doi.org/10.1007/s00603-011-0204-z.

Roshan, H., Rahman, S.S., 2013. The effect of water content on stress changes around a wellbore drilled in a chemically active elastoplastic formation. Pet. Sci. Technol. https://doi.org/10.1080/10916466.2011.553649.

Roshan, H., Ehsani, S., Marjo, C.E., Andersen, M.S., Acworth, R.I., 2015. Mechanisms of water adsorption into partially saturated fractured shales: an experimental study. Fuel 159, 628–637. https://doi.org/10.1016/j.fuel.2015.07.015.

Roshan, H., Al-Yaseri, A.Z., Sarmadivaleh, M., Iglauer, S., 2016a. On wettability of shale rocks. J. Colloid Interface Sci. 475, 104–111. https://doi.org/10.1016/j.jcis.2016.04.041.

Roshan, H., Andersen, M.S., Rutlidge, H., Marjo, C.E., Acworth, R.I., 2016b. Investigation of the kinetics of water uptake into partially saturated shales. Water Resour. Res. 52, 2420–2438. https://doi.org/10.1002/2015WR017786.

Roshan, H., Sari, M., Arandiyan, H., Hu, Y., Mostaghimi, P., Sarmadivaleh, M., Masoumi, H., Veveakis, M., Iglauer, S., Regenauer-Lieb, K., 2016c. Total porosity of tight rocks: a welcome to the heat transfer technique. Energy Fuels 30, 10072–10079. https://doi.org/10.1021/acs.energyfuels.6b01339.

Roshan, H., Chen, X., Pirzada, M.A., Regenauer-Lieb, K., 2019. Permeability measurements during triaxial and direct shear loading using a novel X-ray transparent apparatus: fractured shale examples from Beetaloo basin, Australia. NDT E Int. 107, 102129. https://doi.org/10.1016/j.ndteint.2019.102129.

Roychaudhuri, B., Tsotsis, T.T., Jessen, K., 2013. An experimental investigation of spontaneous imbibition in gas shales. J. Pet. Sci. Eng. 111, 87–97. https://doi.org/10.1016/j.petrol.2013.10.002.

Saraji, S., Piri, M., Goual, L., 2014. The effects of SO_2 contamination, brine salinity, pressure, and temperature on dynamic contact angles and interfacial tension of supercritical CO_2/brine/quartz systems. Int. J. Greenh. Gas Control 28, 147–155. https://doi.org/10.1016/j.ijggc.2014.06.024.

Satterfield, C.N., Katzer, J.R., 1971. Counterdiffusion of Liquid Hydrocarbons in Type Y Zeolites., https://doi.org/10.1021/ba-1971-0102.ch055.

Satterfield, C.N., Van Eek, M.W., Bliss, G.S., 1978. Liquid-solid mass transfer in packed beds with downward concurrent gas-liquid flow. AIChE J. https://doi.org/10.1002/aic.690240421.

Settari, A., Sullivan, R.B., Bachman, R.C., 2002. The modeling of the effect of water blockage and geomechanics in waterfracs. In: SPE Annual Technical Conference and Exhibition., https://doi.org/10.2118/77600-MS.

Siddiqui, M.A.Q., Gajbhiye, R.N., 2017. Stability and texture of CO_2/N_2 foam in sandstone. Colloids Surf. A Physicochem. Eng. Asp. https://doi.org/10.1016/j.colsurfa.2017.04.049.

Siddiqui, M.A.Q., Roshan, H., 2022. Thermodynamic characterization of chemical damage in variably saturated water-active shales. Rock Mech. Rock Eng. https://doi.org/10.1007/s00603-022-02916-6.

Siddiqui, M.A.Q., Al-Nuaim, S., Khan, R.A., 2014. Stochastic optimization of gas cycling in gas condensate reservoirs. In: Soc. Pet. Eng.—30th Abu Dhabi Int. Pet. Exhib. Conf. ADIPEC 2014 Challenges Oppor. Next 30 Years 6, pp. 4331–4341. https://doi.org/10.2118/172107-ms.

Siddiqui, M.A.Q., Gajbhiye, R.N., Sultan, A.S., Abu-Khamsin, S., 2015. Effect of Temperature on Stability of sc-CO_2-Foam With New Fluorosurfactant., https://doi.org/10.2118/178039-ms.

Siddiqui, M.A.Q., Ali, S., Fei, H., Roshan, H., 2018. Current understanding of shale wettability: a review on contact angle measurements. Earth Sci. Rev. https://doi.org/10.1016/j.earscirev.2018.04.002.

Siddiqui, M.A.Q., Chen, X., Iglauer, S., Roshan, H., 2019. A multiscale study on shale wettability: spontaneous imbibition vs contact angle. Water Resour. Res. 0–3. https://doi.org/10.1029/2019WR024893.

Siddiqui, M.A.Q., Lv, A., Regenauer-Lieb, K., Roshan, H., 2020a. A novel experimental system for measurement of coupled multi-physics-induced surface alteration processes in geomaterials. Meas. J. Int. Meas. Confed. 166, 108211. https://doi.org/10.1016/j.measurement.2020.108211.

Siddiqui, M.A.Q., Ueda, K., Komatsu, H., Shimamoto, T., Roshan, H., 2020b. Caveats of using fractal analysis for clay rich pore systems. J. Pet. Sci. Eng. https://doi.org/10.1016/j.petrol.2020.107622.

Siddiqui, M.A.Q., Salvemini, F., Ramandi, H.L., Fitzgerald, P., Roshan, H., 2021. Configurational diffusion transport of water and oil in dual continuum shales. Sci. Rep. https://doi.org/10.1038/s41598-021-81004-1.

Singh, H., 2016. A critical review of water uptake by shales. J. Nat. Gas Sci. Eng. 34, 751–766. https://doi.org/10.1016/j.jngse.2016.07.003.

Wall, J., Vile, M., Cousins, K., 2013. Shale gas development in Australia: potential impacts and risks to ecological systems. Aust. Counc. Learn. Acad. Ecol. Aust., 67.

Wei, Z., Zhang, D., 2010. Coupled fluid-flow and geomechanics for triple-porosity/dual-permeability modeling of coalbed methane recovery. Int. J. Rock Mech. Min. Sci. https://doi.org/10.1016/j.ijrmms.2010.08.020.

Wu, K., Chen, Z., Li, X., Dong, X., 2016. Methane storage in nanoporous material at supercritical temperature over a wide range of pressures. Sci. Rep. https://doi.org/10.1038/srep33461.

Xiao, J., Wei, J., 1992. Diffusion mechanism of hydrocarbons in zeolites-I. Theory Chem. Eng. Sci. https://doi.org/10.1016/0009-2509(92)80236-6.

Xie, Q., Saeedi, A., Pooryousefy, E., Liu, Y., 2016. Extended DLVO-based estimates of surface force in low salinity water flooding. J. Mol. Liq. 221, 658–665. https://doi.org/10.1016/j.molliq.2016.06.004.

Xie, Q., Chen, Y., You, L., Hossain, M., Saeedi, A., 2018. Drivers of wettability alteration for oil/brine/kaolinite system: implications for hydraulic fracturing fluids uptake in shale rocks. Energies 11, 1666. https://doi.org/10.3390/en11071666.

Xu, M., Dehghanpour, H., 2014. Advances in understanding wettability of gas shales. Energy Fuels 4362–4375. https://doi.org/10.1021/ef500428y.

Yang, X., Liang, Y., Chen, W., 2020. Anomalous imbibition of non-Newtonian fluids in porous media. Chem. Eng. Sci. 211. https://doi.org/10.1016/j.ces.2019.115265.

Yassin, M.R., Begum, M., Dehghanpour, H., 2017. Organic shale wettability and its relationship to other petrophysical properties: a Duvernay case study. Int. J. Coal Geol. 169, 74–91. https://doi.org/10.1016/j.coal.2016.11.015.

Yuan, Q., Zhu, X., Lin, K., Zhao, Y.P., 2015. Molecular dynamics simulations of the enhanced recovery of confined methane with carbon dioxide. Phys. Chem. Chem. Phys. https://doi.org/10.1039/c5cp06649b.

Yucel Akkutlu, I., Fathi, E., 2012. Multiscale gas transport in shales with local kerogen heterogeneities. SPE J. https://doi.org/10.2118/146422-pa.

Zhang, J., Standifird, W.B., Roegiers, J.C., Zhang, Y., 2007. Stress-dependent fluid flow and permeability in fractured media: from lab experiments to engineering applications. Rock Mech. Rock Eng. https://doi.org/10.1007/s00603-006-0103-x.

Zhang, J., Al-Bazali, T.M., Chenevert, M.E., Sharma, M.M., 2008. Factors controlling the membrane efficiency of shales when interacting with water-based and oil-based muds. SPE Drill. Complet. https://doi.org/10.2118/100735-ms.

Zhang, C., Gamage, R.P., Perera, M.S.A., Zhao, J., 2017. Characteristics of clay-abundant shale formations: use of CO_2 for production enhancement. Energies. https://doi.org/10.3390/en10111887.

Zhou, Z., Abass, H., Li, X., Bearinger, D., Frank, W., 2016. Mechanisms of imbibition during hydraulic fracturing in shale formations. J. Pet. Sci. Eng. 141. https://doi.org/10.1016/j.petrol.2016.01.021.

Zolfaghari, A., Dehghanpour, H., Noel, M., Bearinger, D., 2016. Laboratory and field analysis of flowback water from gas shales. J. Unconv. Oil Gas Resour. 14, 113–127. https://doi.org/10.1016/j.juogr.2016.03.004.

Zou, C., Zhu, R., Liu, K., Su, L., Bai, B., Zhang, X., Yuan, X., Wang, J., 2012. Tight gas sandstone reservoirs in China: characteristics and recognition criteria. J. Pet. Sci. Eng. 88–89, 82–91. https://doi.org/10.1016/j.petrol.2012.02.001.

Chapter 7

Interactions during various enhanced oil recovery operations

Anas M. Hassan[a], Mysara Eissa Mohyaldinn[b], Mohammed A. Ayoub[b], and Emad W. Al-Shalabi[a]

[a]*Petroleum Engineering Department, Khalifa University of Science and Technology, Abu Dhabi, United Arab Emirates*

[b]*Petroleum Engineering Department, Universiti Teknologi Petronas (UTP), Seri Iskandar, Malaysia*

7.1 Introduction

After the primary oil recovery methods, such as natural depletion and artificial lift, have been exhausted, the improved/enhanced oil recovery (IOR/EOR) techniques are subsequently applied during the secondary and tertiary stages. The water flooding method is the most well-established secondary oil recovery technique, followed by gas injection. For oil recovery exceeding the water flooding and gas flooding levels, tertiary methods of oil recovery, such as chemical enhanced oil recovery (cEOR), are implemented. In the cEOR techniques, chemical additives, such as polymers, surfactants, and alkali, are mixed with water to augment the microscopic and macroscopic displacement (areal and volumetric sweep efficiencies). The latter is achieved through changes in viscosities of water and oil, reduction in the water/oil and gas/oil interfacial tension (IFT) as well as wettability alteration of the reservoir rock, resulting in incremental oil recovery (Lwisa, 2021). During the tertiary oil recovery stage, these cEOR methods are either applied in stand-alone mode or hybrid mode along with other EOR methods, such as low-salinity water flooding (LSWF) and CO_2 foam flooding. One of the promising novel hybrid synergistic techniques is the smart-water-assisted foam (SWAF) flooding, which imparts the dual-improvement effects of low-salinity water (LSW) effect combined with gas-induced foam flooding effects, resulting in exceptional incremental oil recovery (Hassan et al., 2021, 2022a,b,c).

Developments in Petroleum Science, Vol. 78. https://doi.org/10.1016/B978-0-323-99285-5.00008-9

Estimates of EOR field applications worldwide show that the lithology of most applications is in sandstone reservoirs. However, in carbonates, applying EOR methods is more challenging due to the harsh conditions such as heterogeneity, low permeability, high temperature, and high salinity. Therefore, understanding the fluid-fluid and fluid-rock interactions (i.e., crude oil/brine/rock or COBR system) is of great importance to maximize the oil recovery methods for field applications in carbonates as well as sandstone reservoirs. Also, with a better understanding of the COBR system, it enables the maximization of oil recovery to cater to the needs of the global energy order. According to the International Energy Agency (IEA, 2019), there are currently nearly 375 EOR projects in the world that either have been completed or are currently undergoing. The reports show a steady increase in the number of projects for the sake of meeting the increasing oil demand. It is projected that the total oil production would increase to 4.5 million barrels/day (mb/d) by 2040 (IEA, 2021; Newell et al., 2021). This underscores the significance of EOR and the development of EOR techniques in the forthcoming decades.

In the EOR process, the implementation of EOR technology begins with the screening tests as the initial stage of a project, which recognizes the best EOR method for the candidate reservoir in terms of the possible incremental oil recovery and the economics of the project (Manrique et al., 2010). The screening criteria are based on both reservoir rock properties (e.g., formation type, reservoir thickness, depth, pH, and temperature), and fluids properties (e.g., oil API gravity, oil viscosity, oil composition, remaining target oil saturation), which is shown in Table 7.1 and further discussed in ensuing chapters.

Therefore, it is critical to define the underlying EOR mechanisms comprehensively, which are comprised of complex rock-fluid interactions, to facilitate improvement in the ultimate recovery factor by EOR. Although the established mechanisms behind EOR methods such as water flooding are well understood, the mechanisms underlying some of the newer EOR techniques are still under speculation, and therefore, studies are in progress, which aim to acquire an improved understanding of the rock/fluid interactions associated with these recent EOR techniques.

This work highlights some of the secondary and tertiary EOR techniques, hybrid EOR processes, and the different injected fluids and their effects on the COBR system. The different underlying mechanisms associated with each technique are discussed from the perspective of fluid-fluid and fluid-rock interactions that either augment or disrupt the efficiency of EOR schemes.

7.2 Water injection

Water injection is the most prevalent oil recovery technique, in which water is injected into an oil reservoir to force the oil toward the producer wells. Water injection creates microscopic oil displacement located within the pore spaces and sweeps it toward the production wells. The efficiency of water flood

TABLE 7.1 A summary of screening criteria for chemical EOR methods

EOR method	Oil gravity (°API)	Viscosity (cP)	Oil saturation (%)	Formation type	Reservoir thickness (ft)	Average permeability (mD)	Depth (ft)	Temperature (°F)
Polymer flooding	>15	<150 and >10	>50	Sandstone is preferred	NC	>10	<9000	>200
Micellar/ polymer, ASP and alkaline flooding	>20	<35	>35	Sandstone is preferred	NC	>10	<9000	>200

depends on the effects of oil viscosity and rock-fluid properties. Thus, to acquire an estimate of the ultimate recovery factor, the water-oil flow measurements are quantified by considering the effects of geology, mineral composition, gravity, and the well-spacing. Generally, the water viscosity at reservoir temperatures is either much lower or equal to that of the reservoir oil, causing the water/oil viscosity ratio to be mostly <1. The efficacy of the water/oil displacement is a function of the water-oil mobility ratio, where the recovery efficacy is inversely proportional to the water/oil mobility ratio. According to the definition of fluid mobility in a porous medium, a low viscosity ($\leq$ 1 cP) indicates a highly mobile fluid, except when relative permeability is extremely low, whereas a low-API crude oil ($\leq$ 20°API) indicates a high viscosity oil with extremely low mobility, except when high temperatures are concerned.

To better understand the effect of reservoir rock (lithology) on the incremental oil recovery, the fluid/rock interactions are studied according to their role in wettability alteration, residual oil saturation to water injection, and the relative oil permeability when water saturation is high. Wettability is an immiscible liquid's proclivity to spread and attach to a solid surface in the presence of another immiscible liquid (Ahmed, 2018). This proclivity of spreading in a liquid is determined by simply measuring the angle of contact at the liquid-solid surface, which is called the contact angle (θ) and is measured through the liquid to the solid by convention at all times. When the contact angle decreases, it indicates an increase in the wetting qualities of the liquid. A zero-contact angle denotes absolute wettability, and a contact angle of 180° denotes absolute nonwetting (Ahmed, 2018). Thus, we understand that wettability is the tendency of a solid to adhere to one type of fluid, in another fluid's presence. The wetting fluid clings to the solid surface creating a contact angle (θ) between the wetting liquid and the surface of the solid. Thus, the wettability of a reservoir is determined by taking the (θ) measurement of the tangent from the rock surface to the droplet of the wetting fluid through the denser fluid. We see this in Fig. 7.1, where the contact angle for two different systems of oil-wet and water-wet states is shown.

Moreover, to effectively gauge the water injection flow and its effects on fluid-solid interactions, the rock properties must be considered. The reservoir stratum holding the hydrocarbon is either sandstone or carbonates rocks. Heterogeneous rock formations are composed of either shale, anhydrite, or other impermeable layers that partially or completely insulate the porous and permeable reservoir layers. Considering this, the efficacy of water injection corresponds to the relationship between the bed angle, the gravity, and the oil/brine density difference at the reservoir, which cause relative vertical and horizontal flow behaviors according to the reservoir structure (Craig, 1971; Rose et al., 1989; Willhite, 1986). Fig. 7.2 portrays the water injection effect through permeable layers, where one can note that the remaining oil is trapped in narrow pores (i.e., low permeability) and the residual oil film is left behind.

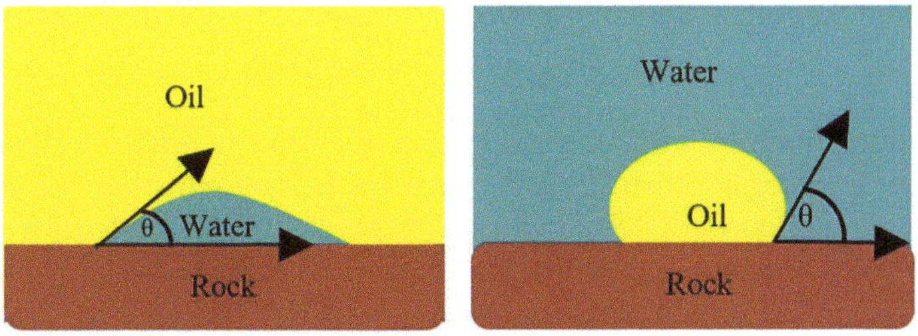

FIG. 7.1 The contact angle of systems of oil-wet and water-wet state.

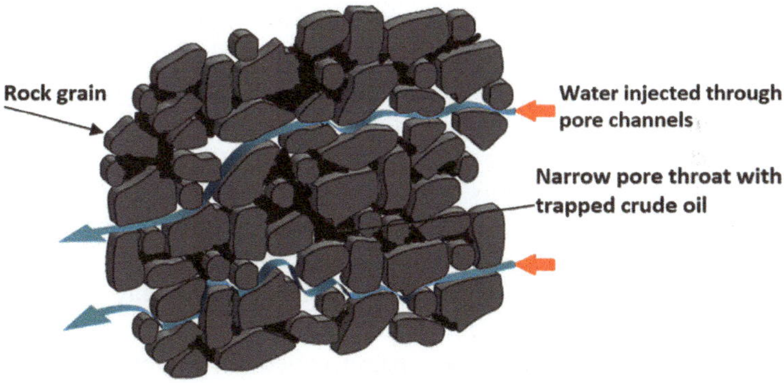

FIG. 7.2 Schematic representation of water injection through pore channels and the effect of narrow pore throat.

Furthermore, based on the reservoir specifications, mathematical calculations or numerical simulations can be used to predict and assess water injection (Ebanks and Cook, 1993; Scheihing et al., 2002). The flow dynamics of a water flood are regulated by the injected water flow, rock properties (composition), and reservoir conditions. Therefore, adjustments that correspond to the reservoir conditions to modify the dynamics of COBR systems can maximize the oil production and minimize water production (Ayoub et al., 2015).

7.3 Gas flooding

Gas flooding has been a well-established viable method for secondary recovery stages for decades (Lake et al., 2014; Taber et al., 1997). Gas flooding employs either nitrogen (N_2) or carbon dioxide (CO_2) gases for injection, which mix with the crude oil to reduce the oil viscosity and displace it, resulting in improved oil mobility. The gas is injected into the reservoir formation until an adequately high downhole pressure is reached, which forms a miscible zone of gas, displacing the trapped oil from the pores.

In order to curtail the contribution of greenhouse gases, N_2 gas derived from the air is regarded as a more environmentally acceptable candidate than CO_2 gas. On the other hand, the CO_2 flooding technique has wider application and is mainly applied in deeper reservoirs of >2000 ft, where oil gravity is equivalent to or higher than 22–25°, wherein CO_2 can remain in a supercritical form. Unlike other EOR methods, the lithology of the reservoir is not relevant to CO_2 flooding, due to its low impact on the fluid-solid interaction. However, other reservoir characteristics can affect the COBR system and the outcome of CO_2 flooding. For instance, in reservoirs where the initial pressure has been diminished due to primary production processes (Orr et al., 1982), CO_2 injections are most feasible. To achieve this, initially, the reservoir pressure is restored to a specific level by water injection, following which CO_2 is injected through the same injection wells into the reservoir. As carbon dioxide permeates into the reservoir and contacts oil, a miscible CO_2 zone with light hydrocarbons is created. This miscible oil zone has good oil solubility, which supports easier oil flow toward the production wells. At the miscible front, an oil bank is created, and the miscible fluids are collected through the production wells, while the CO_2 returns to its gaseous state. This provides a "gas lift" effect, like the initial natural gas pressure in the reservoir (Speight, 2016). The CO_2 gas and water injections may be alternated, which enables the fluids to reach further into unswept areas and improve the sweep efficiency of oil, in a scheme called the water alternating gas (WAG) technique (i.e., miscible CO_2-WAG flooding). Fig. 7.3 shows the

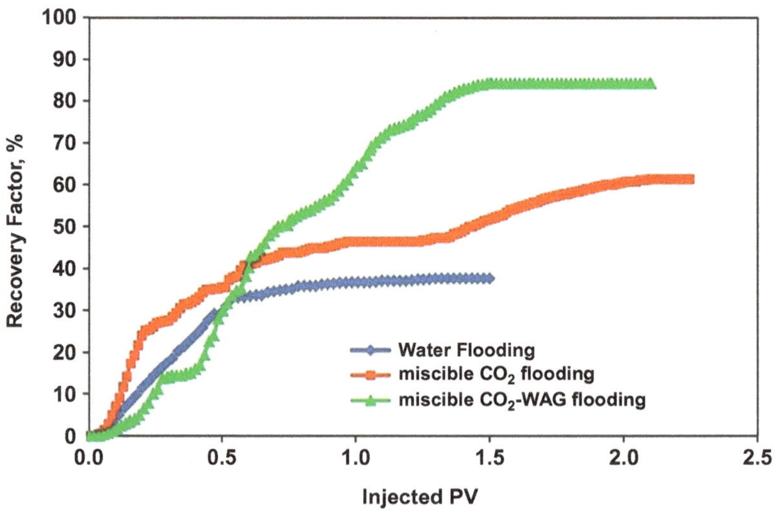

FIG. 7.3 Comparison of oil recovery factors for water flooding, miscible continuous CO_2 flooding, and miscible CO_2-WAG flooding. *The information was taken from data points of Rahimi, V., Bidarigh, M., Bahrami, P., 2017. Experimental study and performance investigation of miscible water-alternating-CO2 flooding for enhancing oil recovery in the Sarvak formation. Oil Gas Sci. Technol. 72 (6), 35.*

graphic comparison of oil recovery factors for water flooding with miscible continuous CO_2 flooding, and miscible CO_2-WAG flooding (Rahimi et al., 2017).

Additionally, in order to evaluate the effects of CO_2-rich or supercritical CO_2, on the fluid-rock interactions in the COBR system, investigations were conducted by Wang et al., at 85°C reservoir temperature on sandstone samples possessing calcite and carbonate cement. Then simulations using PHREEQC assessed the significance of the surface area of the sample, partial pressure, i.e., PCO_2 (at 10 and 20 MPa), and formation of water composition (3 M NaCl or 1 M $CaCl_2$–2 M NaCl) on the water-rock geochemistry. After CO_2 injection, a considerable decrease in pH of formation water was observed, resulting in the dissolution of alkaline minerals, viz. carbonate areas, and K-feldspar dissolution. The formation water type had a great influence on carbonate minerals resulting in calcite dissolution, which was considerably lessened when formation water had $CaCl_2$ present. Moreover, the sandstone dissolution was proportionate to the sample surface area and PCO_2 increase. This phenomenon of carbonate dissolution signifies that CO_2 injection into formation strata enhances the porosity and permeability of the reservoir, allowing for further propagation of the injected CO_2 (Wang et al., 2017). Also, CO_2 can be derived from the atmosphere through the carbon capture, utilization, and storage (CCUS) technologies, which are useful in reducing greenhouse gases. However, during storage CO_2 and formation water leakage must be prevented, as it could pollute shallow freshwater aquifers, to ensure robust CO_2 sequestration (Wang et al., 2017).

7.3.1 Crude-oil effect on supercritical CO_2 flooding

During supercritical CO_2 flooding, miscible solvents injected with crude oil can improve highly viscous and immobile heavy oil recovery. The inherent mechanisms of miscible fluid interactions affect several physical processes in both natural and engineered schemes. The fluids' interactions under micro-constrained geological geometries, like the porous media, exhibit the occurrence of spontaneous fingering between the crude oil having the complex composition and the solvent. Fractal-like fingers form when the applied external pressure gradient is nil (i.e., imposed Peclet number is zero). This is due to the crude-oil composition complexity where the light crude-oil components are observed to be mobile and to interchange with the solvent phase, whereas heavy crude-oil components are not as moveable and stay behind. Fractal-like fingering occurs when diffusion-driven mass exchange continues, causing instabilities at the interface that leads to variations of dynamic interfacial tension (DIFT) along with the interface. The Marangoni flow that results from this, which exacerbates the mass exchange, persists, until a local equilibrium is reached between crude-oil phase components and the solvent phase (Mehraban et al., 2022).

7.3.2 Carbonate dissolution

During CO_2 flooding, the COBR system and CO_2 come in contact to have several physical and chemical interactions in the reservoir (Perrin et al., 2009; Zhang et al., 2013). The CO_2 dissolves in the brine forming carbonic acid, and this carbonate saline water can dissolve the carbonate and silicate minerals. In this manner, the reservoir formation matrix disintegrates, which creates new flow channels to increase the rock layers' permeability (Alam et al., 2014; Gaus, 2010; Lin et al., 2008; Zhang et al., 2013). Moreover, the formation water contains carbonate ions including metal ions (Ca^{2+}, Mg^{2+}) that form precipitates, which settle on rock surfaces as deposits. Also, with the dissolved clay particles through the carbonate cement, kaolinite and intermediate products are formed, which travel toward pore throats and choke the smaller ones to cause permeability reduction (Izgec et al., 2008a,b; Shiraki and Dunn, 2000; Vetter and Farone, 1987; Yu et al., 2012).

Against this background, Zhao and Yin (2021) examined the before and after flooding effects of CO_2 injection on the rock mineral composition (in tight sandstone reservoirs), in terms of micromorphology, change in composition, and redistribution of rock pores and throats. The investigative methods such as the EDS (energy dispersive X-ray spectroscopy), SEM (scanning electron microscopy), and NMR (nuclear magnetic resonance spectroscopy) analyses, were conducted. It was revealed that in the pores, which expand and interlink, higher mineral dissolution was typically observed in contrast to the particle migration effect and sedimentation on plugged pores. It was also observed that due to CO_2 flooding, the permeability of CO_2 increased. Furthermore, pressure increase improved the CO_2 solubility in water, and a lower solution pH intensifies CO_2-water-rock interactions (Zhao and Yin, 2021).

The graphical representations in Figs. 7.4 and 7.5 show the respective increase in permeability and porosity of cores saturated with simulated formation water (no scale-forming ions water) after CO_2 flooding. Fig. 7.6 displays the pH of formation water at different pressures during CO_2 injection at a temperature of 50°C, showing that the decrease in pH is proportional to the increase in flooding pressure, whereas in Fig. 7.7, one can observe the probability distribution of small, medium, and large pores in the cores saturated with no scale-forming ions water under various pressures before and after CO_2 flooding. It is observed that the probability of small and large pore distribution increased with pressure, whereas for medium pores, the probability distribution decreased with pressure.

Additionally, through SEM analysis, higher mineral dissolution was observed, which is evident from the images below that feldspar and carbonate dissolutions are the most severe among all, contributing to secondary pores creation on the surface as shown in Figs. 7.8 and 7.9 (Zhao and Yin, 2021).

In another study on this subject by Wang et al., SEM analysis was applied to examine the fluid-rock-CO_2 interactions that transformed the morphology

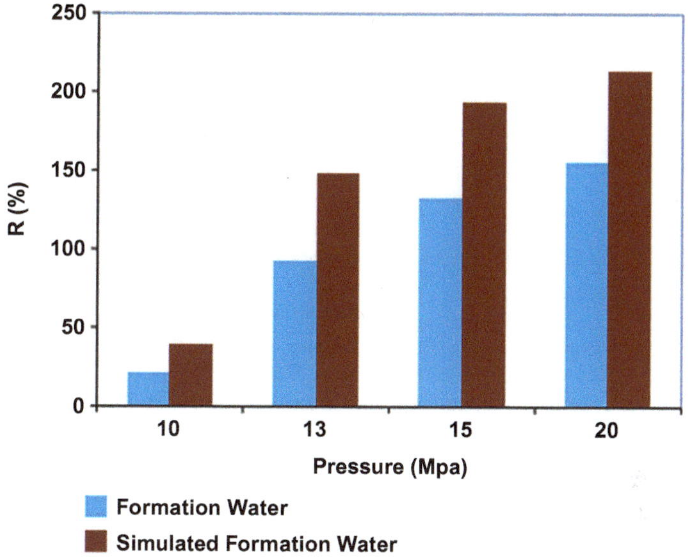

FIG. 7.4 Permeability change in accordance with pressure after CO_2 flooding. *The information was taken from data points of Zhao, D., Yin, D., 2021. Effect of fluid/rock interactions on physical character of tight sandstone reservoirs during CO2 flooding. Geofluids 2021 (6), 1–14.*

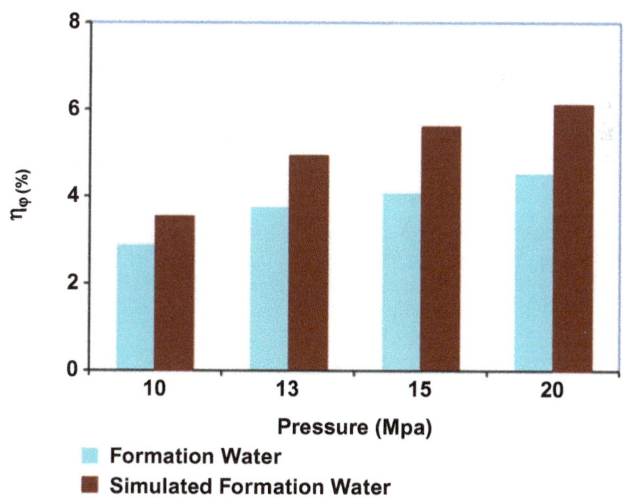

FIG. 7.5 Porosity changes with pressure after CO_2 flooding. *The information was taken from data points of Zhao, D., Yin, D., 2021. Effect of fluid/rock interactions on physical character of tight sandstone reservoirs during CO2 flooding. Geofluids 2021 (6), 1–14.*

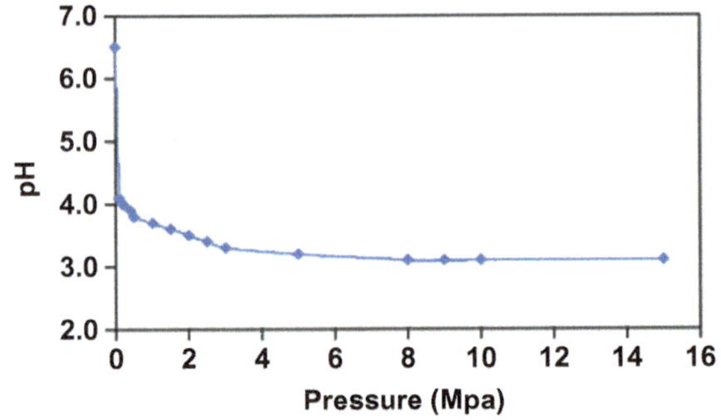

FIG. 7.6 pH of formation water at different pressures during CO_2 injection at a temperature of 50°C. *The information was taken from data points of Zhao, D., Yin, D., 2021. Effect of fluid/rock interactions on physical character of tight sandstone reservoirs during CO2 flooding. Geofluids 2021 (6), 1–14.*

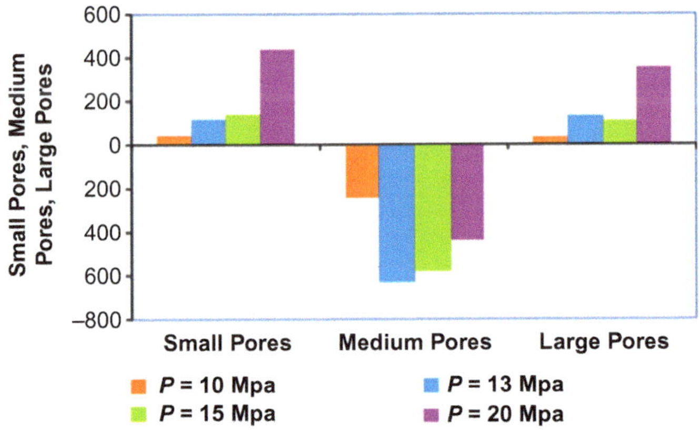

FIG. 7.7 Probability distribution of cores saturated with no scale-forming ions water under different pressures before and after CO_2 flooding. *The information was taken from data points of Zhao, D., Yin, D., 2021. Effect of fluid/rock interactions on physical character of tight sandstone reservoirs during CO2 flooding. Geofluids 2021 (6), 1–14.*

of the rock surface. The images reveal that certain precipitates crystallize (salt fragments) unevenly around the pore and do not occupy the big pore space, leaving the large pore throats open. Thus, this goes to show that good connectivity between the throats and permeability are not affected by the pore morphology of rock. In conclusion, the size and morphology of some pores in the core sample were significantly altered following the CO_2 flooding process. This is clearly seen in Fig. 7.10 where the core pore distribution radius is observed to completely change at the core scale (Wang et al., 2017).

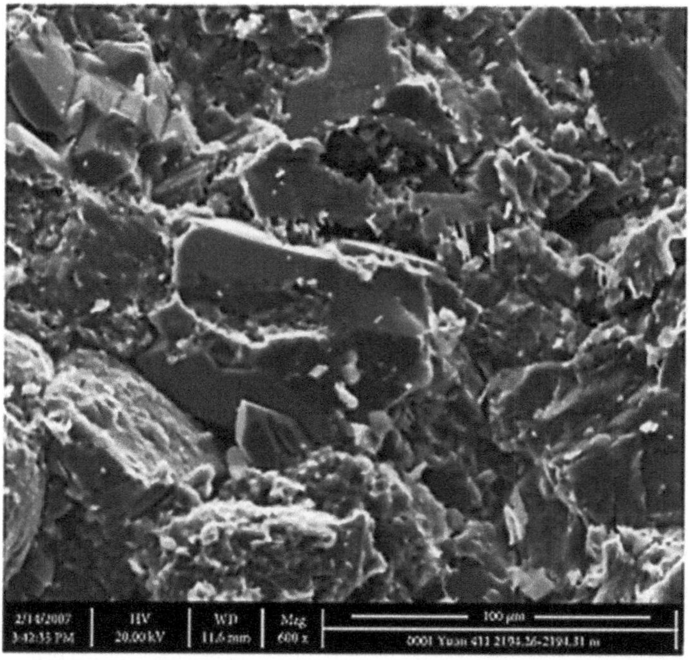

FIG. 7.8 SEM pictures of rocks before interactions with CO_2, no scale-forming ions water. *Reproduced with permission from Zhao, D., Yin, D., 2021. Effect of fluid/rock interactions on physical character of tight sandstone reservoirs during CO2 flooding. Geofluids 2021 (6), 1–14, Copyright (2021), Geofluids.*

Furthermore, in a related study, Cai et al. (2021) focused on reducing carbon emissions via the CCUS technology. They observed that when CO_2 flooding was performed in unconventional reservoirs (with nanopores), the observations of the physical and chemical interactions of fluid-fluid and fluid-solid systems, and the visualization of nanofluids revealed some unique phenomena of nanoscale phase transition. With the aid of molecular dynamic (MD) simulation technology, the adsorption and desorption of CO_2 in micro- and nanoscale pores, CO_2 and components of heavy oil interaction, salinity and temperature effects, and the interface properties of the brine/rock/CO_2 system were studied. They revealed that certain phase behavior caused IFT reduction and minimum miscible pressure (MMP) change in the oil-bearing nanopores. They also observed that mineral dissolution and precipitation could have varying levels of impact on connectivity and effective porosity. Besides, the process of CO_2 extraction involves a high chance of asphaltene deposition, which may obstruct pore throats or lead to rock wettability modification. During CO_2 pressurization, the chemical reactions can alter the mechanical rock properties, and the stress variations can damage

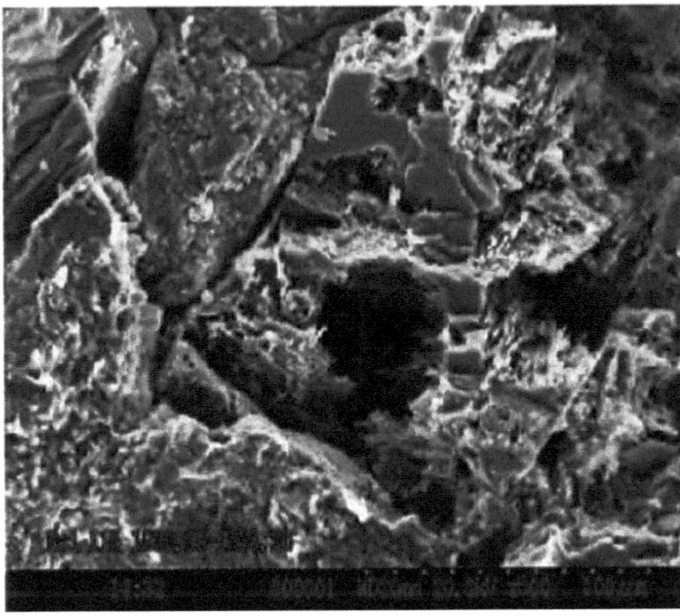

FIG. 7.9 SEM pictures of rocks after interactions with CO_2, no scale-forming ions water. *Reproduced with permission from Zhao, D., Yin, D., 2021. Effect of fluid/rock interactions on physical character of tight sandstone reservoirs during CO2 flooding. Geofluids 2021 (6), 1–14, Copyright (2021), Geofluids.*

the rock or create microseismic formation hazards, expose microfractures, restart faults, or cause nonnatural fractures (Cai et al., 2021).

Thus, because the reactive nature of carbonates is the cause of dissolution, the fluid-solid transport turns out to be complicated. The process of formation stimulation (i.e., acidizing) and storage of CO_2 requires acidic fluids to move within the carbonate pore space, leading to reactive interactions that change the surface and pore space properties, and consequently, pore geometry is modified, which affects the multiphase flow. Studies on the influence of carbonates dissolution under acidic flows for safe CO_2 storage projects found a new mechanism. When the reaction product, i.e., CO_2 is wetted on the susceptible grain surface, it prevents further dissolution of the grain at varying temperatures, pressures, and acidities associated with surface and subsurface settings.

7.3.2.1 Asphaltene precipitation

It is also noteworthy that, gas injections alter the innate dispersal structure of colloids and asphaltenes within crude oil, as the injected gases cause asphaltene instability. This leads to asphaltene adsorption on the reservoir rock

FIG. 7.10 SEM images of calcite before and after dissolution in the 3M NaCl and 1M $CaCl_2$-2M NaCl solutions when PCO_2 is 20MPa. (A, B) Calcite cement before the reaction, (C) and (D) are calcites reacted in the 3M NaCl solution for 25 and 72H, respectively, and (E) and (F) are the calcites reacted in the 1M $CaCl_2$-2M NaCl solution for 21 and 114H, respectively. Inserted are the EDS spectra for calcite. *Reproduced with permission from Wang, X., Wang, X., Hu, W., Wan, Y., Cao, J., Lv, C., Wang, R., Cui, M., 2017. Supercritical CO2-involved water–rock interactions at 85°C and partial pressures of 10–20 MPa: sequestration and enhanced oil recovery. Energy Explor. Exploit. 35 (2), 237–258. doi:10.1177/0144598716687933, Copyright (2017), Energy Exploration & Exploitation.*

surface, blocked pore throats, possible formation wettability alteration, and reduction in mobility of hydrocarbon fluids (Leontaritis, 1996). Furthermore, asphaltene deposition may lead to severe reservoir impairment, particularly near production wells (accompanied by the highest pressure drop). Moreover, the asphaltene's flocculation on the surface of rock formations, can break during detachment and block the reservoir pore openings (Zendehboudi et al., 2014). Thus, all the injection gases, viz. CH_4, CO_2, and N_2 lead to asphaltene precipitation, with N_2 and CH_4 having the greatest effect, in that order (Gonzalez et al., 2012; Jamaluddin et al., 2002; Moradi et al., 2012).

7.4 Fluid-solid interactions in water-based EOR methods

Water flooding is usually ineffective in reservoirs with mixed-wet or oil-wet and heterogeneous rocks. Heterogeneity and gravity forces reduce the sweep efficiency, and the vertical permeability variation in the reservoir causes water movement into the highly permeable layers. Thus, reservoir discontinuity prevents the water flood from reaching a greater pore volume amount in the reservoir. Moreover, huge amounts of oil produced using water flooding could be accompanied by large quantities of produced water, which requires treatment before disposal (Zitha et al., 2011).

Furthermore, the hydrocarbons can be subjected to unfavorable viscosity ratios or different capillary forces, which keep them trapped in pores. In water flooding or gas injection, on certain occasions, the oil viscosity is high, and the required mobility ratio between the displacing fluid and displaced oil is unlikely to be achieved, which leaves huge sections of oil untouched by the injected fluid, and the contacted oil is displaced ineffectively (Adibhatla and Mohanty, 2006). The fluid-rock interactions lead to fluids being attached to the porous medium of either oil-wet, water-wet, or mixed-wet conditions. The mixed or oil-wet rocks hold the oil in the pores, as the oil is attracted to the rock surface and the capillary action impedes the easy mobilization of oil globules over macroscopic distances (Zitha et al., 2011). Although water flooding is the appropriate technique for light oil reservoirs, in the case of reservoirs with high viscosity oils, oil recovery is challenging, and other EOR (cEOR or hybrid EOR) approaches may be required (Zitha et al., 2011).

7.4.1 Low-salinity water flooding

LSWF is a fairly recent EOR technique in which the interactive chemistry of fluid and rock components plays a vital role. For instance, the salt ions concentration in the injection brine is directly contributory to IOR. At the same time, the mechanisms that define the process of LSWF in affecting oil production are not yet thoroughly understood. Properties such as surface chemistry and fluid-rock interactions, which cause sensitivity of reservoir minerals to solution properties and have a significant effect on oil recovery, have not been fully perceived so far (Buckley et al., 1989; Tang and Morrow, 1999). However, many studies have shed light on the interactions at the pore scale of fluid-solid interfaces to characterize the fundamental mechanisms of LSWF.

It is understood that, during LSWF in sandstones (Al-Shalabi and Sepehrnoori, 2016; Al Shalabi and Sepehrnoori, 2017), the oil molecules that are attached to clay particles due to divalent cation linkages (Ca^{2+} and Mg^{2+}) can be released and yielded (Buckley et al., 1989). In reservoirs with high-salinity formation water, the electrical forces compress these linkages onto the clay surface, causing stagnation of a sizeable number of oil molecules from flowing freely. With LSWF, the water salinity is reduced, making these linkages expand and loosen up allowing the unlinked monovalent ions, viz. Na^+, K^+ to contact and substitute the divalent ions, which dislodges the oil molecules to be swept easily into the producing wells. LSWF can be used for both secondary and tertiary EOR; however, under certain conditions, tertiary LSWF is inadequate to achieve incremental oil recovery, and therefore, the utility of LSWF is mostly reserved for the secondary stage (Alotaibi et al., 2011).

Correspondingly, certain preconditions are suggested for a successful LSWF based on the fluid-solid interactive tendencies, such as the presence of clay in the porous media (this has been refuted by some in more recent works), the oil must possess polar components (i.e., acids and bases) (Tang and Morrow, 1999), and initial formation water (FW) with divalent cations (Ca^{2+}, Mg^{2+}) which must be present, and the LSW salinity should be in the range of 1000–2000 ppm (Bernard, 1967). As well, the formation should be oil-wet if a low-salinity effect is to be achieved since the trials on weak water-wet samples have not shown effective additional oil recovery (Mahani et al., 2013).

A mechanism of the workings of LSWF was given by Tang and Morrow (1999). They stated that during low-salinity injection, mixed-wet particles are released, which move to the core, and the oil drops attached to them are transported, leading to IOR. The separation of these mixed-wet particles from the pore surface exposes the water-wet underlying surface to the fluids even more, which causes greater wetness of rock (Tang and Morrow, 1999).

Another mechanism was proposed by McGuire et al. (2005), where the pH increase due to LSWF was put forth as the principal cause of IOR. According to them, the injected LSW reacts with native minerals of the reservoir to produce hydroxyl ions, which lead to an increase in pH, i.e., from 7 to 8 and even to 9 or more pH value in certain cases. Moreover, when LSW behavior was compared with alkaline flooding, it was observed that LSW caused slight IFT reduction between the residual oil and water and high pH, rendering the rock more water-wet and modifying the crude-oil characteristics (McGuire et al., 2005).

Additionally, it has been observed that the incremental oil recovery from LSWF is governed mainly by wettability alteration, where the rock is observed to change toward a more water-wetting state (Fjelde et al., 2012; Nasralla et al., 2013). The starting and concluding wettability are the main indicators, even though there are still differences in opinion on the precise causative effect of this. Some researchers have argued that the salinity type and level of LSW injection lead to reservoir wettability change, while others have cited the presence of plagioclase minerals and a high salinity as the cause behind reduced pH (to <7), which creates a mixed-wet condition that impedes LSW effect. However, contrary to this, the minerals in moderate LSW injection resulted in a pH of 7, leading to an undesirable water-wet condition (Buckley and Liu, 1998). Alternatively, the ultimately favorable wettability still contends, where some views state that LSWF alters the wettability toward a neutral state instead of a water-wet state (Alotaibi et al., 2011; Ashraf et al., 2010; Fjelde et al., 2012).

Buckley et al. (1989) proposed a different mechanism of LSWF. They suggested that when the high pH LSW contacts oil, it causes saponification of the acid or polar components in the oil, to behave like an in-situ surfactant. In this mechanism, the negatively charged oil is presumed to have linkages to the clay minerals via multivalent cation bonds (Buckley et al., 1989). The mechanism asserts that there occurs multicomponent ion exchange (MIE) while LSWF is injected, which leads to EOR. This MIE between rock/oil/brine occurs when the oil separates from the rock surface through a certain process, causing incremental oil recovery. The polar components interact with high pH LSW to become saponified, and like surfactants, create in-situ soaps. Apart from this, the hardness of certain ions, viz. Ca^{2+} and Mg^{2+}, generates surfactant precipitates in high-salinity water injection, which cease any further oil recovery increment.

In a later work by Buckley and Liu (1998), the wettability alteration mechanism involves has been explained in four principal processes related to the COBR mineral interactions as given below:

- Polar interaction between polar functional group of oil and polar surface sites that have minerals occurs when clean and moisture-free surface conditions exist.
- In the case of oils having a low level of solubility for the asphaltenes, the likelihood of the system going through wettability alterations and surface precipitation is greater.
- In the presence of water, polar functional groups of both mineral and crude-oil phases act like acids and bases to generate charges at both solid and oil interfaces. In such cases, pH is the leading cause of the surface charge level.
- Due to the ion-binding interactions the divalent ions, i.e., Ca^{2+}, Mg^{2+} behave as a link to likely create three kinds of interactions, viz. between oil-Ca-oil, mineral-Ca-mineral, and oil-Ca-mineral phases. In the final type of interaction, a significant wettability alteration can occur. In this case, instead of pH, the governing factor is the temperature that significantly affects the interactions (Buckley and Liu, 1998).

In another study, Hamouda and Valderhaug (2014) studied, at length, the effects of formation composition and brine imbibition and the crude-oil type by conducting LSWF imbibition tests. They reported that the presence of Ca^{2+}

and Mg^{2+} in formation brine led to alteration of sample wettability toward oil-wet, based on their concentration, which reduced recovery from conventional oil reservoirs. In addition, there is low surface precipitation because of the water present in the pore space. Moreover, regarding the effect of oil properties, the API gravity (solvent quality of the oil), and total acid and base numbers (TAN and TBN) are the key features that determine the wetting alteration.

Similarly, Kakati and Sangwai investigated the effect of monovalent and divalent salts on IFT reduction through LSWF at ambient temperature and atmospheric pressure. Results indicated that a low-salinity concentration is influential whenever a minimum value of the oil-brine IFT was exhibited. It was also noted that the salt type had an important effect on the IFT of crude oils that were aliphatic and aromatic. Monovalent salt was active in the IFT reduction of aliphatic crude oils, whereas, divalent salts affected the aromatic hydrocarbons' IFT reduction (Kakati and Sangwai, 2017).

On the same subject, Patil et al. (2008) investigated different injection brines with salinities ranging from reservoir brine salinity of 22,000 to 5500 TDS, which were applied in conducting corefloods. The study included ultralow-salinity water along with the characterization of COBR system wettability using the Amott-Harvey wettability indices ($I_{(AH)}$). In this wettability test, the value of (+1) of the $I_{(AH)}$ specifies a highly water-wet system, while an $I_{(AH)}$ value of (−1) denotes a highly oil-wet system (Chen et al., 2006; Cuiec, 1984), which has been tabulated in Table 7.2. The results from the coreflooding experiments exhibited a steady residual oil saturation reduction of around 20%, with a small surge in the Amott-Harvey wettability index, and injection brine salinity decrease at reservoir temperature. This indicated the potential of LSWF in EOR compared to high-salinity water floods (Patil et al., 2008).

TABLE 7.2 Wettability classification

$I_{(AH)}$ range	Wettability state
+0.3 to +1.0	Strongly water-wet
+0.1 to +0.3	Slightly water-wet
−0.1 to +0.1	Natural
−0.3 to −0.1	Slightly oil-wet
−1.0 to −0.3	Strongly oil-wet

7.4.1.1 Effect of oil chemistry

The chemistry of oil affects all aspects of the LSWF performance. It has been identified that the polar components present in crude oil (nitrogen-, sulfur-, oxygen-bearing compounds, i.e., NSOs), termed surface-active components (SACs) (not found in synthetic oil) regulate the surface wettability in a COBR system (Denekas et al., 1959; Tang and Morrow, 1999; Tawfik et al., 2019). The SACs are adsorbed either directly onto the rock surface or through cationic bridging (Buckley et al., 1989; Dubey and Doe, 1993; Fjelde et al., 2014; Standnes and Austad, 2003). These NSOs are therefore considered the prime factor in the failure of cEOR methods (Feng and Liang, 2015; Samanta et al., 2011; Shabib-Asl et al., 2015). The molecular-scale interactions in the presence of various SACs types are dependent on the degree of their affinity to the rock surface, adsorption strength, solubility in brine, and their unique pore-scale wetting abilities. Hence, on the Darcy-scale of oil recovery, significant variations are observed. Furthermore, the length of the carboxylic acid chain affects the number of SACs adsorbed onto carbonate rocks during the aging process, and affects the strength of adsorption, affecting the extent of wettability change during LSWF. Segregation of SACs in brine may affect the rock-oil-brine system interactions, and create in-situ carboxylate salts (soaps), which have been observed in the presence of certain kinds of oil-brine pairs. The solubility of SACs in brine is another factor that results in water-wetness (Tawfik et al., 2020). Additionally, the total acid number (TAN) value has a negligible effect on monoprotic acidic SAC types in the oil phase, in case the molar concentration of TAN and SACs in the oil is the same.

In Table 7.3, the adsorbed amount of each SAC type (in % of the original number of SACs present in oil through aging) is provided. By observing the adsorption and desorption with temperature, the degree of adsorption of the different SACs on the rock surface can be determined. At high temperatures,

TABLE 7.3 SAC amount adsorbed onto the rock sample postaging

SAC	Amount adsorbed (% of initial SAC mass available during aging)
Acetic acid	11.9
Decanoic acid	2.9
Stearic acid	1.9
Quinoline	5

loss of weight occurs due to strong chemisorption, where more heat energy is consumed for releasing SACs, which raises the temperature during the thermal gravimetric analysis measurement. The adsorption and desorption of the diverse SACs on the rock surface is the key marker that defines the initial rock wettability state before chemically tuned water flooding (CTWF) or LSWF, and the ease of changing rock sample wettability by CTWF/LSWF (Tawfik et al., 2020).

Furthermore, at high brine salinity, the contact angle and area of oil contact arrive at an even equilibrium with the substratum, while being receptive to low-salinity brine with a lower oil wetting state, which eventually causes the clay to detach from the substratum. At the outset, the clay particles stick at the oil-water-solid interface at high salinity. However, with contact to low-salinity brine, this affinity decreases, resulting in the contact area and contact angle reduction, indicating the wettability change closer to water-wet conditions. When the contact angle reaches a critical value of around 40–50°, it initiates the detachment of oil drops from the clay. This detachment phenomenon causes a larger portion of oil to be freed; however, tiny oil residue pockets may remain in the clay layers. Although it may take days to complete, the investigations of various salinity levels have deduced that the wettability change kinetics are associated with buoyancy rather than the binding force equilibrium (Mahani et al., 2013).

7.4.1.2 Mechanisms of LSWF in sandstone

In sandstone reservoirs, the main proposed mechanisms behind LSWF are fines migration. pH increase, multicomponent ion exchange (MIE), salting-in, and wettability alteration (Al Shalabi and Sepehrnoori, 2017). The fines migrations mechanism was suggested by Tang and Morrow (1999), as they observed that LSW could release clay fragments called fines which are mostly kaolinite, to make clay minerals alter to water-wet (Tang and Morrow, 1999). Nevertheless, the actual reason for enhanced microscopic displacement efficiency was the diversion of water into unswept pores as the released clay fines blocked pore throats and not wettability change. More subsequent studies categorized fines migration as an auxiliary mechanism rather than the main mechanism of LSWF (Lager et al., 2008).

McGuire et al. suggested that LSW caused increasing pH of the system via an alkaline flood, to be the leading mechanism of LSWF (McGuire et al., 2005). This was however denied by Lager et al., stating that for in-situ surfactants to form, the acid number (AN) had to be greater than 0.2 mg KOH/g. Yet, in some cases of AN = 0.5 mg KOH/g, the LSW effect was noticed. Moreover, the literature showed that the IFT reduction was not very high, and that, in many cases, the solution was on the basic side with pH increases of not more 1 unit. The proposal by Lager et al. stated that the LSW effect resulting in IOR was caused by wettability alteration toward water-wetness due to LSW injection. It was the presence of divalent ions such as Ca^{2+} and Mg^{2+} which affects ion exchange between clay particles and surface-active components in crude oil.

Additionally, the salting-in and salting-out mechanisms were proposed by Doust et al., who stated that the polar solubility of the organic component is altered by LSW injection, which disrupts the equilibrium of the water/oil/rock phases (RezaeiDoust et al., 2009). Salting-in occurs as the salinity of the system decreases due to salt elimination from solvent fluid (water), leading to enhanced solubility increase of organic components. Salting-out occurs with more salt added to the system, resulting in a solubility decrease of organic material. The presence of divalent ions (Ca^{2+}, $Mg^{2+,}$ and Na^{+}) can rupture the water structure, leading to a decrease in the solubility of organic materials. Salting-in happens with the salinity of the system decreasing to lower than critical ionic strength, which causes enhanced organic solubility (Al Shalabi and Sepehrnoori, 2017). Furthermore, the mechanism of wettability alteration was studied in depth by several authors. The prerequisites are the clay presence in the porous medium, polar components in oil, and the presence of divalent ions. Wettability alteration occurs through fines migration, pH increases which caused a reduction in IFT, MIE, and the expansion of the electrical double layer, causing greater oil recovery (Al Shalabi and Sepehrnoori, 2017; Lager et al., 2007, 2008; McGuire et al., 2005; Nasralla and Nasr-El-Din, 2011; Shehata and Nasr-El-Din, 2015; Suijkerbuijk et al., 2012; Tang and Morrow, 1997; Zhang et al., 2007).

In a recent study by Mehraban et al., in a clay-rich sandstone using two contrasting crude-oil types, microdispersion (formation of separated water phases within the crude-oil phase) was described as the cause of EOR through a change in wettability and swelling up of crude oil. Using Fourier-transform infrared spectroscopy and water content measurements, the propensities of crude oils toward microdispersion formation were studied. It was noted that the mechanism underlying LSW is related to the formation of microdispersion by acidic materials, viz. carboxylic acids, and/or carboxylic functional groups of asphaltenes. Moreover, they also stated that the presence of clay is not necessary for the low-salinity effect, as was believed in earlier literature, and the electrical double layer expansion and multiion exchange (MIE) mechanisms have been observed to contribute in minor terms to low-salinity effect (Mehraban et al., 2022).

7.4.1.3 Mechanism of LSWF in carbonate reservoirs

In carbonates, most studies agree that wettability alteration is the main mechanism of EOR in carbonates (Al Shalabi and Sepehrnoori, 2017). This effect is created by the modified ionic composition of the injection water, which alters the thermodynamic equilibrium of the initial COBR system (Austad et al., 2007, 2009; Fathi et al., 2010a,b; RezaeiDoust et al., 2009; Shariatpanahi et al., 2016; Zhang and Austad, 2006). Moreover, dissolution or desorption of organic components alters the charge of the surface and leads to wettability change (Al Shalabi and Sepehrnoori, 2017), and at a higher temperature of >90°C, injection of water with SO_4^{2-} or Ca^{2+} or Mg^{2+} may be used to achieve this.

Furthermore, in calcites, LSW improves the oil recovery by inducing favorable fluid-fluid and rock-fluid interactions. The crucial role of these interactions on LSWF efficacy was observed by Mehraban et al. when investigating pure calcites that were originally water-wet or oil-wet. In initially water-wet cores, the oil recovery efficiency is higher in high-salinity water, owing to dynamic IFT, dynamic contact angle, oil/brine and rock/brine interactions, zeta-potentials, and pH of effluent. In addition, the electric charge variations of the crude-oil/brine and calcite/brine interfaces result in short-term IFT reduction and long-term wettability changes toward neutral-wetness. In contrast, the oil-wet cores behave in a completely dissimilar manner to water-wet cores, yet they still show wettability dependency.

Additionally, in the works of Meherban et al., using the contact angle measurements, it was noted that diluted seawater (dSW) is more effective in changing wettability toward water-wetness. With the decline in brine salinity, the crude oil and rock surfaces tend to become negatively charged (evident from zeta-potential investigations). The wettability alteration that ensues is the result of massive repulsion between the two surfaces when salinity is low. In dSW brine, the CO_3^{2-} ions can turn the calcite surface more negatively charged. Thus, the original wetting state is highly influential in oil recovery. When using seawater brine in an originally oil-wet core, it showed earlier breakthrough, than in an initially water-wet core, and the resultant oil recovery was low. The dSW yielded higher oil recovery with a similar breakthrough time in an originally oil-wet core than in an originally water-wet core. Therefore, with the wettability alteration toward a more water-wet state, a piston-like movement occurs, resulting in better performance LSW in an oil-wet core (Mehraban et al., 2022).

7.4.2 Chemical EOR

During the secondary and tertiary EOR, the cEOR techniques, in which chemical additives are mixed with water to adjust viscosity or reduce surface/IFT between oil and water interfaces or alter rock wettability or combinations, have been technically highly effective (Ayoub et al., 2015).

7.4.2.1 Polymer flooding

Polymer flooding (PF) is a cEOR technique in which a highly viscous polymer solution is injected into the rock formation to adequately reduce the mobility ratio, causing swept volume expansion, which leads to EOR. Lately, it has been suggested that polymer flooding may replace water flooding as the main method in secondary and tertiary EOR stages, because polymer flooding has a higher potential for water cut reduction during low water period, than water flooding (Sheng, 2015

Polymer properties impart robustness and viscoelasticity, which is attributed to their long-chain structure, and high molecular weight. This makes them able to stretch the oil droplets and the oil film through the polymer solution movement, which imparts considerable carrying capacity to them (Li et al., 2021). Polymer behavior is sensitive to temperature, formation water salinity, rock permeabilities, divalent contents, clay contents, oil viscosity, and heterogeneity distinctions (i.e., minerals) (Musa et al., 2021; Shakeel et al., 2020). During PF, the nanocomposite (clay) samples yielded the most oil and lowered the water cut ratio. However, polymer adsorption and high polymer slug size are some challenges of PF that need alleviation (Salehi et al., 2017).

Hydrolyzed polyacrylamide (HPAM) is the most extensively used polymer in all cEOR techniques worldwide. During PF in a reservoir formation where oxygen is present, Fe^{2+} ion is oxidized to Fe^{3+}, releasing the free radical O^{2-}, which reacts with HPAM to generate peroxide leading to rupture of HPAM backbones. Also, O^{2-} reacts with Fe^{3+} ion-producing Fe^{2+}, and further oxidization creates Fe^{3+} ions and O^{2-} free radicals. The ensuing chain reaction considerably diminishes polymer viscosity. In this case, Fe^{2+} plays the role of a catalyst and thus far is considered the solitary compound to which polymer viscosity reduction can be attributed. This is a rapid reduction where the solution loses the viscosifying power to almost the water viscosity level in a matter of seconds; hence, the Fe^{2+} concentrations should be below 0.5 mg/L (Luo et al., 2006).

HPAM has negative carboxyl groups which are highly sensitive to divalent metal cations (e.g., Ca^{2+} and Mg^{2+}) on the rock surface, which also lead to solution viscosity reduction due to reactive injection-fluid-rock tendencies that form gels (or precipitates) (Sheng et al., 2015). Therefore, it is clear that salinity affects the viscosifying power of the polymer negatively. Moreover, the impact of divalent or multivalent ions is much greater in viscosity reduction than that of monovalent ions; thus, divalent concentration in mixing water for polymers must be limited in the reservoir formation water (Sheng, 2014).

Thus, the polymers can modify the rock-brine equilibrium by attracting divalent cations present on the surface of the rock. According to studies, in comparison to the equilibrated brine, the concentration of the divalent cation could increase up to 5 times due to PF. The anticipated viscosity loss ranges in PF and LSP can be up to 30% loss. Hence, it is the fluid-rock system, which determines the cation exchange level and viscosity loss. In this system, the important parameters are the charge ratio of divalent cations' positive charges to the polymer's negative charges, the cation mass-action ratio (MAR) of the divalent to monovalent, and the polymer's viscosifying power which is sensitive to divalent cations concentration (Ca^{2+} and Mg^{2+}) in the formation water (Divers et al., 2017).

Concerning the interfacial interaction during PF with the cations on the clay surface, cations are exchanged within the pore fluids (Farajzadeh et al., 2017; Pope and Nelson, 1978), and the degree of the exchange is mostly affected by relative concentration, size, and ion valences on the rock and in the solution. The cation exchange capacity (CEC) specifies the number of cation exchange sites in a specific weight (or pore space) of rock, with which the ions in solution can interact. CEC is the sum of the transferrable positive charges that ionically bind to the anionic charges on the rock exterior. In reservoir rocks, there is an equilibrium between the mono- and divalent cations on the negatively charged sites in the clays and the formation of water ions, corresponding to their respective MAR as given in Eq. (7.1) (Hirasaki, 1982; Pingo-Almada et al., 2013):

$$\mathrm{MAR} = \frac{[\mathrm{X}^{+}]^{2}}{[\mathrm{X}^{2+}]}, \tag{7.1}$$

where $[X^+]$ and $[X^{2+}]$ denote the monovalent and divalent cation concentrations in equivalents/liter (eq./L), respectively. In case a dissimilar ionic composition of water is injected, the cations in the injected water exchange with cations on the clay locations to create a new equilibrium, and the brine slug's ionic composition changes (Hirasaki, 1982; Pingo-Almada et al., 2013). Correspondingly, in the work of Unsal et al. (2018), the divalent cations increase leads to around 30% bulk viscosity loss (Unsal et al., 2018).

Similarly, in the experiments by Berge et al. (2018), it was observed that the polymer solutions altered due to cation exchange, which affected the viscosity. The MAR is a key factor in the cation exchange process. Berge et al., designed a low-salinity (LS) brine that had 10 times reduced MAR than the formation brine (of high salinity, i.e., HS), and a second low-salinity brine (LS_{MAR}) was designed with MAR reservoir brine MAR of HS. The aim was to realize if regulating the MAR could reduce the cation exchange process. They prepared the $LS_{MAR}P$ (low-salinity MAR polymer) solution, which was added with extra cations in amounts equivalent to the increases noted in produced fluid during their preceding experimental steps, i.e., the produced $LS_{MAR}P$ and HSP (high-salinity polymer) fluids.

It was observed that the anionic polymer groups either behaved as cation exchange spots, where divalent cations attach favorably and conforming with the type of rock-polymer-fluid system, the viscosity increased (~20%) or decreased (~30%) (Berge et al., 2018). For instance, the viscosity loss in the solution was 9% when Na^+ was added in their experimental study. On the other hand, there was a 32% viscosity loss in the solution when Ca^{2+} was added, although it was not entirely due to the increase in divalent cations, it was to a certain extent (~9%) due to the total salinity increase. Moreover, as seen in Table 7.4 below, the total viscosity loss (>20% at $8\,s^{-1}$) is high due to Ca^{2+} increase, which is affected by the shear rate, and it increases when shear rates are lower (Berge et al., 2018).

TABLE 7.4 Relative viscosity losses at $8 s^{-1}$ and 50°C of the $LS_{MAR}P$ solutions to which either 20 ppm Na^+ or 20 ppm Ca^{2+} was added

Polymer solution	Viscosity loss (%)
$LS_{MAR}P + 20$ ppm Na^+	9
$LS_{MAR}P + 20$ ppm $[Ca^{2+}]$	32
$LS_{MAR}P$	—

Furthermore, to conduct a successful polymer flood scheme, the appropriate approximation of polymer retention is mandatory. The three main causes of retention are adsorption, mechanical entrapment, and hydrodynamic retention of polymers (Al-Hajri et al., 2018). The phenomenon of chemical adsorption is when a portion of the mass of a chemical component of a fluid (adsorbate) inhabits the voids, or portions of it amasses on the solid surface matrix (adsorbent), at the interface of liquid-solid (Sun and Zhang, 2020). Polymers have high molecular weights and a structure with long chains containing numerous polar clusters, which tend to bind to the polar elements accessible on the surface of the rock (Manichand and Seright, 2014). The polymer occupies the adsorption sites at the rock surface so that the larger available surface area in relation to the polymer molecules has a higher level of adsorption (Theng, 2012). Polymer adsorption is irreversible and therefore can be a serious issue (Green and Willhite, 1998; Lee and Fuller, 1985), if the adsorbed polymer layers settle into huge pore areas, it would obstruct the recovery of a considerable amount of oil.

Thus, the polymer molecules attach to the solid rock surface mainly because of physical adsorption, i.e., the van der Waal's, hydrogen bonding, and hydrophobicity (Sheng, 2010). It is unlikely that complete chemical bonds are formed between the polymer molecules and rock surface by chemisorption (Sorbie, 2013). Generally, due to a large low free energy, the polymer molecules are adsorbed onto the rock surface. Owing to this large low free energy, the ensuing polymer adsorption leads to the release of water molecules from the polymer solution or rock surface causing the entropy to increase (Kolodziej, 1988). Additionally, when adsorption of the bulk polymer solution concentration decreases, it adds to entropy increase. Contrary to this, these entropy increases are countered by entropy loss, which is a result of the polymer losing its' structural autonomy when it is adsorbed on the surface of the rock (Dang et al., 2014b). For ionic polymers, enthalpy leads to low free energy due to polymer adsorption. Because of the polymer electrostatic forces, which either repel or attract, depending on the net ionic charge of the surface, the enthalpy occurs.

Moreover, it is to be expected that in the case of sandstones there is less polymer adsorption due to the alike negative charge of the polymer chain and the sandstone surface, whereas, in carbonate rocks, the polymer adsorption is observed to be of greater extent, because of the positively charged carbonate surface attracting the negatively charged polymer ions. In nonionic polymers, e.g., PAM, the adsorption can be mainly associated with hydrogen bonds, and in ionic polymers, e.g., HPAM, adsorption can be mostly attributed to electrostatic repulsion. As the temperature increases, it facilitates the hydrogen bond to ease up and break, which decreases the adsorption of PAM. The temperature increase affects the negative charge of the rock surface to surge, creating greater electrostatic repulsion reducing adsorption (Sheng, 2010).

Therefore, polymer adsorption is an important factor to be estimated accurately for designing polymer floods. For this purpose, numerical simulation in commercial software such as CMG-STARS (Computer Modelling Group's Thermal & Advanced Processes Simulator) has been used to calculate polymer adsorption, based on the Langmuir isotherm equation (Goudarzi et al., 2013). Eq. (7.2) for polyadsorption has a nonlinear salinity relation and polymer mole fraction when in an aqueous phase, as given in Eq. (7.2) (Goudarzi et al., 2013):

$$\mathrm{ad} = \frac{(\mathrm{tad1} + \mathrm{tad2} * \mathrm{xnal})\mathrm{ca}}{(1 + \mathrm{tad3} * \mathrm{ca})} \tag{7.2}$$

where ad denotes polymer adsorption, xnal denotes the salinity, ca stands for polymer mole fraction in an aqueous phase, and tad1, tad2, and tad3 are input fitting parameters. The results show that after PF, according to the parameters, such as rock and fluid properties, absolute and relative permeability, a saturation of fluid, porosity, and water effective permeability, the velocity decreases afterward. Polymer adsorption on rock surface after polymer flooding reduces the permeability depending on the type of polymer (cationic, anionic, or nonionic), the molecular weight of the polymer, shear effects, rock properties, and reservoir environment (temperature, pressure, and salinity) (Goudarzi et al., 2013).

Furthermore, formation permeability is very important for a successful polymer injection, as low permeability formations stop the polymer molecules in the injection slug from easily moving through rock pores (Taber et al., 1997). Fig. 7.11 shows the various polymer retention types including adsorption, mechanical entrapment, and hydrodynamic entrapment, ordered from the most significant to the least.

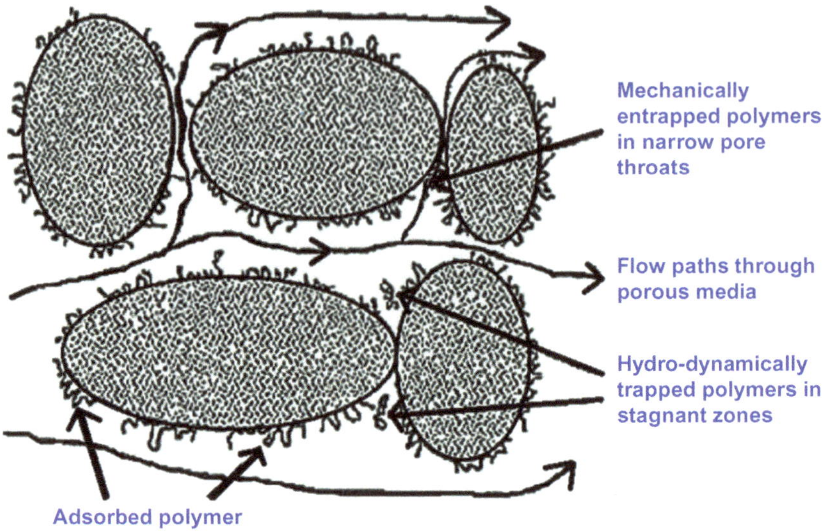

FIG. 7.11 Polymer retention mechanisms in porous media. *Reproduced with permission from Firozjaii, A.M., Saghafi, H.R., 2020. Review on chemical enhanced oil recovery using polymer flooding: fundamentals, experimental and numerical simulation. Petroleum 6 (2), 115–122, Copyright (2020), Petroleum.*

To date, most PF projects have been performed in sandstone reservoirs, while few have been implemented in carbonate reservoirs due to the high adsorption of anionic polymers (HPAM) in carbonates. Also, the heterogeneous carbonate reservoirs have a low-permeability matrix, which results in poor polymer fluid injectivity due to obstruction in the movement of large polymer molecules from entering and dislodging the trapped crude oil from the pores. Due to years of water flooding, the reservoir-water salinity tends to be closer to the injected water salinity; thus, screening of salinity and divalent ions of reservoir water before polymer injection is necessary. We now know that the salinity of formation water and the divalent ion concentration can affect the polymer viscosity and performance when mixed with the injected fluid. To mitigate this, a preflush wherein the water chemistry has been modified can lessen the negative effect of the salinity of formation water and divalent ions on polymer-flooding performance. Thus, merging low salinity with seawater may result in much lower polymer consumption in the carbonate reservoir (Masalmeh et al., 2019). In addition, using salt-tolerant biopolymers in high-salinity reservoirs can prevent polymer degradation, and need not require the use of preflush (Leonhardt et al., 2014; Ogezi et al., 2014).

Furthermore, current reservoirs have matured, and to replenish the global oil supply, newer reservoirs with unconventional and harsh conditions (i.e., high temperature, high salinity, HTHS carbonate formations) are being exploited for oil recovery, which is highly challenging. Novel polymers have been synthesized that are tolerant to HTHS carbonate conditions. For instance, modified-HPAM has greater resistance to degeneration, precipitation, and exhaustion in the reactive fluid-rock or fluid-fluid interactions within HTHS carbonate environments. Furthermore, plant-waste-derived polymers are biodegradable and nontoxic, which have a minimal environmental impact with the least adsorption on rock surfaces. Surfactants are chemical agents that can reduce the IFT between fluid-fluid, and surface tension (ST) between gas-liquid interfaces. Moreover, surfactants can alter the wettability of the rock affecting the fluid-rock interface (Olajire, 2014). A surfactant is structurally comprised of hydrophobic groups in their tails and hydrophilic groups in their heads, imparting dual oil and water solubility (amphiphilic). (Rosen and Kunjappu, 2012).

The amphiphilic property of surfactants that induces IFT reduction between the water-oil interfaces results in the reduction in capillary forces within the porous medium of the reservoir rock (Ragab and Mansour, 2021). Moreover, based on different temperatures and water salinities, cosurfactants may be required in the formulation fluid, to make surfactants more robust. In an experimental study conducted by Keshtkar et al., using oil of viscosity 2.24cP, three varying surfactant concentrations of 0.00, 0.01, and 0.004 were injected to examine their capacity in oil recovery. For initial injection, only water flooding was conducted (0.00 mole fraction of surfactant concentration), followed by surfactant concentrations of 0.01 and 0.004 mole fractions. As given in Fig. 7.12, the oil recovery at 0.01 mole fraction was the highest, since greater surfactant concentration results in greater IFT reduction (Keshtkar et al., 2016). However, since surfactants are costly, the optimum surfactant concentration is essential. When the injection fluid reaches the critical micelle concentration (CMC), we acquire the optimum surfactant concentration.

For residual capillary oil trapped in the reservoir, the movement of oil through the narrow capillary pores requires the oil/water IFT to be extremely low Sheng (2013). The capillary number is described through viscous terms as given in Eq. (7.3) (Sheng, 2013):

$$N_c = \frac{\mu_w \varphi q}{\gamma_{o/w}}, \tag{7.3}$$

where N_c stands for a capillary number, ϕ denotes porosity, μ_w stands for aqueous viscosity, q denotes the flow rate/unit cross-sectional area of water, and $\gamma_{(o/w)}$ stands for IFT between oil and water. Thus, when the capillary number rises, additional residual oil is retrieved (Sheng, 2013). A greater capillary number could be achieved by minimizing the IFT parameters (Donaldson et al., 1989).

Selecting the appropriate surfactant is crucial in designing SF projects. According to Sheng (2010), surfactant types are characterized based on the

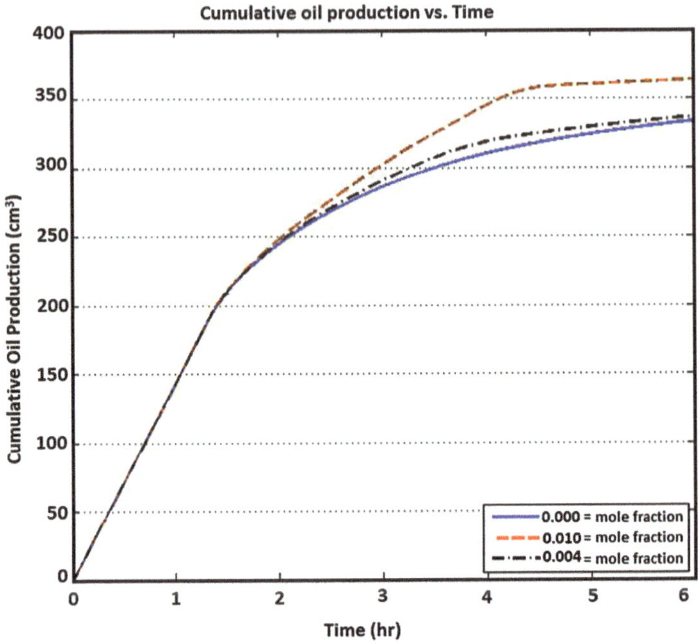

FIG. 7.12 The cumulative oil production at various surfactant mole fractions of 0.00, 0.01, and 0.004. *The information was taken from data points of Keshtkar, S., Sabeti, M., Mohammadi, A.H., 2016. Numerical approach for enhanced oil recovery with surfactant flooding. Petroleum 2 (1), 98–107.*

nature of the ionic head portion of the surfactant. Anionic surfactants have a negative charge on their polar sections, which makes them ideal for sandstone reservoirs as there is minimal internal pore surface adsorption, whereas the cationic surfactants have a positively charged head part, which makes them ideal for surfactant flood in the positively charged carbonate reservoirs. Nonetheless, in general, anionic surfactants are ideally chosen, owing to their low adsorption at neutral-to-high pH on both sandstones and carbonates, and because they can be modified for use in a wide range of conditions while being of low cost (in certain cases) (Sheng, 2015). In carbonates generally, the anionic surfactant is employed as IFT reduction inducers owing to HTHS tolerance, while the cationic surfactants are employed for wettability alteration owing to low adsorption on carbonates. However, the nonionic surfactants have so far been inefficient in EOR trials. In carbonate reservoirs, sulfates are suitable for low-temperature applications, while sulfonates are suitable for high-temperature applications, whereas cationic surfactants can be more useful as cosurfactants.

The pH and potential determining ions (PDIs), viz. Ca^{2+}, Mg^{2+}, CO_3^{2-}, SO_4^{2-} are the leading factor in the wettability alteration via surfactants. These interactions are governed by the surface charges (zeta-potential) at fluid-fluid and rock-fluid interfaces, and the optimal salinity is defined by the types and characteristics of crude oil. At extreme magnitudes of zeta-potentials, the carbonates change to highly oil-wet or water-wet (Boampong et al., 2022; Hassan et al., 2020a,b,c). The surface charges of minerals cause varying dissolution behavior. The metal oxides usually have a point of zero charges (PZCs) greater than minerals contained in carbonate rocks, which makes oxides and carbonates of opposing polarities at certain pH levels (Kosmulski, 2002; Stumm and Morgan, 2012). Thus, there is continuous and irregular surface potential (Na et al., 2007), which creates the zeta-potential of carbonate residues (Chen et al., 2014; Kasha et al., 2015; Mahani et al., 2017; Moulin and Roques, 2003). Even though zeta-potential does not entirely describe the adsorption of surfactants due to surface charge, nonuniform adsorption (e.g., charge-regulated vs hydrogen-bonded) occurs, and this dynamic can lead to patch-wise adsorption (Nwidee et al., 2017). As the surfactant solution contacts the rock surface, mineral dissolution changes the ionic composition from the initial state (Cui et al., 2015). In addition, as pH increases, anionic surfactants adsorption decreases consequently, and a small ionic strength increment and divalent cations concentration increase occur (Budhathoki et al., 2016; Pham et al., 2015; Seethepalli et al., 2004; ShamsiJazeyi et al., 2014; Sharma et al., 2015; Somasundaran and Fuerstenau, 1966; Southwick et al., 2016).

7.4.2.1.1 Surfactant adsorption mechanisms

Surfactant adsorption behavior may be of either physical or chemical character (Sheng, 2010). Normally, the adsorption type is categorized according to the charge transfer. When there is no sharing of an electron during the adsorption process, it is categorized as physical adsorption, while the opposite holds for chemical adsorption as it involves electron transfer during the adsorption process. For instance, in a study by Tagavifar et al., they concluded that adsorption of an anionic surfactant on carbonate surface was through two mechanisms independent of each other. The first adsorption mechanism involves a monodentate (carboxylate or sulfate) inner-sphere complex, which is charge-regulated, along with a weak bonding or hydrogen bonding. The ionic surfactant head groups of either carboxylate or sulfate are associated with the monodentate inner-sphere complexes and are charge-regulated, while the surfactant's ethoxy and propoxy groups are associated with weak adsorption. Irrespective of surface load, for low pH conditions, the principal adsorption on calcite and clay is by charge-regulated means, while for pH values of $\sim$10, the principal adsorption on calcite is using hydrogen bonding (Tagavifar et al., 2018).

Furthermore, the surfactant molecule adsorption onto the internal pore surfaces depends on the petrophysical properties and environment of the reservoir. The surfactant is inclined to latch onto pore surfaces at lower temperatures, and this tendency is reduced significantly at very high temperatures

(Hamid and Onur, 2013; Ziegler and Handy, 1981). In addition, based on the principal PZC, rock surface charge turns from negative to positive or vice versa, due to the pH variations. Sandstone has PZC of around 2.5 and carbonates have PZC of around 9.0. Under usual reservoir environments where pH ranges from 6 and 8, sandstone is found to be negatively charged while carbonates have a positive charge (Srisuriyachai, 2008). This means that if the pH of the reservoir is altered, surfactant adsorption behavior may be changed. It has been observed that adsorption decreases with pH increase when anionic surfactants are used, which is due to the electrostatic repulsion of like negative charges between anionic surfactant polar molecules and the pore surface (Hamid and Onur, 2013). Furthermore, when NaCl is added to the surfactant solution, the electrostatic repulsion of surfactant molecules wanes, leading to higher surfactant adsorption with a higher concentration of NaCl (Azam et al., 2013). That is, higher surfactant concentration leads to greater surfactant adsorption (Hamid and Onur, 2013). In addition, surfactant adsorption is affected by clay and rock surface chemistry, water chemistry, surfactant characteristics, and functional groups (Chandar et al., 1987; Hu and Bard, 1997; Levitt and Bourrel, 2016; Ma et al., 2013; Somasundaran and Krishnakumar, 1997; Yu et al., 2003; Zhong and Claverie, 2013; Zhu & Gu, 1991). In addition, the formation of gels and viscous microemulsions could occur in the surfactant-oil-water solution, depending on the surfactant structure at certain temperatures (Skauge and Palmgren, 1989).

7.4.2.1.2 Surfactant adsorption in carbonates

The carbonate rocks (Levitt and Bourrel, 2016) and soil (Ishiguro and Koopal, 2016) have heterogeneous mineral makeup and rock surface chemistries, on which cationic surfactants have low adsorption on synthetic calcite, but high adsorption on natural limestone. The causative reason is deduced to be the aluminum and silicon in the rock, and the minerals in carbonates, viz. magnesite, Fe-dolomite, clay, anhydrite, and metal oxides present (Folk, 1959; Levitt et al., 2015; Ma et al., 2013). However, cationic surfactants are more expensive than anionic surfactants. Hence, anionic surfactants may be injected in the presence of a sacrificial agent that curtails the anionic surfactants' adsorption due to the negative charges of the surfactant and positive charges on the carbonate surface interactions. Through the steric effect of an adsorbed layer of polymer, even at low concentrations, coinjection additives that act as scaling and corrosion inhibitors can reduce anionic surfactants adsorption in carbonate reservoirs by acting as a sacrificial agent (Gutierrez et al., 2019). Static adsorption of an anionic surfactant (sodium olefin sulfonate) on carbonate rock can be reduced using scaling and corrosion inhibitors, viz. poly sodium methacrylate, and sacrificial agent, viz. diethanolamine (Gutierrez et al., 2019).

In addition, water chemistry also affects the size and dynamics of adsorption in a nonuniform style, owing to the polymorphic tendency of oxide-like locations on the clay platelets extremities. Levitt and Bourrel observed that in montmorillonite clay and another sample of oxidized outcrop rock with high clay and iron content, simply rinsing with sodium chloride resulted in less adsorption. The underlying mechanism is deemed a consequence of decreased surface charges due to divalent cations exchange from the clays, because of the rinse (Levitt and Bourrel, 2016). Furthermore, Klimenko et al. observed that even minute calcium pockets on the surface retained the carboxylate surfactant, which had been used for a coreflood trial (Klimenko et al., 2020).

Still, surfactant adsorption is reversible (desorption) through water floods; although it does not significantly affect incremental oil recovery except in heterogeneous reservoirs. The reversibility behavior is sensitive to forces of attraction between the monomer heads to the internal rock surface, and the self-interaction of the monomer's tail section (Ngo et al., 2017). As an alternative to conventional surfactants, plant-based surfactants have been developed, which are nonhazardous, biodegradable, and nontoxic, and hence, have negligible environmental impact, i.e., their adsorption on the rock surface is minimal and reversible.

7.4.2.2 Surfactant polymer flooding

The synergy of polymer with surfactant flooding enhances the water/oil IFT reduction, both microscopic and macroscopic sweep efficiencies, and stimulates the water viscosity. Surfactants are amphiphilic that renders them capable to decrease the IFT of water-oil phases (fluid-fluid) and limit the reservoir rock capillary forces, resulting in tertiary oil movement by the viscous forces, whereas the polymers affect the mobility ratio (fluid-rock). In other words, water-soluble electrolytes help the micellar slug with greater viscosity control of the solution and the polymer slug circulates the micellar solution slug for mobility control.

Following the injection procedure, decreased oil-water IFT is established, and mobility control is enhanced by the creation of a large oil bank (Cockcroft et al., 1988). In other words, simultaneous adsorption of the auxiliary surfactant, ionized acid, and unionized acid on the interface forms a compound adsorption layer, and the DIFT behavior occurs (Bera et al., 2011; Karović-Maričić et al., 2014; Mandal, 2015). This is described by the creation of thermodynamically formed stable phases, i.e., microemulsions, when surfactants, oil, water, and electrolytes mix under certain conditions (Bourrel and Schechter, 2010; Winsor, 1954). When polymer solution is injected, these microemulsions dissolve in oil or water to create an oil bank that improves the sweep efficiency, given that a stable pressure gradient exists (Bera et al., 2011; Hill et al., 1973).

Thus, the combined effects of fluid-rock interactions encourage residual oil recovery (Ragab and Mansour, 2021). Moreover, when surfactants are added to a solution, micelles are created. That is, at the interface, surfactant concentration reaches a particular critical concentration (i.e., CMC), and the surfactants begin accumulating to form micelles. The micelles solutions have an immiscible solubilized phase with microemulsions (Green and Willhite, 1998). This microemulsion phase solubilizes both oil and water to create a three-phase region of microemulsion that has ultralow IFT against both water and oil, and hence, is crucial in surfactant-based cEOR (Green and Willhite, 1998). These microemulsion phase changes can be dependent on variations in salinity, temperature, surfactant type, cosolvent (alcohol) molecular constructions, oil compositions, and the solids dissolved in the aqueous phase (Bera et al., 2011; Bourrel and Schechter, 2010; Hill et al., 1973; Healy and Reed, 1977; Winsor, 1954).

Ineffective surfactant-polymer-based flooding, the mechanisms behind IFT reduction between crude oil and water have been investigated by several researchers (Adibhatla and Mohanty, 2006; Hirasaki and Zhang, 2003; Lu et al., 2014; Mandal, 2015; Seethepalli et al., 2004). The imbibition experiments distinguish the low IFT imbibition mechanism to be the resultant of the wettability changes and buoyancy. For the anionic surfactants tested, the ultralow IFT reduces the capillary pressure to a near-zero value, allowing gravity and viscous forces to induce imbibition. During the dynamic imbibition process, simulations show that the transverse pressure gradients of the fractures and matrix propagates the surfactant further into the matrix (Abbasi-Asl et al., 2010).

As discussed earlier, the wettability of a fluid is its ability to flow and wet a solid surface in the presence of another immiscible fluid (Crain, 2002). Eq. (7.4) describes wettability in general (Anderson, 1986):

$$\sigma_{\mathrm{os}} - \sigma_{\mathrm{ws}} + \sigma_{\mathrm{ow}} \cos\theta_{\mathrm{c}} = 0 \tag{7.4}$$

where σ_{os} is the interfacial energy between oil and solid, σ_{ws} is the interfacial energy between water and solid, σ_{ow} is the IFT between oil and water, and θ_{c} is the contact angle. To determine wettability, a contact angle measurement or the product of the contact angle and the IFT (the last term in Eq. 7.4, Young's equation) can be conducted. Wettability is also measured by relative quantities of oil and water displaced in similar conditions, and by observing the displacement and/or surface mechanisms of water/oil phases (Anderson, 1986).

As earlier mentioned, the wettability changes by surfactants have been attributed to the main pH and PDI, viz. Ca^{2+}, Mg^{2+}, $CO_3{}^{2-}$, $SO_4{}^{2-}$. These reactions are controlled by the surface charges (zeta-potential), produced at the interfaces of oil-brine and rock-brine convergences. In addition, the optimal salinity is governed by the characteristics and varieties of crude oil. Dissolved salt content and concentration of the wettability changing elements in the injection water are also influential in enhancing the oil recovery factor (Austad et al., 2012; Qiao et al., 2016). The correlation between rock-brine-oil interactions in the reservoir shows that, in terms of ionic composition and salinity, the wettability alteration is principally dependent on pH and PDI, and the total oil recovery is subject to the proximity of these ions (Austad et al., 2012; Hiorth et al., 2010; Lager et al., 2008; Mahani et al., 2015, 2017; Puntervold et al., 2007; RezaeiDoust et al., 2011; Zhang and Austad, 2006; Zhang and Sarma, 2012). However, when ionic concentration decreases, wettability alteration occurs, which is not a direct consequence of PDI activity (Romanuka et al., 2012).

7.4.2.2.1 Fluid-solid interactions during SP flooding in carbonate

The oil recovery mechanisms for SP-based EOR in carbonate reservoirs are mainly influenced by gravity, capillary forces, viscous forces, and injection water composition. During SP flooding in carbonates, chromatographic separation of chemicals, surfactant adsorption on carbonate rock surface (Ragab and Mansour, 2021), and instability at higher pressure, temperature, and salinity conditions may happen (Mansour et al., 2016). In carbonate reservoirs, typically, when wettability changes from oil or mixed-wet to water-wet, an improvement in oil recovery is observed. This is because polymer and surfactant molecules compete for adsorption sites on the rock surface, and in this way, polymer additives can lower the adsorption of surfactant, or vice versa (Sheng, 2013), as can be seen in Fig. 7.13.

7.4.2.2.2 Kinetics of surfactant adsorption

The kinetics of surfactant adsorption is an important parameter because it provides the definitions of the diffusive transport of surfactant molecules from the bulk solution to the liquid-solid interface and the specifics of kinetics occurring at these interfaces. Existing adsorption kinetics can detail helpful

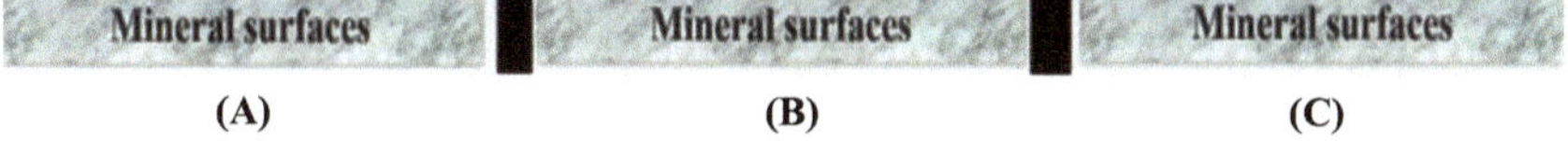

FIG. 7.13 (A) Graphic representation of the anionic surfactant adsorption on mineral surfaces through cation bridging. (B) Polymer polyelectrolytes compete with anionic surfactants as the sacrificial agent vying for the rock surface binding locations, reducing the adsorption of anionic surfactant. (C) Polymers as chelating agents increase the negative potential on the rock surfaces, and hence, the surfactant adsorption is reduced. *Reproduced with permission from Liu, Z., Zhao, G., Brewer, M., Lv, Q., Sudhölter, E.J.R., 2021. Comprehensive review on surfactant adsorption on mineral surfaces in chemical enhanced oil recovery. Adv. Colloid Interface Sci. 102467, Copyright (2021), Advances in Colloid and Interface Science.*

data about mechanisms of the adsorption, and define the adsorption rate against time. Table 7.5 below provides the four types of proposed adsorption kinetics (Zhang et al., 2014).

7.4.2.2.3 The kinetics of polymer adsorption

The kinetic adsorption process of polymers could be characterized by three main mechanisms. The first is the conveyance of polymer molecules in the direction of the surface from the bulk solution. The second is the surface attachment of the polymer, and the third is the unraveling of the polymer chain. The degree of adsorption could also be affected by other factors such as hydrodynamic settings, polymer type, polymer molar mass, polymer concentration, surface coverage, the strength of the binding, and the quality of the solvent (Dijt, 1993). Furthermore, following cEOR process, the wettability state of the sample reservoir rock has to be restored by reducing fluids that eliminate the iron oxides formed at the surface. Even though there is a lack of comprehensive knowledge of such reducing fluids on minerals and pore structure, these reducing fluids affect the rock mineralogy causing calcite cement dissolution. In addition, based on the presence or absence of a chelating agent, precipitates may reform in the core (iron precipitating with sulfur under pyrrhotite form) of the iron and calcite solubilized earlier to the reducing fluid. These dissolutions and precipitations affect the pore structure and may cause a slight decrease in the adsorption of surfactants from the reducing procedure (Guillon et al., 2017).

7.4.2.3 Alkaline-surfactant polymer flooding

On mixing high pH alkali additives with surfactants, negatively charged rock surfaces are generated, having enormous repulsion terms, which leads to decreased surfactant adsorption of anionic sulfonates (Delshad et al., 2011). The fluid-fluid interactions via alkalis include IFT reduction between oil and water interfaces, while the effective fluid-rock interactions assist in trapped oil displacement. In alkaline-surfactant polymer (ASP) flooding, the alkali, such as sodium carbonate, sodium hydroxide, and organic acids, undergoes saponification with crude oil and acts as in-situ surfactants, which causes IFT reduction. The greatest IFT reduction has been noted to occur at the lowest alkali concentrations (Jenning, 1975; Radke and Somerton, 1978; Reisberg and Doscher, 1956). During the ASP flooding, alkali is consumed while chemically reacting with the reservoir rock minerals, and then additional alkali is needed for maintaining the alkali concentration at the required amount. The mechanisms behind alkali consumption are reversible cation exchange, mineral dissolution, and insoluble hydroxides precipitation (Mohammadi, 2008). Also, the pH of the aqueous solution strongly affects the IFT reduction of the oil-water interface (Hirasaki and Zhang, 2003).

TABLE 7.5 Details of four proposed adsorption kinetics types

Kinetic model	Equation	Application conditions	Examples
Pseudo-first-order (PFO)	$dq_t/dt = K_1(q_e - q_t)$ where q_t = biosorption quantity (mg/g) at time t, q_e = biosorption capacity at equilibrium (mg/g); k_1 = kinetic rate constant (1/min), k_2 = pseudo-second kinetic rate constant (mg/g min); A and B are the constants relating to fraction of the surface covered and chemisorption activation energy	PFO model is valid only under either of these two sets of conditions (i) reaction control and Henry regime adsorption, or (ii) reaction control and high adsorbent dose (Tan and Hameed, 2017). For many adsorption processes, the PFO model is found suitable only for the initial 20–30 min of interaction (McKay et al., 1999)	PFO model can best predict the kinetic process of Congo red adsorption from aqueous solutions using cationic surfactant-modified wheat straw (Zhang et al., 2014)
Pseudo-second-order (PSO)	$dq_t/dt = K_2(q_e - q_t)^2$	Most environmental kinetic adsorption can be modeled well by PSO, when the solute concentration is not too high (Lagergren, 1898). In most cases, PSO model shows a wide applicability over PFO	The kinetics of saponin surfactant adsorption on the shale sandstone were persuasively estimated with the PSO model (Ahmadi and Shadizadeh, 2015)
Intra particle diffusion (IPD)	$q_t = K_i\, t^{\frac{1}{2}} + C$	Multilinearity nature in adsorption of the surfactant is emerged, indicating multiple adsorption mechanisms, such as mass transfer, film diffusion, surface diffusion, and pore diffusion (Ahmadi and Shadizadeh, 2015)	The results matched well with PSO and IPD model suggests that the adsorption process proceeds by surface sorption and intraparticle diffusion (Ibrahim et al., 2016)
Elovich	$dq_t/dt = \alpha e^{-\gamma_1 q_t}$	Assuming strong heterogeneity at the adsorbent surface, it is suitable for kinetics far from equilibrium and describes chemisorption as well (Gupta and Bhattacharyya, 2011)	When the dodecylamine is mainly in the form of micelle or precipitation, the initial rapid stage is best fitted by the PFO model, while the second stage is best fitted by the Elovich model (Kou and Xu, 2016)

In oil-wet reservoirs, ASP flooding alters the rock wettability from oil-wet to water-wet, which improves the oil's effective permeability. Also, combining IFT reduction with effective viscous water-in-oil emulsions reduces residual oil saturation, by forming a high capillary number (N_c), which is defined by Eq. (7.5) (Sheng, 2010):

$$N_c = \frac{(\mu_w u_t)}{\sigma}, \tag{7.5}$$

where μ_w is the viscosity of water, u_t is the total velocity of water and oil phases, and σ denotes the IFT (Sheng, 2010).

The mechanisms of alkaline flooding include the emulsification of crude oil in the reservoir as IFT reduction occurs along with entrainment, based on alkali and water solution movement. This requires high pH, a high acid number of crude oils (for low acid number of crude oils, ASP floods at the initial stage are preferred), low salinity, and for the size of the crude oil-water emulsion to be smaller than pore throat, otherwise, the droplets of emulsified oil formed can block the tighter pore throats (Lwisa, 2021).

Generally, during the tertiary cEOR stage, when implementing ASP flooding, the COBR interactions form emulsions inside the reservoir. These emulsions have very low IFT values, and they are produced due to the shearing action of the moving fluid within the porous media (Li et al., 2020). The emulsification process occurs due to the entrainment and entrapment mechanisms to enhance oil recovery. The entrainment mechanism is when emulsification of crude-oil droplets occurs within the aqueous phase owing to surface energy reduction. These droplets move to the aqueous phase, forming oil masses as minute particles, while the entrainment is when these emulsion droplets' sizes are either equal to or lesser than the size of the pore opening. Moreover, when surfactant flooding fails to achieve ultralow IFT (or low IFT), large emulsions with diameters exceeding the pore throat diameters form, leading to emulsification entrapment. These larger emulsion droplets block the pore throats and water channels, resulting in a reduction of the oil-to-water mobility ratio, and transmitting the surfactant solution to upswept zones. This effect improves the areal and vertical sweep efficiencies and is mainly significant in viscous (heavy) oil, for which the volumetric sweep efficiency of water flooding is ineffective (Bryan and Kantzas, 2009; Chowdhury et al., 2022; Kakati and Sangwai, 2017; Shi et al., 2015; Türksoy and Bağci, 2000).

Because the clays cause significant alkali consumption due to chemical reactions and ion exchanges, the total clay content must be within 15%–25% to limit this consumption effect. Also, the conducive formation permeability for ASP flooding is $>$100 mD, the oil viscosity $<$100 cP, the formation water salinity $<$20 wt%, and injection water containing divalent $<$0.4 wt%, which are prerequisites (Johnson, 1976; Sheng, 2010).

With the increase in salinity of the solution, the microemulsion phase behavior shifts from the lower to the middle to the upper phase. For some surfactants, the temperature rise may result in an optimal salinity increase. In the case of greater pressure, higher optimal salinity is needed, and with pressure increase, the system leans toward Type I phase behavior (Skauge and Palmgren, 1989). Typically, higher equivalent alkane carbon number results in higher optimal salinity (Solairaj, 2011). The cosolvents, e.g., diethylene glycol monobutyl ether, help to form clear, transparent, and aqueous stable surfactant mixtures at optimum settings, as cosolvents can encourage a quick equilibration and prevent viscous emulsions, gels, and liquid crystals formation (Sahni et al., 2010). Alkaline-cosolvent-polymer flooding has also been reported to be as effective as alkaline-surfactant-polymer flooding, if not better (Fortenberry et al., 2015). Moreover, the cosolvents with branched structures are more hydrophilic than linear structure cosolvents, which transforms the optimum salinity into a higher value (Shah et al., 1976).

Upon comparing ASP flooding through several simulation scenarios with other tertiary recovery methods namely, high-salinity (HS) water flooding, polymer flooding, alkaline-surfactant (AS) flooding, and LSWF, for a 10-year-old traditional high-salinity water-flooded reservoir. In this particular scenario, the underlying fluid-rock interactions prove to be most effective during ASP flooding. Favorable COBR interactions owing to surfactant and alkali effects lead to ultralow IFT reduction between oil and displacing fluid phases and improved volumetric sweep efficiency due to polymer interactions (Dang et al., 2017).

7.4.2.3.1 Pore structure modification through fluid-solid interaction on sandstones

When strong alkaline ASP flooding is performed, the pore structures of sandstones undergo substantial alteration at different stages of flooding. In the quantitative analysis of pores, the better ASP flood stages were determined by using factors such as pore width (PW), mean pore depth (PD), and fractal dimensions (FD1, FD2). Through testing the best pore structure in correspondence to the ASP flooding from 10 to 500 PV, the optimum displacement effect was observed at 50–100 PV for the ASP flooding stage. Moreover, it was reported that the factors controlling the modifications in the pore structure of sandstone include the intergranular particles migration, the clay, feldspar, and quartz corrosions, and the creation of new intergranular substances (Bai et al., 2016). Thus, as the ASP flood continues, the pore width further degenerates, as illustrated in Figs. 7.14 and 7.15. Fig. 7.16 illustrates feldspar and quartz corrosion, i.e., rock degeneration in relation to ASP flooding. With a persistent increase in ASP flooding, there occurs a steady decline in potassium feldspar and substantial differences in quartz. This occurs due to the rock components of silica acid radicals reacting with alkaline hydroxyl in ASP agents in the injected fluid to form aluminum hydroxide leading to a reduction in feldspar. On the other hand, the quartz on the rock converts into fragile crystalline or into amorphous SiO_2 that have robust silane bonds active, and fluid (H_2O), with Na^+, K^+, and OH^- ions enters the matrix, leading to the production of soluble Si $(OH)^{3-}$ (Bai et al., 2016).

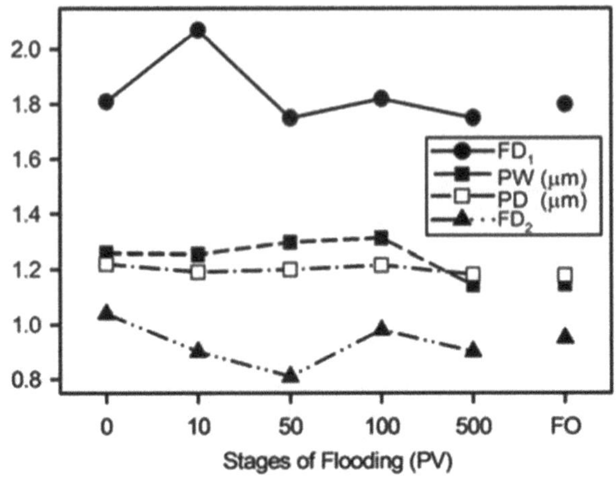

FIG. 7.14 Comparison of mean pore width (PW), mean pore depth (PD), and fractal dimensions (FD1, FD2) at different ASP flooding stages. *The information was taken from data points of Bai, Y., Chunmei, Y., Mei, L., Zhenxue, J., 2016. Variations of micropores in oil reservoir before and after strong alkaline alkaline-surfactant-polymer flooding. Open Pet. Eng. J. 9 (1).*

7.4.3 Hybrid EOR techniques

To promote EOR with the added benefits of synergic effect, low environmental impact, and easier applications and economic gains, hybrid EOR techniques are being developed. These methods combine two or more individual EOR methods, either synergistically or technically for an overall enhanced recovery performance. Following descriptions of the fluid-solid interactions defines the prominent mechanisms of some promising hybrid EOR methods.

7.4.3.1 CO_2 foam flooding

The CO_2 foam flooding technique provides better sweep efficiency than polymer flooding and evades the potential formation impairment due to polymer adsorption. During CO_2 foam flooding, carbon sequestration occurs in the formations underground, and since the CO_2 foam barely contains any water, the possible formation damage of water-reactive sections in the formation rocks is avoided, which enables a swift cleanup afterward (Sagir et al., 2018).

When a surfactant is incorporated to interact with CO_2, it generates foam that supplements water flooding process and causes oil displacement from the reservoir, owing to its greater sweep efficiency. Furthermore, the CO_2 mixes with oil to make it less viscous, swelling up the oil, which reduces the high IFT between oil and rock, leading to the mobilization of trapped oil

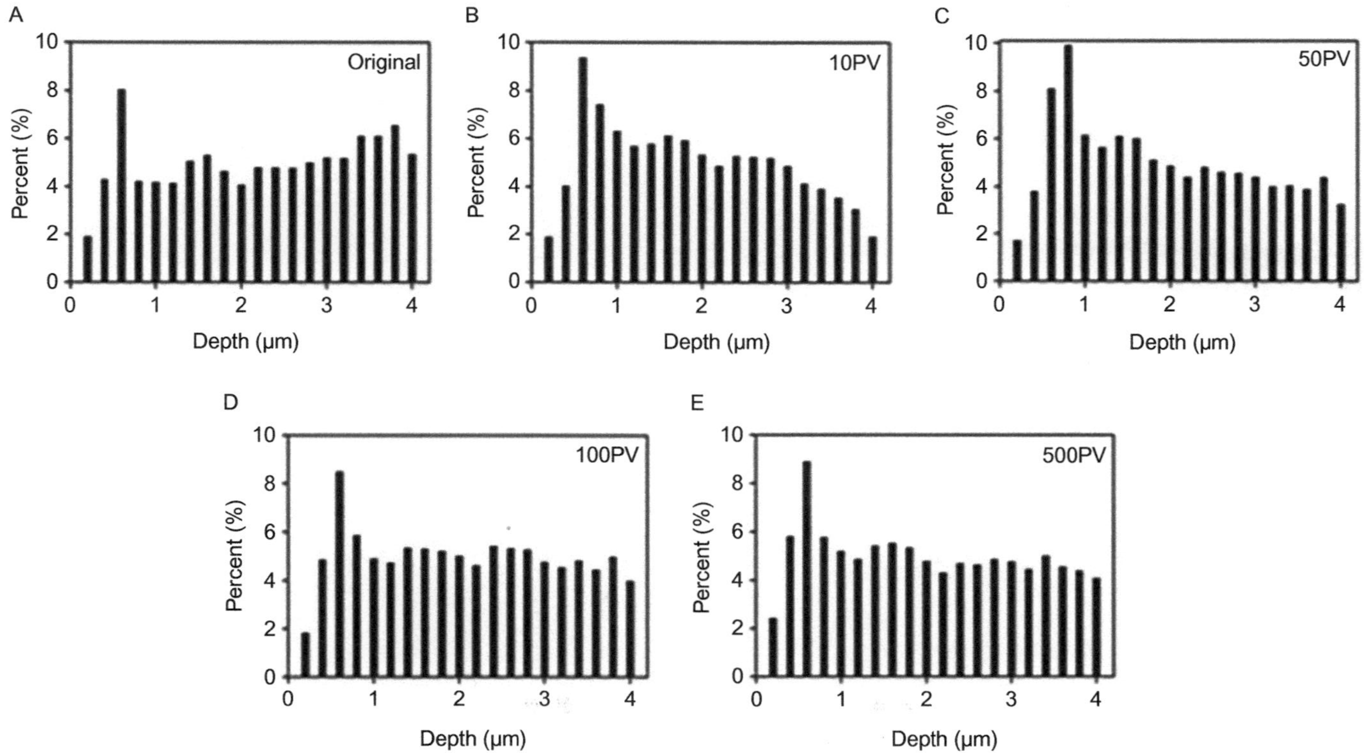

FIG. 7.15 Depth distributions of pores at different flooding stages. *The information was taken from data points of Bai, Y., Chunmei, Y., Mei, L., Zhenxue, J., 2016. Variations of micropores in oil reservoir before and after strong alkaline alkaline-surfactant-polymer flooding. Open Pet. Eng. J. 9 (1).*

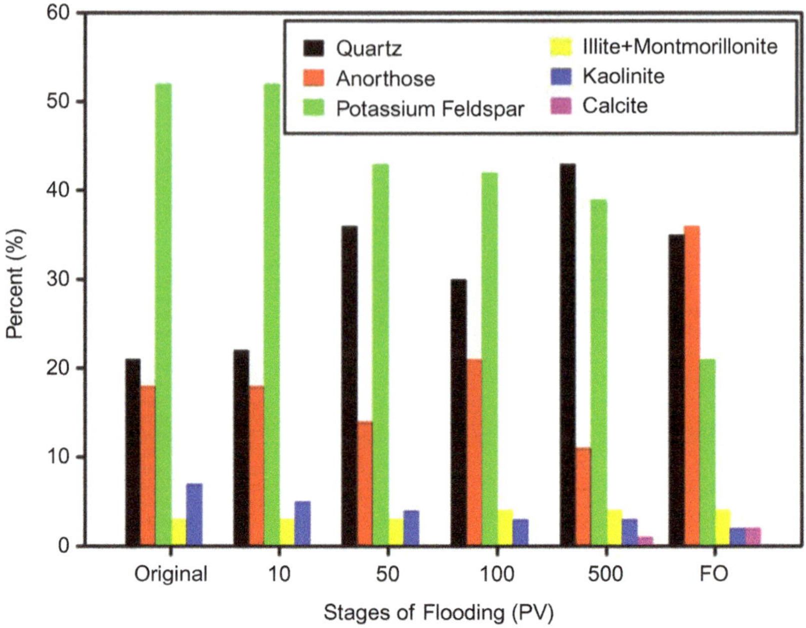

FIG. 7.16 Quality of minerals in sandstones (percentage) at different ASP flooding stages. *The information was taken from data points of Bai, Y., Chunmei, Y., Mei, L., Zhenxue, J., 2016. Variations of micropores in oil reservoir before and after strong alkaline alkaline-surfactant-polymer flooding. Open Pet. Eng. J. 9 (1).*

(Sagir et al., 2018). However, foams are highly susceptible to high temperature and high-salinity (HTHS) conditions as well as the presence of crude oil (Ibrahim and Nasr-El-Din, 2019).

In a typical reservoir with normal conditions, the CO_2 viscosity is about 10–50 times lower than the oil viscosity, resulting in channeling of CO_2 through the oil phase, and preferential flow through high permeability rock zones (Bernard et al., 1980). Surfactant additives that create in-situ foams with CO_2, can assuage this issue (Bernard et al., 1980; Cummings et al., 2012; Duda et al., 2010; Enick and Olsen, 2012; Hirasaki et al., 2011; Li et al., 2010; Sagir et al., 2014; Tadros, 2013; Xing et al., 2012). As foams have a greater viscosity than both surfactants and CO_2, they become more viscous with larger pore diameters, which reduces preferential flow through high permeability rock sections. Thus, a foam front with enhanced uniformity forces the oil further along the cavities of reservoir geological formations as the fingering is abated (Clark and Santiso, 2018).

Sagir et al. synthesized and tested a new surfactant to examine its CO_2 compatibility and evaluated it by inspecting the fluid-fluid interactions at different temperatures, salinity, and hardness, along with the foaming characteristics. Without a surfactant, the IFT between CO_2 and brine decreased as the CO_2 neared critical pressure and remained at the same level even when higher pressures were applied. Contrariwise, when the surfactant was added, an ultralowest IFT at critical pressure conditions was attained. Likewise, foam stability increased with the addition of surfactants, and similarly, the mobility reduction factor also increased with adding surfactants. Thus, adding a suitable surfactant in CO_2 foam flooding can yield OOIP with very low surfactant adsorption, due to the consistent IFT reduction between the gas and brine interface (Sagir et al., 2018).

The key challenge of CO_2 foam flooding occurs when foam film is exposed to harsh conditions (, i.e., HTHS). Moreover, the polar oil components (i.e., resin and asphaltenes) act in dividing the pressure of brine film between mineral and oil interfaces (Basu and Sharma, 1997). The presence of resin and asphaltenes can also reduce the film stability, and with stronger brine salinity, the pH decreases, whereas when resin and asphaltenes are absent, the increase in pH and salinity results in a stable film. However, the later outcome did not concur with DLVO theory, i.e., Derjaguin-Landau-Verwey-Overbeek (DLVO) theory of colloid stability quantifications, which was associated with variation of hydrophobic force with salinity (Basu and Sharma, 1997).

7.4.3.2 *Hybrid low-salinity with gas injection (i.e., LSW/gas)*

LSW flooding is combined with gas injection (i.e., LSW/gas) to improve oil recovery through mobility control, wettability change, and IFT reduction mechanisms. Furthermore, hybrid LSW/gas can be optimized for various injection scenarios to achieve the best potential recovery. As a result, the oil and gas industry has become highly interested in hybrid LSW/gas applications recently.

Various injection scenarios of LSW/gas have been explored through laboratory and simulation studies. The following injection scenarios have been tested: (i) LSW injected with WAG, also known as surfactant water alternating gas (SWAG), (ii) LSW injected into the WAG technique, and (iii) LSW flooding followed by gas injections, all of which result in IOR. Generally, carbon dioxide (CO_2) is used as the injected gas for hybrid gas injection techniques owing to its economic value, low MMP, and environmental acceptability (Pourafshary and Moradpour, 2019).

7.4.3.2.1 Mechanisms of LSW/gas flooding

The LSW/gas flooding affects the oil-gas interaction, by enhancing gas solubility in water, whereby oil recovery improves (Pourafshary and Moradpour, 2019). Based on literature related to laboratory and modeling studies, the solubility of gas (e.g., CO_2) in seawater rises when with salting-out effect decreases the salinity of seawater (Duan et al., 2006; Enick and Klara, 1990; Li and Nghiem, 1986; Pollack et al., 1988). Austad and coworkers explained that the salting-out effect indicates the low solubility of polar components of crude-oil high ionic strength water. The salting-in effect, on the other hand, implies that polar components have a higher water solubility in water with a low ionic strength (Austad, 2013). Furthermore, higher gas (CO_2) solubility leads to decrease in availability of free gas that can interact with the crude oil to induce mobility oil, causing low oil recovery efficiency. Alternatively, higher gas solubility means higher gas diffusion in LSW, the effects of which alter the waterfront, mobility ratio, and influence the stability of the injected LSW. Augmenting these factors lowers fingering in the LSW injection leading to greater contact with bypassed oil droplets in the porous medium. As a result, gas dissolved in water changes the structure of the waterfront, making it easier for it to come into contact with remaining and residual oil in the pores (Pourafshary and Moradpour, 2019).

The LSW/gas injection is effective in enhancing injected fluid's mobility (LSW) in both sandstones and carbonates reservoirs. In relation to this, Aleidan and Mamora (2010) investigated LSW/CO_2 injection with WAG and SWAG scenarios at different salinities, to evaluate the oil recovery effect in carbonate reservoirs, wherein results showed both injection schemes yielded high oil recovery when LSW was used. They discovered that injecting LSW alone in a nonhybrid scenario did not result in effective recovery, demonstrating that the incremental oil recovery was related to the synergy effect of combined LSW/gas injection.

In a similar study using sandstone rock samples, Jiang et al. (2010) examined the effects of LSW on the efficiency of WAG during miscible flooding on highly viscous crude oil. Higher gas solubility in LSW controls injected-LSW movement and lowers the mobility ratio of LSW and viscous crude oil, resulting in better oil recovery. In a related project, Teklu et al. investigated hybrid LSW/gas method that resulted in 20% incremental oil recovery (Teklu et al., 2016).

Moreover, Yang et al. demonstrated that when pressure and temperature are steady, the dissolved CO_2 present leads to the LSW-oil IFT reduction (Yang et al., 2005). As a result, the reduction in LSW/oil IFT might be viewed as a significant mechanism for incremental oil recovery using the hybrid LSW/gas technique. However, Ramanathan et al. (2015, 2016) and Teklu et al. (2016) found a minor reductions in IFT of less than 10 dynes/cm. Alternatively, the study of Kumar et al. revealed that as solubility of injected gas (CO_2) increased in the LSW, it led to the reduction in IFT of LSW and the injected gas (CO_2), resulting in altered wettability (Kumar et al., 2016). Therefore, based on the substantiations, wettability modification and mobility control are the main mechanisms behind hybrid LSW/gas methods. In case the rock is water-wet initially, applying LSW flooding only (nonhybrid) is ineffective. In such conditions, the mobility control (i.e., salting-out effect) is the main mechanism that has the greatest effect in oil recovery improvement. Hence, Teklu et al. noted the important impact of initial rock wettability (Teklu et al., 2016).

Moreover, for a successful hybrid LSW/gas injection, the efficiency of continued gas injection (CGI) is a vital factor. In situations where high oil recovery is achieved from individual gas injection, there isn't much scope for hybrid LSW/gas approach, as no noticeable incremental oil recovery is obtained. For instance, Al-Shalabi et al. modeled the hybrid LSW/gas (CO_2) technique, in a carbonate formation utilizing the UTCOMP-simulator application (Al-Shalabi et al., 2016). The outcomes indicated the recovery factors by CGI at 98.9%, while the outcome for hybrid LSW/gas technique was at 99.7% recovery factor (Al-Shalabi et al., 2016). Thus, it is clear that, in similar circumstances, which have gas miscibility as the leading mechanism, hybrid LSW/gas application is not endorsed.

Al-Shalabi et al. also used simulations to compare the geochemical analysis findings of LSW, CGI, and hybrid approaches on carbonate rocks with various properties (Al-Shalabi et al., 2016). According to the findings, hybrid approaches may speed up the dissolution process, especially in locations with high dolomite concentrations. However, more research into the geochemical parameters that influence oil recovery during hybrid LSW/gas injection is required to fully understand the oil recovery mechanisms of the process. Because the majority of the evaluations of the LSW/gas injection method have been laboratory experiments, and only a few were simulations and field implementations. In one of the simulation studies, Dang et al. modeled an injection of hybrid LSW-alternating miscible CO_2 for a (1-D) heterogeneous core sample, which was then upscaled to simulate the LSW/gas injection process at field scale (Dang et al., 2014a). In this study, the delayed production problem that occurs in WAG was mitigated by LSWF/gas method. Also, Pourafshary and Moradpour simulated a scaled-up model, founded on experimental studies that comprehensively evaluated field-scale application, and recommended that hybrid LSW/gas method(s) can be considered for field applications (Pourafshary and Moradpour, 2019).

7.4.3.3 Low-salinity polymer flooding

A recent but noteworthy hybrid EOR method is the low-salinity polymer (LSP) flooding, in which the synergies within this method can initiate both displacement efficiency and sweep efficiency improvements (Al-Shalabi et al., 2022). The function of polymers in EOR has been as mobility controllers of injected fluid to lead to improved sweep efficiency (Sorbie, 1991, 2013). In instances where polymers are combined with low-salinity aqueous solutions, it alters the wettability of the formation and causes the water-cut and water/oil ratio to be reduced (Karimov et al., 2020). We know from previous Section 7.4.2.1 that polymers solution including LSP fluid may lose up to 30% of the viscosity due to the polymers' attraction to divalent ions on the rock surface. It is the properties of this fluid-rock system that determines the extent of cation exchange and viscosity loss. Berge et al. (2018), investigated

LSP flooding, where the makeup water salinity was much lower than the formation water. Due to the exchange of cations on the rock surface, high viscosity loss of the fluid occurs. The partially HPAM polymer was studied to determine the effects of the rock-fluid interactions by scrutinizing the fluids produced from many different coreflooding experiments. Higher levels of divalent cations such as Ca^{2+} and Mg^{2+} were found in all the samples of produced fluids having polymer in comparison to the produced fluids of injected slugs of brine that had no polymers. This demonstrates the proclivity of the polymer in the injection fluid to be attracted to divalent ions on the rock surface and its detrimental effect on the viscosity of the solution. Al-Shalabi et al. (2022) modeled LSP flooding, using a coupled numerical model simulator to study the geochemical interactions of the polymer-brine-rock system during an LSP flooding, and to analyze the flow dynamics, they used the MATLAB Reservoir Simulation Toolbox and coupled it with the geochemical software IPhreeqc, which is the interface module of Phreeqc (pH-Redox-Equilibrium in C programming language). The parameters along with the physical effects that were considered in the model are inaccessible pore volume, the effect of polymer adsorption, permeability reduction, the effect of polymer viscosity, the effect of shear rate, and the effect of salinity. The stable complexes are formed, when the divalent ions bind with the acrylic acid groups of the polymer, resulting in salinity reduction. To restore the equilibrium lost during these reactions, the rocks discharge supplementary divalent ions into the solution and equilibrate it. Thus, polymer viscosity rheology is strongly influenced by the divalent ions while the monovalent ions in the solution limit the salinity and have very little effect on polymer viscosity. In another study by Khorsandi et al. (2017), a mechanistic model of wettability alteration centered around cation exchange reactions was performed using LSP flood in sandstones and analytical solutions were derived. They concluded that their proposed model for cation exchange was likely the mechanism for the wettability alteration. They noted the cation exchange front moved sluggishly in contrast to the salinity shock or the anion front, and that surface composition and wettability modification were limited to the cation exchange front only. Also, the recovery of oil was slow which required many pore volumes of LSW to be injected, and a larger quantity of the oil was recovered after the breakthrough of the wettability front. It is suggested that high-salinity shock traveled quicker than the wettability front, which made the small slugs of LSW to be unproductive, without any wettability alteration, implying that the low-salinity slugs must have adequate size to spread the wettability front toward the production wells (Khorsandi et al., 2017).

7.4.3.4 Hybrid low-salinity water and surfactant (LSW/surfactant)

The hybrid LSW flood combined with surfactant flooding (i.e., LSW/surfactant recovery) (Fathi et al., 2012), can improve the oil recovery through synergic effects of mobility control, wettability alteration, and IFT reduction.

As already stated, the efficiency of LSW is primarily attributed to the mechanism of reservoir rock's wettability alteration (in both carbonates and sandstones) from oil-wet toward more water-wet state, resulting in the separation of crude-oil films (adsorption) from the surface of the rock. Surfactant flooding reduces IFT across crude oil-water (brine) interfaces and modifies rock wettability, resulting in a reduction in capillary forces that keep crude oil trapped inside porous media, as detailed in earlier sections (Salager et al., 1979; Zhang, 2005). Hence, in comparison to stand-alone LSW injection or surfactant flooding, the hybrid LSW/surfactant approach provides greater oil recovery due to the synergy of both mechanisms.

Moreover, the effects of LSW include reservoir acclimatization, or conditioning, in order to prepare the environment to respond favorably to the effects of surfactant flooding, in which the SAS causes the solubilization of remaining and residual oil defined by Winsor Type-II microemulsion behavior (Mohyaldinn et al., 2019). It is evident from several studies that have reported excellent (tertiary) oil recovery when surfactant flooding was applied following a LSW flooding in both sandstones and in carbonate reservoirs. Enge reported 5%–7% incremental oil recovery achieved by injection of sodium dodecyl-benzene sulfonate surfactant solution following the LSW into a core sample of sandstone (Enge, 2014). Similarly, in a work by Alameri et al., it was determined that incremental oil recovery $\leq$10% was achieved through LSW-surfactant (LSS) injection following a LSW flooding using a core sample of carbonate (Alameri et al., 2015).

Thus, in the hybrid LSW/surfactant method, the resulting wettability alteration that leads to EOR is affected by LSW flooding and the IFT reduction between fluid-fluid (i.e., crude oil-brine) interfaces is affected by surfactant aqueous solution (SAS). Teklu et al. and Spildo et al. demonstrated that, in cases where LSW injection is not effective, the SAS can induce IFT reduction and alter the rock wettability, which results in IOR (Spildo et al., 2012; Teklu et al., 2016).

In addition, the advantage of hybrid LSW/surfactant method includes prevention of surfactant consumption, low-cost operations (OpEx), and ease of operations, for which, the water chemistry aspects, such as ionic composition of the brine solution, are vital factors in surfactant-flooding designs in order to mitigate surfactant molecule retention (i.e., adsorption) onto the porous medium. Reduction in surfactant retention on to the rock heightens the performance of surfactant-flooding project, in terms of cost-effectiveness and operational advantages, wherein applying LSW flooding has resulted in decrease of surfactant adsorption (Alagic and Skauge, 2010; Alagic et al., 2011; Araz and Kamyabi, 2021; Glover et al., 1979; Shaddel and Tabatabae-Nejad, 2015; Teklu et al., 2017).

Pourafshary and Moradpour discovered that injecting LSW makes it easier to achieve low IFT during surfactant flooding, and low surfactant retention occurs as a result of these conditions, making surfactants more stable and soluble (Pourafshary and Moradpour, 2019). More accurately, the results of their assessments showed that the divalent ions contained in brines lead to precipitation of surfactant in sandstones. Also, the calcium cations affected the surfactant adsorption and surfactant precipitation due to calcium interactions (i.e., negative effect), and the maintenance of micelles (i.e., positive effect) (Pourafshary and Moradpour, 2019). Consequently, in sandstones, divalent cation can be present in a limited concentration and must be taken into account during the LSS design.

It is to be noted that, in a hybrid LSW/surfactant flooding, the movement of stability front is vital for an effective oil recovery project. Tavassoli et al. identified that surfactant-flood fronts, which are unstable owing to excess velocity, result in a sluggish oil recovery process (Tavassoli et al., 2016). To alleviate this issue LSW with surfactant/polymer (SP) injection can be useful (Tavassoli et al., 2016; Wang et al., 2018).

Overall, the majority of research on hybrid LSW/surfactant injection has been experimental. On this subject, there is a clear lack of modeling research and procedure optimization through simulation evaluations (Pourafshary and Moradpour, 2019). More research, such as computer studies that estimate the efficacy of hybrid LSW/surfactant-flooding schemes, is required (Tavassoli et al., 2016).

7.4.3.5 Smart-water-assisted foam flooding

In carbonate rocks, water injection is ineffective, owing to the oil-wet state of the rock (Kokal and Al-Kaabi, 2010; Lake, 1989; Lake et al., 2014; Taber et al., 1997). Based on previous literature, the oil-wetting of rock is considered a vital factor affecting the stability of foam and circulation of foam fluid in oil-wet carbonates. Crude oil has an antifoam effect and weakens the foam on contact (Farajzadeh et al., 2012; Lee et al., 2013; Rossen, 1996; Simjoo et al., 2013). Thus, a new EOR method that combines the effects of smart water (SW) and foam flooding termed SWAF flooding was developed by Hassan et al. to target the residual oil recovery in carbonate reservoirs (Hassan et al., 2020a,b,c, 2021, 2022a,b,c). This novel smart water is designed to coerce a dual-improvement effect that enhances oil displacement, by affecting both the fluid-rock and fluid-fluid interactions. The fluid-rock interaction alters the surface charge at the fluid-rock interface, which leads to the modification of the reservoir rock wettability state toward a more water-wet state (Hassan et al., 2022a,b,c). Then, the smart water changes the fluid-fluid interaction through shielding effect minimization of the electrical double layer (EDL), which stabilizes the foam film (Hassan et al., 2022a,b,c). Due to the surface charge alterations at both interfaces of fluid-rock and fluid-fluid (Morrow, 1990), the rock-water interface and the water-oil interface develop the same sign, creating stable water and foam films, which augments oil recovery enhancement.

7.4.3.5.1 Mechanisms of SWAF

The primary mechanisms of SWAF process are considered to be the wettability modification of the carbonate rock from oil-wet state to more water-wet state, which is brought about by the surface charge modification and multicomponent ionic exchange (MIE) when PDI (SO_4^{2-}, Ca^{2+}, and Mg^{2+}) is present. These PDIs (SO_4^{2-}, Ca^{2+}, and Mg^{2+}) are adsorbed onto the carbonate rock surface, which causes desorption, causing the charged organic constituent (i.e., carboxylic acid) of crude oil to be discharged from the rock surface of carbonates (Hassan et al., 2022a,b,c; Lee and Lee, 2019; Sheng, 2013). In the early stages, the negatively charged carboxylic acid crude-oil components attach to the positively charged carbonate rock surface (i.e., calcite rock sample). Typically, the connate/formation water in a carbonate reservoir has negligible SO_4^{2-} ion concentration, whereas the seawater has a significant SO_4^{2-} concentration. Therefore, when seawater and the smart water designed by Hassan et al. (2022a,b,c) are injected together into the oil-wet calcite rock, the negatively charged SO_4^{2-} ions contained in seawater injection possibly adsorb onto the positively charged water-wet sites on the calcite rock, i.e., Indiana Limestone sample. SO_4^{2-} attaches to the calcite rock surface, causing a reduction in the density of positively charged ions on the chalk rock surface. The positive charge density on the calcite surface diminishes, resulting in a decrease in the electrostatic repulsive force between the cation (i.e., Ca^{2+} and Mg^{2+}) and the calcite surface. The smaller the electrostatic repulsion, the greater the adsorption of positively charged Ca^{2+} ions onto the calcite rock surface (i.e., the more Ca^{2+} remains close to the calcite rock surface). Adsorption of positively charged Ca^{2+} also speeds up SO_4^{2-} coadsorption on the calcite rock surface. The adsorbed negatively charged organic carboxylic groups of crude oil may react with the Ca^{2+}, resulting in neutralization of the carboxylic groups of crude oil. Thus, the organic carboxylic groups on the calcite surface are eliminated during the process (Hassan et al., 2022a,b,c; Lee and Lee, 2019; Sheng, 2013).

On the other hand, with increase in temperature (e.g., >70°C), a different mechanism takes place, i.e., the substitution of Ca^{2+} by Mg^{2+} onto the calcite rock surface occurs, which is called dolomitization. The exchange potentially extricates the carboxylic materials which had reacted with Ca^{2+} on the surface of calcite rock and creates further increase in the water-wetness of the calcite rock surface (Hassan et al., 2022a,b,c; Lee and Lee, 2019; Sheng, 2013).

Moreover, the coadsorption of SO_4^{2-} reduces electrostatic repulsion, which promotes the ionic interaction between Mg^{2+} and SO_4^{2-}, resulting in an increase in Mg^{2+} concentration at the calcite rock's surface. The increase in Mg^{2+} concentration at the surface may contribute to more Mg^{2+} replacement of Ca^{2+}. According to the mechanism proposed by Zhang et al. (2007), in both low and high temperature settings, neither Mg^{2+} nor Ca^{2+} can remove the negatively charged carboxylic acid groups from the positively charged surface of calcite rock without the presence of SO_4^{2-} (Hassan et al., 2022a,b,c; Lee and Lee, 2019; Sheng, 2013).

Hence, the smart water produces dual-improvement effect to yield compounded oil displacement, through wettability alteration of carbonate rock and foam film stabilization during the foam flooding stage. Thus, the COBR system is affected by the designed smart water solutions in two ways: first, by improving the fluid-rock interaction by altering the surface charge at the fluid-rock interface, causing reservoir rock wettability to change to a more water-wet state (Hassan et al., 2022a,b,c). The second enhancement impact of smart water is the modification of fluid-fluid interaction via reduction in the shielding effect of the electrical double layer (EDL), which improves foam film stability (Akbar et al., 2000; Hassan et al., 2022a,b,c; Roehl and Choquette, 2012; Sheng, 2013). The rock-water interface and the water-oil interface develop the same sign (i.e., either negative or positive charge) as a result of these surface charge modifications at both fluid-rock and fluid-fluid interfaces, resulting in stable water and foam films and increased EOR (i.e., observed in better recovery factor or RF) (Akbar et al., 2000; Hassan et al., 2022a,b,c; Lee and Kam, 2013; Roehl and Choquette, 2012; Sheng, 2013).

Furthermore, when the smart water solution is merged with foam flooding by alternating surfactant aqueous solution (SAS) and gas (i.e., SAG) injection, a synergy effect is achieved. Adding surfactant to smart water solutions to generate foam creates microemulsions at the oil-brine interface, resulting in a synergistic effect that leads to increased wettability alteration and IFT reduction, which improves displacement efficiency (oil recovery) (Hassan et al., 2021, 2022a,b,c).

Consequently, SWAF process is conducted by first conditioning the carbonate-reservoir rock with the designed smart water during the secondary phase, and then foam (SAS and gas) is injected as a tertiary phase to improve the carbonate rock wettability further. It has been discovered that improved oil displacement is achieved (Hassan et al., 2021, 2022c) as a result of the effect of SAS and IFT reduction, as well as the microemulsion generated between the aqueous and oleic phases (Anderson, 1986). Hence, Fig. 7.17 shows the illustration of different SWAF technology injection scheme scenarios for enhancing favorable COBR interactions (Hassan et al., 2022a,b,c).

Accordingly, Hassan et al. (2022a,b,c) evaluated the effect of the designed smart water on COBR system, to detect contact angle reduction, accompanied by lowered salinity, which altered the wettability of limestone core sample toward a more water-wet state (Hassan et al., 2021). Afterward, the designed smart water had been mixed with selected surfactants (forming SAS) and injected with CO_2 and N_2 gases, which produced a robust foam and lead to further wettability alteration, and IFT reduction to result in high oil recovery, i.e., cumulative oil recovery of ~92%, as shown in Figs. 7.18 and 7.19 (Hassan et al., 2022a,b,c).

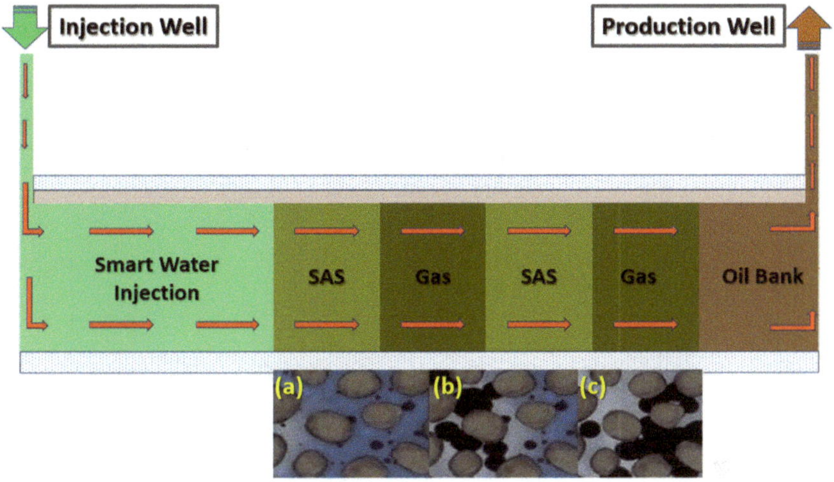

FIG. 7.17 Schematic diagram of SWAF technology, which displays the injected scheme scenarios of SWAF technology, which starts with smart water (SW) injection as a secondary mode to change the reservoir wettability (conditioning the reservoir), then the injection of SAG (SAS alternating gas), i.e., surfactant aqueous solution (SAS) followed by gas injection (i.e., carbon dioxide or CO_2 and nitrogen or N_2) slugs injections as a tertiary mode or stage. (A) Injection of SWAF solution in presence of crude oil, (B) SWAF solution interacts with crude oil, and (C) oil detaches from rock and moves toward production well (Hassan et al., 2022a,b,c).

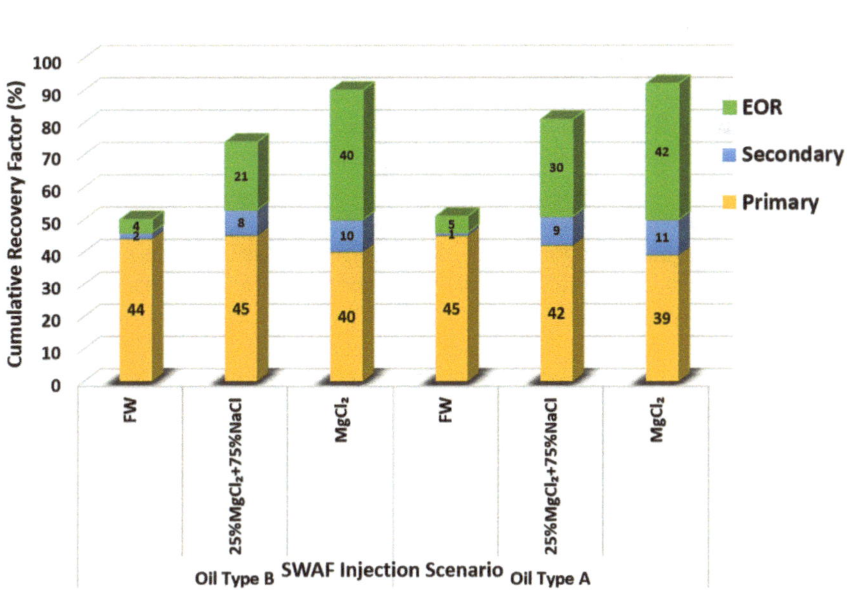

FIG. 7.18 Smart-water-assisted foam (SWAF) injection scenarios; the effect of crude-oil type (Type-A and Type-B) with different smart water (SW) solutions (i.e., 3500 ppm) and at surfactant concentration of 1000 (ppm) (Hassan et al., 2022a,b,c).

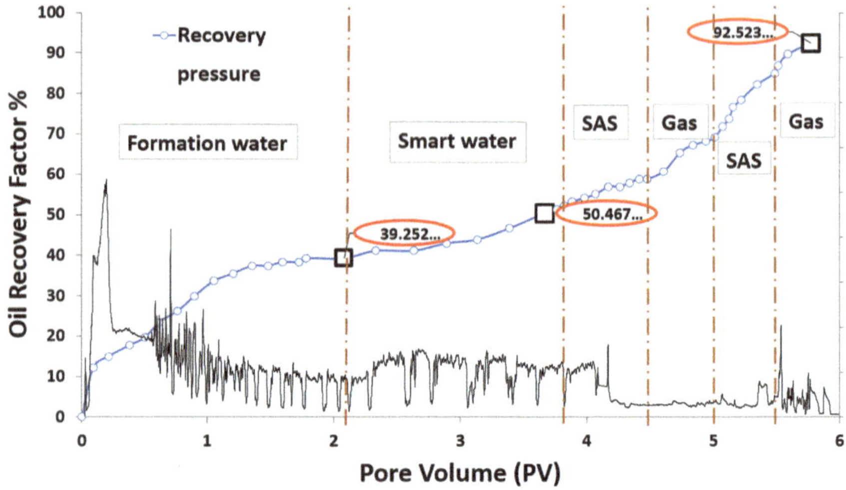

FIG. 7.19 Smart-water-assisted foam (SWAF) injection process for the best case scenario (i.e., $MgCl_2$), The method works as an improved/enhanced (IOR/EOR) recovery method for the ionically modified brine, which goes up to 39% RF for the primary stage (i.e., formation water), 11% RF (i.e., 50% cumulative OIIC) for the secondary stage (i.e., smart water injection), and 92% cumulative OIIC for the tertiary stage (SAG injection, i.e., SAS, gas, SAS, and gas) (Hassan et al., 2022a,b,c).

As a result, the SWAF process's positive performance implies lower circulation costs and a low CO_2 footprint (as defined by the principle of exergy return on exergy investment), and therefore a low environment impact (Hassan et al., 2019). As a result, the innovative hybrid SWAF technique has been developed as a possible option for successful EOR applications, particularly in HTHS carbonates, as the more cost-effective and environmentally friendly alternative (Hassan et al., 2022a,b,c).

It is worth mentioning that some other alternatives as foam stabilizers have been provided in literature, such as AVS polymers and clay nanoparticles. AVS is a terpolymer of acrylamide (AM), AMPS with one synthesized functional monomer, which is well suited for HTHS conditions. When AVS are incorporated with alpha olefin sulfonate, it generates highly viscous foam, without affecting the foam fluid, which makes them viable alternative for fortifying foams in HTHS reservoirs (Xu et al., 2015). Nonetheless, in HTHS reservoir conditions, most polymers have limited function as foam stabilizer due to the degradation of larger polymer molecules. Therefore, foam flooding using nanoparticles has been proposed by some as substitute for polymers as foam stabilizers. Clay nanoparticles that were evaluated have shown remarkable drainage of the foam system half-life. The results have exhibited greater generated foam density in the foam fluid, which had been incorporated with nanoparticle stabilizer. Thus, along with good injectivity in cores having permeability over >200 mD, the nanoparticle-stabilized foam has decent mobility control. The estimated yield from foam flooding is approximately 17%–21% of OOIP incremental oil recovery (under optimum conditions) (Zhu et al., 2017). Nonetheless, during experiments in carbonate cores at 85°C, the innovative SWAF technology achieved exceptionally high cumulative oil recovery (92%), as shown in Figs. 7.18 and 7.19. Moreover, with the expected increase in the number of carbonate oil fields in the coming years, SWAF technology can be a great candidate for HTHS carbonate reservoirs due to dual-improvement effects for maximum recovery at low cost and with minimal environmental impact (Hassan et al., 2019, 2021, 2022a,b,c).

7.5 Conclusion and recommendation

It is anticipated that the global energy demand will increase steadily in the forthcoming decades, with hydrocarbon energy as a significant contributor. To sustain the global energy supply, the oil production has expanded beyond the existing aged reservoirs to explore the newer unconventional and harsh-conditioned reservoirs (e.g., high-temperature, high-salinity, and HTHS carbonate formations). Consequently, IOR/EOR can be a powerful tool to unlock a large amount of residual and trapped oil from these existing and new hydrocarbon resources. This chapter discusses the various EOR techniques in both carbonates and sandstone reservoirs with regard to water-based EOR, cEOR, and hybrid EOR techniques, from the perspective of fluid-fluid and fluid-solid interactions. Each of these EOR methods has its unique challenges and proclivities to the environment it is applied. To refine the performance of these EOR techniques, recent studies have afforded new insights into fluid-rock interactions (i.e., COBR system). An accurate understanding of COBR systems can help in better implementation of EOR methods in the field to maximize oil recovery. In addition, the correct knowledge of COBR systems can aid in precise predictions of chemical adsorption and degeneration, mineral dissolution and precipitations, pore-clogging, reservoir damage, and aquifer pollution. Apart from that, a complete understanding of the COBR system allows for cost-effective projects and minimizes environmental impact (CO_2 footprint). Accordingly, the key purpose of this chapter is to enhance the reader's knowledge of the scope of fluid-rock interactions, to enable efficient EOR project designs that capture the accurate requirement of candidate reservoirs for upstream oil and gas applications.

In summary, for the future of EOR applications, improved understandings of the fluid-fluid and fluid-rock interactions that affect the outcome of EOR application have progressively expanded. As EOR methods become more advanced, to maximize oil recovery from challenging reservoirs, the key to the successful field applications of EOR methods, lies in the comprehensive knowledge of fluid-fluid and fluid-rock interactions.

Following key insights and recommendations can be highlighted in this chapter.

- The EOR techniques are applied to maximize oil production from both matured reservoirs and unconventional and harsh-conditioned carbonate reservoirs.
- An accurate understanding of the geochemistry of fluid-rock interactions is paramount, to achieving successful field implementations of EOR thus maximizing oil recovery. In addition, an improved understanding of the COBR system facilitates economic projects and minimizes environmental impact (CO_2 footprint).
- During LSWF, the surface charge alteration due to PDIs (Ca^{2+}, Mg^{2+}, SO_4^{2-}) accelerates the wettability alteration (brine-rock interface) in harsh-conditioned carbonate reservoirs. These divalent ions (Ca^{2+}, Mg^{2+}) form linkages such as oil-Ca-mineral phases, which can produce substantial wettability alteration. In sandstone reservoirs, the LSWF mechanisms that cause wettability alteration are fines migration, pH increase, multiion exchange (MIE), and salting-in effect.
- Polymers improve the viscosity of the solution and impart mobility control, to enhance the volumetric sweep efficiency of the injected fluid, by freeing trapped oil from pores.
- In recent years, many have suggested that water flooding may need to be replaced by polymer flooding as a replacement for secondary and tertiary EOR, since polymer flooding is more likely to be effective in terms of reducing water cut during low water periods, unlike water flooding (Sheng, 2015).
- When the polymer is added to LSW injection, the negatively charged polymer species may attract the positively charged divalent ions on the rock, which might lead to mineral dissolution in carbonate reservoirs and thus increase the polymer viscosity loss.
- Surfactant flooding interrupts the fluid-fluid interactions by inducing the water-oil IFT reduction. Also, SF influences fluid-rock interactions and causes wettability alterations.
- The synergy of surfactant-polymer flooding increases water viscosity, decreases the IFT of the water-oil interfaces, and reduces the capillary forces of the reservoir rock, enhancing oil recovery.
- As more reservoirs mature and the current reservoirs are mostly carbonate formations with harsh conditions (high temperature, high salinity, i.e., HTHS and low permeability), the chemicals used in cEOR such as polymers and surfactants have to be redesigned and innovated to be stable in HTHS conditions of carbonates.
- During ASP flooding, the alkali interacts with crude oil to generate in-situ soaps that reduce the IFT between the fluid-fluid interfaces, whereas ASP flooding alters the rock wettability from oil-wet to water-wet, in oil-wet

reservoirs. ASP flooding forms crude-oil emulsions in the reservoir, that have very low IFT. These emulsions are formed by the shearing action of the fluid movement inside the pores.

- Furthermore, alkali loss is significant in clays owing to the chemical interactions between the fluid-rock interfaces. Moreover, potent ASP flooding in sandstones leads to rock degeneration due to the feldspar and quartz corrosion, which modifies the pore structures.
- The SWAF flooding combines smart water, surfactant, and gas injection to result in maximized oil recovery. The smart water is designed to have dual-improvement effects of wettability alteration in carbonate rocks toward the water-wet state and foams stability enhancement. In this way, smart water affects both rock-fluid and fluid-fluid interactions. The rock-fluid interaction is enhanced when the surface charges at the interface are altered, which modifies the carbonate rock wettability toward a more water-wet state. Besides that, smart water enhances the fluid-fluid interaction by decreasing the shielding effect of the electrical double layer (EDL) that augments foam film stability (Hassan et al., 2022a,b,c).
- Moreover, mixing surfactant with smart water generates foam, wherein the microemulsions created at the oil-brine interface further supplement wettability alteration and IFT reduction, leading to a compounded increase in displacement efficiency. Thus, through the synergy effect, oil recovery improves (Hassan et al., 2022a,b,c).
- Moreover, as discussed in the above sections, the novel hybrid EOR technique called SWAF flooding utilizes a unique designed smart water, i.e., ionically modified brine which produces a dual-improvement effect of both IFT reduction and foam film stabilization. This leads to wettability alteration of the carbonate rocks from oil-wet to water-wet (i.e., which improves the rock-fluid interaction), resulting in greater oil recovery. Thus, under optimum conditions, the SWAF technology can be effective EOR method for successful field applications, particularly in HTHS carbonate reservoirs (Hassan et al., 2022a,b,c).

References

Abbasi-Asl, Y., Pope, G.A., Delshad, M., 2010. Mechanistic modeling of chemical transport in naturally fractured oil reservoirs. In: SPE Improved Oil Recovery Symposium.

Adibhatla, B., Mohanty, K.K., 2006. Oil recovery from fractured carbonates by surfactant-aided gravity drainage: laboratory experiments and mechanistic simulations. In: SPE/DOE Symposium on Improved Oil Recovery.

Ahmadi, M.A., Shadizadeh, S.R., 2015. Experimental investigation of a natural surfactant adsorption on shale-sandstone reservoir rocks: static and dynamic conditions. Fuel 159, 15–26.

Ahmed, T., 2018. Reservoir Engineering Handbook. Gulf Professional Publishing.

Akbar, M., Vissapragada, B., Alghamdi, A.H., Allen, D., Herron, M., Carnegie, A., Dutta, D., Olesen, J.-R., Chourasiya, R.D., Logan, D., 2000. A snapshot of carbonate reservoir evaluation. Oilfield Rev. 12 (4), 20–21.

Alagic, E., Skauge, A., 2010. Combined low salinity brine injection and surfactant flooding in mixed-wet sandstone cores. Energy Fuels 24 (6), 3551–3559.

Alagic, E., Spildo, K., Skauge, A., Solbakken, J., 2011. Effect of crude oil ageing on low salinity and low salinity surfactant flooding. J. Pet. Sci. Eng. 78 (2), 220–227.

Alam, M.M., Hjuler, M.L., Christensen, H.F., Fabricius, I.L., 2014. Petrophysical and rock-mechanics effects of CO2 injection for enhanced oil recovery: experimental study on chalk from South Arne field, North Sea. J. Pet. Sci. Eng. 122, 468–487.

Alameri, W., Teklu, T., Graves, R., Kazemi, H., AlSumaiti, A., 2015. Low-salinity water-alternate surfactant in low-permeability carbonate reservoirs. In: IOR 2015—18th European Symposium on Improved Oil Recovery. European Association of Geoscientists & Engineers, p. cp-445.

Aleidan, A.A., Mamora, D.D., 2010. Swaco2 and waco2 efficiency improvement in carbonate cores by lowering water salinity. In: Canadian Unconventional Resources and International Petroleum Conference. Society of Petroleum Engineers.

Al-Hajri, S., Mahmood, S.M., Abdulelah, H., Akbari, S., 2018. An overview on polymer retention in porous media. Energies 11 (10), 2751.

Alotaibi, M.B., Nasralla, R.A., Nasr-El-Din, H.A., 2011. Wettability studies using low-salinity water in sandstone reservoirs. SPE Reserv. Eval. Eng. 14 (06), 713–725.

Al-Shalabi, E.W., Sepehrnoori, K., 2016. A comprehensive review of low salinity/engineered water injections and their applications in sandstone and carbonate rocks. J. Pet. Sci. Eng. 139, 137–161.

Al Shalabi, E.W., Sepehrnoori, K., 2017. Low Salinity and Engineered Water Injection for Sandstone and Carbonate Reservoirs. Gulf Professional Publishing.

Al-Shalabi, E.W., Sepehrnoori, K., Pope, G., 2016. Numerical modeling of combined low salinity water and carbon dioxide in carbonate cores. J. Pet. Sci. Eng. 137, 157–171.

Al-Shalabi, E.W., Alameri, W., Hassan, A.M., 2022. Mechanistic modeling of hybrid low salinity polymer flooding: role of geochemistry. J. Pet. Sci. Eng., 110013.

Anderson, W., 1986. Wettability literature survey-part 2: wettability measurement. J. Pet. Technol. 38 (11), 1246–1262.

Araz, A., Kamyabi, F., 2021. Experimental study of combined low salinity and surfactant flooding effect on oil recovery. Oil Gas Sci. Technol. 76, 4.

Ashraf, A., Hadia, N.J., Torsaeter, O., Tweheyo, M.T., 2010. Laboratory investigation of low salinity waterflooding as secondary recovery process: effect of wettability. In: SPE Oil and Gas India Conference and Exhibition.

Austad, T., 2013. Water-based EOR in carbonates and sandstones: new chemical understanding of the EOR potential using "smart water". In: Enhanced Oil Recovery Field Case Studies. Elsevier, pp. 301–335.

Austad, T., Strand, S., Madland, M.V., Puntervold, T., Korsnes, R.I., 2007. Seawater in chalk: an EOR and compaction fluid. In: International Petroleum Technology Conference.

Austad, T., Strand, S., Puntervold, T., 2009. Is wettability alteration of carbonates by seawater caused by rock dissolution. In: Paper SCA2009-43 Presented at the International Symposium of the Society of Core Analysts Held in Noordwijk, The Netherlands, September, 27–30.

Austad, T., Shariatpanahi, S.F., Strand, S., Black, C.J.J., Webb, K.J., 2012. Conditions for a low-salinity enhanced oil recovery (EOR) effect in carbonate oil reservoirs. Energy Fuels 26 (1), 569–575.

Ayoub, M.A., Shabib-Asl, A., AbdellahiZein, A.M., Elraies, K.A., MohdSaaid, I., bin., 2015. Recovery optimization of an oil reservoir by water flooding under different scenarios; a simulation approach. Res. J. Appl. Sci. Eng. Technol. 10 (4), 357–372.

Azam, M.R., Tan, I.M., Ismail, L., Mushtaq, M., Nadeem, M., Sagir, M., 2013. Static adsorption of anionic surfactant onto crushed Berea sandstone. J. Pet. Explor. Prod. Technol. 3 (3), 195–201.

Bai, Y., Chunmei, Y., Mei, L., Zhenxue, J., 2016. Variations of micropores in oil reservoir before and after strong alkaline alkaline-surfactant-polymer flooding. Open Pet. Eng. J. 9 (1).

Basu, S., Sharma, M.M., 1997. Investigating the role of crude-oil components on wettability alteration using atomic force microscopy. In: International Symposium on Oilfield Chemistry.

Bera, A., Ojha, K., Mandal, A., Kumar, T., 2011. Interfacial tension and phase behavior of surfactant-brine–oil system. Colloids Surf. A Physicochem. Eng. Asp. 383 (1–3), 114–119.

Berge, A.B., Lenchenkov, N., Wever, D.A., Farajzadeh, R., Al-Mjeni, R., Glasbergen, G., 2018. The role of synthetic polymer on rock-fluid interactions and the resulting change in ionic composition and viscosity of the polymer slug for a cEOR flood in the Sultanate of Oman. In: SPE EOR Conference at Oil and Gas West Asia.

Bernard, G.G., 1967. Effect of floodwater salinity on recovery of oil from cores containing clays. In: SPE California Regional Meeting.

Bernard, G.C., Holm, L.W., Harvey, C.P., 1980. Use of surfactant to reduce CO2 mobility in oil displacement. Soc. Pet. Eng. J. 20 (04), 281–292.

Boampong, L.O., Rafati, R., Haddad, A.S., 2022. A calibrated surface complexation model for carbonate-oil-brine interactions coupled with reservoir simulation—application to controlled salinity water flooding. J. Pet. Sci. Eng. 208, 109314.

Bourrel, M., Schechter, R.S., 2010. Microemulsions and Related Systems: Formulation, Solvency, and Physical Properties. Editions Technip.

Bryan, J., Kantzas, A., 2009. Potential for alkali-surfactant flooding in heavy oil reservoirs through oil-in-water emulsification. J. Can. Pet. Technol. 48 (02), 37–46.

Buckley, J.S., Liu, Y., 1998. Some mechanisms of crude oil/brine/solid interactions. J. Pet. Sci. Eng. 20 (3–4), 155–160.

Buckley, J.S., Takamura, K., Morrow, N.R., 1989. Influence of electrical surface charges on the wetting properties of crude oils. SPE Reserv. Eng. 4 (03), 332–340.

Budhathoki, M., Barnee, S.H.R., Shiau, B.-J., Harwell, J.H., 2016. Improved oil recovery by reducing surfactant adsorption with polyelectrolyte in high saline brine. Colloids Surf. A Physicochem. Eng. Asp. 498, 66–73.

Cai, M., Su, Y., Li, L., Hao, Y., Gao, X., 2021. CO2-fluid-rock interactions and the coupled geomechanical response during CCUS processes in unconventional reservoirs. Geofluids 2021 (49), 1–22.

Chandar, P., Somasundaran, P., Turro, N.J., 1987. Fluorescence probe studies on the structure of the adsorbed layer of dodecyl sulfate at the alumina–water interface. J. Colloid Interface Sci. 117 (1), 31–46.

Chen, J., Hirasaki, G.J., Flaum, M., 2006. NMR wettability indices: effect of OBM on wettability and NMR responses. J. Pet. Sci. Eng. 52 (1–4), 161–171.

Chen, L., Zhang, G., Wang, L., Wu, W., Ge, J., 2014. Zeta potential of limestone in a large range of salinity. Colloids Surf. A Physicochem. Eng. Asp. 450, 1–8.

Chowdhury, S., Shrivastava, S., Kakati, A., Sangwai, J.S., 2022. Comprehensive review on the role of surfactants in the chemical enhanced oil recovery process. Ind. Eng. Chem. Res. 61 (1), 21–64.

Clark, J.A., Santiso, E.E., 2018. Carbon sequestration through CO2 foam-enhanced oil recovery: a green chemistry perspective. Engineering 4 (3), 336–342.

Cockcroft, P., Anli, J., Duignan, J., 1988. EOR Potential of Indonesian Reservoirs.

Craig, F.C., 1971. The Reservoir Engineering Aspects of Waterflooding. Society of Petroleum Engineers of AIME, Monograph Series.

Crain, E.R., 2002. Crain's Petrophysical Handbook. Spectrum 2000 Mindware Limited.

Cui, L., Ma, K., Abdala, A.A., Lu, L.J., Tanakov, I., Biswal, S.L., Hirasaki, G.J., 2015. Adsorption of a switchable cationic surfactant on natural carbonate minerals. SPE J. 20 (01), 70–78.

Cuiec, L., 1984. Rock/crude-oil interactions and wettability: an attempt to understand their interrelation. In: SPE Annual Technical Conference and Exhibition.

Cummings, S., Enick, R., Rogers, S., Heenan, R., Eastoe, J., 2012. Amphiphiles for supercritical CO2. Biochimie 94 (1), 94–100.

Dang, C.T.Q., Nghiem, L.X., Chen, Z., Nguyen, N.T.B., Nguyen, Q.P., 2014a. CO2 low salinity water alternating gas: a new promising approach for enhanced oil recovery. In: SPE Improved Oil Recovery Symposium. Society of Petroleum Engineers.

Dang, T.Q.C., Chen, Z., Nguyen, T.B.N., Bae, W., 2014b. Investigation of isotherm polymer adsorption in porous media. Pet. Sci. Technol. 32 (13), 1626–1640.

Dang, C., Nghiem, L., Nguyen, N., Chen, Z., Yang, C., Bae, W., 2017. A comprehensive evaluation of alkaline surfactant polymer flooding and hybrid process for enhanced oil recovery. In: SPE Annual Technical Conference and Exhibition.

Delshad, M., Han, C., Veedu, F.K., Pope, G.A., 2011. A simplified model for simulations of alkaline-surfactant-polymer floods. In: SPE Reservoir Simulation Symposium.

Denekas, M.O., Mattax, C.C., Davis, G.T., 1959. Effects of crude oil components on rock wettability. Trans. AIME 216 (01), 330–333.

Dijt, J.C., 1993. Kinetics of Polymer Adsorption, Desorption and Exchange. Wageningen University and Research.

Divers, T., Gaillard, N., Bataille, S., Thomas, A., Favéro, C., 2017. Successful polymer selection for cEOR: brine hardness and mechanical degradation considerations. In: SPE Oil and Gas India Conference and Exhibition.

Donaldson, E.C., Chilingarian, G.V., Yen, T.F., 1989. Enhanced Oil Recovery, II: Processes and Operations. Elsevier.

Duan, Z., Sun, R., Zhu, C., Chou, I.-M., 2006. An improved model for the calculation of CO2 solubility in aqueous solutions containing Na+, K+, Ca2+, Mg2+, Cl−, and SO42−. Mar. Chem. 98 (2–4), 131–139.

Dubey, S.T., Doe, P.H., 1993. Base number and wetting properties of crude oils. SPE Reserv. Eng. 8 (03), 195–200.

Duda, J.R., Kuuskraa, V., Godec, M., van Leeuwen, T., 2010. Modeling exercises assess US CO2-EOR potential. Oil Gas J. 108 (12), 52–54.

Ebanks Jr., W.J., Cook, C.B.P., 1993. Sedimentology and Reservoir Properties of Eocene Ngimbang Clastics Sandstones in Cores of the Pagerungan-5 Well Pagerungan Field, East Java Sea, Indonesia.

Enge, I.B., 2014. The Effect of Brine Composition and Rock Type on Oil Recovery by the Use of Combined Low-Salinity Waterflooding and Surfactant Flooding: A Literature Review and Experimental Study. Master's thesis, Institutt for petroleumsteknologi og anvendt geofysikk.

Enick, R.M., Klara, S.M., 1990. CO2 solubility in water and brine under reservoir conditions. Chem. Eng. Commun. 90 (1), 23–33.

Enick, R.M., Olsen, D.K., 2012. Mobility and conformance control for carbon dioxide enhanced oil recovery (CO2-EOR) via thickeners, foams, and gels—a detailed literature review of 40 years of research. Contract DE-FE0004003, Activity 4003 (01).

Farajzadeh, R., Andrianov, A., Krastev, R., Hirasaki, G.J., Rossen, W.R., 2012. Foam–oil interaction in porous media: implications for foam assisted enhanced oil recovery. Adv. Colloid Interface Sci. 183, 1–13.

Farajzadeh, R., Guo, H., van Winden, J., Bruining, J., 2017. Cation exchange in the presence of oil in porous media. ACS Earth Space Chem. 1 (2), 101–112.

Fathi, S.j., Austad, T., Strand, S., 2012. Water-based enhanced oil recovery EOR by smart water in carbonate reservoirs. In: SPE EOR conference at oil and gas West Asia, OnePetro.

Fathi, S.J., Austad, T., Strand, S., Frank, S., Mogensen, K., 2010a. Evaluation of EOR potentials in an offshore limestone reservoir: a case study. In: Proceedings of the Eleventh International Symposium on Reservoir Wettability, Calgary, AB, Canada, pp. 6–8.

Fathi, S.J., Austad, T., Strand, S., Puntervold, T., 2010b. Wettability alteration in carbonates: the effect of water-soluble carboxylic acids in crude oil. Energy Fuels 24 (5), 2974–2979.

Feng, L., Liang, X., 2015. Implications of shale oil compositions on surfactant efficacy for wettability alteration. In: SPE Middle East Unconventional Resources Conference and Exhibition.

Fjelde, I., Asen, S.M., Omekeh, A., 2012. Low salinity water flooding experiments and interpretation by simulations. In: SPE Improved Oil Recovery Symposium.

Fjelde, I., Omekeh, A.V., Sokama-Neuyam, Y.A., 2014. Low salinity water flooding: effect of crude oil composition. In: SPE Improved Oil Recovery Symposium.

Folk, R.L., 1959. Practical petrographic classification of limestones. AAPG Bull. 43 (1), 1–38.

Fortenberry, R., Kim, D.H., Nizamidin, N., Adkins, S., Pinnawala Arachchilage, G.W., Koh, H., Weerasooriya, U., Pope, G.A., 2015. Use of cosolvents to improve alkaline/polymer flooding. SPE J. 20 (02), 255–266.

Gaus, I., 2010. Role and impact of CO2–rock interactions during CO2 storage in sedimentary rocks. Int. J. Greenh. Gas Control 4 (1), 73–89.

Glover, C.J., Puerto, M.C., Maerker, J.M., Sandvik, E.L., 1979. Surfactant phase behavior and retention in porous media. Soc. Pet. Eng. J. 19 (03), 183–193.

Gonzalez, D.L., Mahmoodaghdam, E., Lim, F., Joshi, N., 2012. Effects of gas additions to deepwater gulf of Mexico reservoir oil: experimental investigation of asphaltene precipitation and deposition. In: SPE Annual Technical Conference and Exhibition.

Goudarzi, A., Delshad, M., Sepehrnoori, K., 2013. A critical assessment of several reservoir simulators for modeling chemical enhanced oil recovery processes. In: SPE Reservoir Simulation Symposium.

Green, D.W., Willhite, G.P., 1998. Enhanced oil recovery. In: Henry L. Doherty Memorial Fund of AIME. vol. 6. Society of Petroleum Engineers.

Guillon, V., Kohler, E., Wartenberg, N., Rousseau, D., 2017. Restoration of reservoir cores to reduced reservoir conditions for chemical EOR studies: impact on mineralogy, pore structure and rock/fluids interactions. In: SPE Europec Featured at 79th EAGE Conference and Exhibition.

Gupta, S.S., Bhattacharyya, K.G., 2011. Kinetics of adsorption of metal ions on inorganic materials: a review. Adv. Colloid Interface Sci. 162 (1–2), 39–58.

Gutierrez, M.E., Gaona, S.J., Calvete, F.E., Botett, J.A., Ferrari, J.V., 2019, March. Effect of scaling and corrosion inhibitors on the static adsorption of an anionic surfactant on a carbonate rock. In: SPE International Conference on Oilfield Chemistry. OnePetro.

Hamid, M.A., Onur, M., 2013. Effects of concentration, temperature, and pH on anionic surfactant adsorption isotherm. Platform, 36–42.

Hamouda, A.A., Valderhaug, O.M., 2014. Investigating enhanced oil recovery from sandstone by low-salinity water and fluid/rock interaction. Energy Fuels 28 (2), 898–908.

Hassan, A.M., Ayoub, M., Eissa, M., Musa, T., Bruining, H., Farajzadeh, R., 2019. Exergy return on exergy investment analysis of natural-polymer (Guar-Arabic gum) enhanced oil recovery process. Energy 181, 162–172. https://doi.org/10.1016/J.ENERGY.2019.05.137.

Hassan, A.M., Ayoub, M., Eissa, M., Bruining, H., Zitha, P., 2020a. Study of surface complexation modeling on a novel hybrid enhanced oil recovery (EOR) method; smart-water assisted foam-flooding. J. Pet. Sci. Eng. 195. https://doi.org/10.1016/J.PETROL.2020.107563.

Hassan, A., Ayoub, M., Eissa, M., Bruining, H., Al-Mansour, A., Al-Guraishi, A., 2020b. A study of combined low salinity foam injection using DLVO theory and surface complexation modeling. In: Offshore Technology Conference Asia. OnePetro. https://doi.org/10.4043/30301-MS.

Hassan, A., Ayoub, M., Eissa, M., Bruining, H., Al-Mansour, A., Al-Guraishi, A., 2020c. A novel hybrid enhanced oil recovery method by smart water-injection and foam-flooding in carbonate reservoirs. In: SPE/IATMI Asia Pacific Oil & Gas Conference and Exhibition. OnePetro. https://doi.org/10.2118/196407-MS.

Hassan, A.M., Ayoub, M., Eissa, M., Al-Shalabi, E.W., Almansour, A., Alquraishi, A., et al., 2022a. Foamability and Foam Stability Screening for Smart Water Assisted Foam Flooding: A New Hybrid EOR Method. International Petroleum Technology Conference (SPE).

Hassan, A.M., Ayoub, M.A., Mohyadinn, M.E., Al-Shalabi, E.W., Alakbari, F.S., 2022b. A new insight into smart water assisted foam SWAF technology in carbonate rocks using artificial neural networks ANNs. In: Offshore Technology Conference Asia.

Hassan, A.M., Ayoub, M., Eissa, M., Al-Shalabi, E.W., Al-Mansour, A., Al-Quraishi, A., 2022c. Increasing reservoir recovery efficiency through laboratory-proven hybrid smart water-assisted foam (SWAF) flooding in carbonate reservoirs. Energies 15, 3058. https://doi.org/10.3390/en15093058.

Hassan, A., Ayoub, M., Eissa, M., Bruining, H., Al-Mansour, A., Al-Quraishi, A., 2021. A new hybrid improved and enhanced oil recovery IOR/EOR process using smart water assisted foam SWAF flooding in carbonate rocks; a laboratory study approach. In: International Petroleum Technology Conference.

Healy, R.N., Reed, R.L., 1977. Immiscible microemulsion flooding. Soc. Pet. Eng. J. 17 (02), 129–139.

Hill, H.J., Reisberg, J., Stegemeier, G.L., 1973. Aqueous surfactant systems for oil recovery. J. Pet. Technol. 25 (02), 186–194.

Hiorth, A., Cathles, L.M., Madland, M.V., 2010. The impact of pore water chemistry on carbonate surface charge and oil wettability. Transp. Porous Media 85 (1), 1–21.

Hirasaki, G., Zhang, D.L., 2003. Surface chemistry of oil recovery from fractured, oil-wet, carbonate formation. In: International Symposium on Oilfield Chemistry.

Hirasaki, G.J., 1982. Ion exchange with clays in the presence of surfactant. Soc. Pet. Eng. J. 22 (02), 181–192.

Hirasaki, G.J., Miller, C.A., Puerto, M., 2011. Recent advances in surfactant EOR. SPE J. 16 (04), 889–907.

Hu, K., Bard, A.J., 1997. Characterization of adsorption of sodium dodecyl sulfate on charge-regulated substrates by atomic force microscopy force measurements. Langmuir 13 (20), 5418–5425.

Ibrahim, A.F., Nasr-El-Din, H.A., 2019. CO2 foam for enhanced oil recovery applications. In: Foams-Emerging Technologies. IntechOpen.

Ibrahim, Q., Ali, I., Kareem, S., 2016. Removal of cetyldimethylbenzylammonium chloride surfactant from aqueous solution by adsorption onto mesoporous silica. Int. J. Sci. Res. 5, 6–391.

IEA, 2019. World Energy Outlook 2019. IEA, Paris. https://www.iea.org/reports/world-energy-outlook-2019.

IEA, 2021. World Energy Outlook 2021. IEA, Paris. https://www.iea.org/reports/world-energy-outlook-2021.

Ishiguro, M., Koopal, L.K., 2016. Surfactant adsorption to soil components and soils. Adv. Colloid Interface Sci. 231, 59–102.

Izgec, O., Demiral, B., Bertin, H., Akin, S., 2008a. CO2 injection into saline carbonate aquifer formations I. Transp. Porous Media 72 (1), 1–24.

Izgec, O., Demiral, B., Bertin, H., Akin, S., 2008b. CO2 injection into saline carbonate aquifer formations II: comparison of numerical simulations to experiments. Transp. Porous Media 73 (1), 57–74.

Jamaluddin, A.K.M., Creek, J., Kabir, C.S., McFadden, J.D., D'Cruz, D., Manakalathil, J., Joshi, N., Ross, B., 2002. Laboratory techniques to measure thermodynamic asphaltene instability. J. Can. Pet. Technol. 41 (07).

Jenning, H.Y., 1975. A study of caustic solution-crude oil interfacial tensions. Soc. Pet. Eng. J. 15 (03), 197–202.

Jiang, H., Nuryaningsih, L., Adidharma, H., 2010. The effect of salinity of injection brine on water alternating gas performance in tertiary miscible carbon dioxide flooding: experimental study. In: SPE Western Regional Meeting. Society of Petroleum Engineers.

Johnson, C.E., 1976. Status of caustic and emulsion methods. J. Pet. Technol. 28 (01), 85–92.

Kakati, A., Sangwai, J.S., 2017. Effect of monovalent and divalent salts on the interfacial tension of pure hydrocarbon-brine systems relevant for low salinity water flooding. J. Pet. Sci. Eng. 157, 1106–1114. https://doi.org/10.1016/j.petrol.2017.08.017.

Karimov, D., Hashmet, M.R., Pourafshary, P., 2020. A laboratory study to optimize ion composition for the hybrid low salinity water/polymer flooding. In: Offshore Technology Conference Asia.

Karović-Maričić, V., Leković, B., Danilović, D., 2014. Factors influencing successful implementation of enhanced oil recovery projects. Podzemni Radovi 22 (25), 41–50.

Kasha, A., Al-Hashim, H., Abdallah, W., Taherian, R., Sauerer, B., 2015. Effect of Ca2+, Mg2+ and SO42− ions on the zeta potential of calcite and dolomite particles aged with stearic acid. Colloids Surf. A Physicochem. Eng. Asp. 482, 290–299.

Keshtkar, S., Sabeti, M., Mohammadi, A.H., 2016. Numerical approach for enhanced oil recovery with surfactant flooding. Petroleum 2 (1), 98–107.

Khorsandi, S., Qiao, C., Johns, R.T., 2017. Displacement efficiency for low-salinity polymer flooding including wettability alteration. SPE J. 22 (02), 417–430.

Klimenko, A., Molinier, V., Dubos, F., Joly, M., Saint-Loubert, M., Jouenne, S., Passade-Boupat, N., Bourrel, M., 2020. Surfactant–polymer flooding at high temperature and high salinity: promising lab scale experiments in challenging conditions. In: Abu Dhabi International Petroleum Exhibition & Conference.

Kokal, S., Al-Kaabi, A., 2010. Enhanced oil recovery: challenges & opportunities. In: World Petroleum Council: Official Publication. vol. 64, pp. 64–69.

Kolodziej, E.J., 1988. Transport mechanisms of Xanthan biopolymer solutions in porous media. In: SPE Annual Technical Conference and Exhibition.

Kosmulski, M., 2002. The significance of the difference in the point of zero charge between rutile and anatase. Adv. Colloid Interface Sci. 99 (3), 255–264.

Kou, J., Xu, S., 2016. In situ kinetics and conformation studies of dodecylamine adsorption onto zinc sulfide using a quartz crystal microbalance with dissipation (QCM-D). Colloids Surf. A Physicochem. Eng. Asp. 490, 110–120.

Kumar, H.T., Shehata, A.M., Nasr-El-Din, A.H., 2016. Effectiveness of low-salinity and CO2 flooding hybrid approaches in low-permeability sandstone reservoirs. In: SPE Trinidad and Tobago Section Energy Resources Conference, OnePetro.

Lager, A., Webb, K.J., Black, C.J.J., 2007. Impact of brine chemistry on oil recovery. In: IOR 2007—14th European Symposium on Improved Oil Recovery, cp-24.

Lager, A., Webb, K.J., Black, C.J.J., Singleton, M., Sorbie, K.S., 2008. Low salinity oil recovery—an experimental investigation. Petrophysics 49 (01).

Lagergren, S., 1898. Zur theorie der sogenannten adsorption geloster stoffe.

Lake, L.W., 1989. Enhanced Oil Recovery.

Lake, L.W., Johns, R., Rossen, B., Pope, G.A., 2014. Fundamentals of Enhanced Oil Recovery. vol. 1 Society of Petroleum Engineers, Richardson, TX.

Lee, J.-J., Fuller, G.G., 1985. Adsorption and desorption of flexible polymer chains in flowing systems. J. Colloid Interface Sci. 103 (2), 569–577.

Lee, S., Kam, S.I., 2013. Enhanced oil recovery by using CO2 foams: fundamentals and field applications. In: Enhanced Oil Recovery Field Case Studies. Elsevier Inc. Chapter 2.

Lee, K.S., Lee, J.H., 2019. Hybrid Enhanced Oil Recovery Using Smart Waterflooding. Gulf Professional Publishing.

Lee, J., Nikolov, A., Wasan, D., 2013. Stability of aqueous foams in the presence of oil: on the importance of dispersed vs solubilized oil. Ind. Eng. Chem. Res. 52 (1), 66.

Leonhardt, B., Ernst, B., Reimann, S., Steigerwald, A., Lehr, F., 2014. Field testing the polysaccharide schizophyllan: results of the first year. In: SPE Improved Oil Recovery Symposium.

Leontaritis, K.J., 1996. The asphaltene and wax deposition envelopes. Fuel Sci. Technol. Int. 14 (1–2), 13–39.

Levitt, D., Bourrel, M., 2016. Adsorption of EOR chemicals under laboratory and reservoir conditions, part III: chemical treatment methods. In: SPE Improved Oil Recovery Conference.

Levitt, D.B., Weatherl, R.K., Harris, H.W., McNeil, R.I., Didier, M., Loriau, M., Gaucher, E.C., Bourrel, M., 2015. Adsorption of EOR chemicals under laboratory and reservoir conditions, part 1—iron abundance and oxidation state. In: IOR 2015—18th European Symposium on Improved Oil Recovery, cp-445.

Li, Y.-K., Nghiem, L.X., 1986. Phase equilibria of oil, gas and water/brine mixtures from a cubic equation of state and Henry's law. Can. J. Chem. Eng. 64 (3), 486–496.

Li, R.F., Yan, W., Liu, S., Hirasaki, G.J., Miller, C.A., 2010. Foam mobility control for surfactant enhanced oil recovery. SPE J. 15 (04), 928–942.

Li, Z., Xu, D., Yuan, Y., Wu, H., Hou, J., Kang, W., Bai, B., 2020. Advances of spontaneous emulsification and its important applications in enhanced oil recovery process. Adv. Colloid Interface Sci. 277, 102119. https://doi.org/10.1016/j.cis.2020.102119.

Li, X., Zhang, F., Liu, G., 2021. Review on polymer flooding technology. IOP Conf. Ser. Earth Environ. Sci. 675 (1), 012199.

Lin, H., Fujii, T., Takisawa, R., Takahashi, T., Hashida, T., 2008. Experimental evaluation of interactions in supercritical CO2/water/rock minerals system under geologic CO2 sequestration conditions. J. Mater. Sci. 43 (7), 2307–2315.

Lu, J., Goudarzi, A., Chen, P., Kim, D.H., Delshad, M., Mohanty, K.K., Sepehrnoori, K., Weerasooriya,, U.P., Pope, G.A., 2014. Enhanced oil recovery from high-temperature, high-salinity naturally fractured carbonate reservoirs by surfactant flood. J. Pet. Sci. Eng. 124, 122–131.

Luo, H.Q., Li, L.N., Li, N.B., 2006. Second chemiluminescence behavior of Fe2+, Fe3+ and Cr3+ in the luminal-KMnO4 system. Anal. Sci. 22 (4), 621–625.

Lwisa, E., 2021. Chemical enhanced oil recovery. Int. J. Innov. Educ. Res. 9, 160–172. https://doi.org/10.31686/ijier.vol9.iss6.3160.

Ma, K., Cui, L., Dong, Y., Wang, T., Da, C., Hirasaki, G.J., Biswal, S.L., 2013. Adsorption of cationic and anionic surfactants on natural and synthetic carbonate materials. J. Colloid Interface Sci. 408, 164–172.

Mahani, H., Berg, S., Ilic, D., Bartels, W.-B., Joekar-Niasar, V., 2013. Kinetics of the low salinity waterflooding effect studied in a model system. In: SPE Enhanced Oil Recovery Conference.

Mahani, H., Keya, A.L., Berg, S., Bartels, W.-B., Nasralla, R., Rossen, W.R., 2015. Insights into the mechanism of wettability alteration by low-salinity flooding (LSF) in carbonates. Energy Fuels 29 (3), 1352–1367.

Mahani, H., Menezes, R., Berg, S., Fadili, A., Nasralla, R., Voskov, D., Joekar-Niasar, V., 2017. Insights into the impact of temperature on the wettability alteration by low salinity in carbonate rocks. Energy Fuels 31 (8), 7839–7853.

Mandal, A., 2015. Chemical flood enhanced oil recovery: a review. Int. J. Oil Gas Coal Technol. 9 (3), 241–264.

Manichand, R.N., Seright, R.S., 2014. Field vs. laboratory polymer-retention values for a polymer flood in the Tambaredjo field. SPE Reserv. Eval. Eng. 17 (03), 314–325.

Manrique, E., Thomas, C., Ravikiran, R., Izadi, M., Lantz, M., Romero, J., Alvarado, V., 2010. EOR: current status and opportunities. In: SPE Improved Oil Recovery Symposium.

Mansour, E.M., Al-Sabagh, A.M., Desouky, S.M., Zawawy, F.M., Ramzi, M.R., 2016. Experimental approach of minimum miscibility pressure for CO2 miscible flooding: application to Egyptian oil fields. Int. J. New Technol. Res. 2 (5).

Masalmeh, S., AlSumaiti, A., Gaillard, N., Daguerre, F., Skauge, T., Skuage, A., 2019. Extending polymer flooding towards high-temperature and high-salinity carbonate reservoirs. In: Abu Dhabi International Petroleum Exhibition & Conference.

McGuire, P.L., Chatham, J.R., Paskvan, F.K., Sommer, D.M., Carini, F.H., 2005. Low salinity oil recovery: an exciting new EOR opportunity for Alaska's North Slope. In: SPE Western Regional Meeting.

McKay, G., Ho, Y.S., Ng, J.C.Y., 1999. Biosorption of copper from waste waters: a review. Sep. Purif. Methods 28 (1), 87–125.

Mehraban, M.F., Farzaneh, S.A., Sohrabi, M., Sisson, A., 2022. Fluid-fluid interactions inducing additional oil recovery during low salinity water injection in inefficacious presence of clay minerals. Fuel 308, 121922.

Mohammadi, H., 2008. Mechanistic Modeling, Design, and Optimization of Alkaline/Surfactant/Polymer Flooding. The University of Texas at Austin.

Mohyaldinn, M.E., Hassan, A.M., Ayoub, M.A., 2019. Application of emulsions and microemulsions in enhanced oil recovery and well stimulation. In: Microemulsion—A Chemical Nanoreactor. IntechOpen.

Moradi, S., Dabiri, M., Dabir, B., Rashtchian, D., Emadi, M.A., 2012. Investigation of asphaltene precipitation in miscible gas injection processes: experimental study and modeling. Braz. J. Chem. Eng. 29 (3), 665–676.

Morrow, N.R., 1990. Wettability and its effect on oil recovery. J. Pet. Technol. 42 (12), 1476–1484.

Moulin, P., Roques, H., 2003. Zeta potential measurement of calcium carbonate. J. Colloid Interface Sci. 261 (1), 115–126.

Musa, T.A., Ibrahim, A.F., Nasr-El-Din, H.A., Hassan, A., 2021. New insights into guar gum as environmentally friendly polymer for enhanced oil recovery in high-salinity and high-temperature sandstone reservoirs. J. Pet. Explor. Prod. 11 (4), 1905–1913.

Nasralla, R.A., Nasr-El-Din, H.A., 2011. Coreflood study of low salinity water injection in sandstone reservoirs. In: SPE/DGS Saudi Arabia Section Technical Symposium and Exhibition.

Na, S., Park, Y., Hwang, A., Ha, J., Kim, Y., Khim, J., 2007. Effect of ultrasound on surfactant-aided soil washing. Jpn. J. Appl. Phys. 46 (7S), 4775.

Nasralla, R.A., Bataweel, M.A., Nasr-El-Din, H.A., 2013. Investigation of wettability alteration and oil-recovery improvement by low-salinity water in sandstone rock. J. Can. Pet. Technol. 52 (02), 144–154.

Newell, R., Raimi, D., Villanueva, S., Prest, B., 2021. Global Energy Outlook 2021: Pathways From Paris.

Ngo, I., Srisuriyachai, F., Sugai, Y., Sasaki, K., 2017. Study of heterogeneous reservoir effects on surfactant flooding in consideration of surfactant adsorption reversibility. In: SPWLA 23rd Formation Evaluation Symposium of Japan.

Nwidee, L.N., Lebedev, M., Barifcani, A., Sarmadivaleh, M., Iglauer, S., 2017. Wettability alteration of oil-wet limestone using surfactant-nanoparticle formulation. J. Colloid Interface Sci. 504, 334–345.

Ogezi, O., Strobel, J., Egbuniwe, D., Leonhardt, B., 2014. Operational aspects of a biopolymer flood in a mature oilfield. In: SPE Improved Oil Recovery Symposium.

Olajire, A.A., 2014. Review of ASP EOR (alkaline surfactant polymer enhanced oil recovery) technology in the petroleum industry: prospects and challenges. Energy 77, 963–982.

Orr, F.M., Silva, M.K., Lien, C.L., Pelletier, M.T., 1982. Laboratory experiments to evaluate field prospects for CO2 flooding. J. Pet. Technol. 34 (04), 888–898.

Patil, S.B., Dandekar, A.Y., Patil, S., Khataniar, S., 2008. Low salinity brine injection for EOR on Alaska North Slope (ANS). In: International Petroleum Technology Conference.

Perrin, J.-C., Krause, M., Kuo, C.-W., Miljkovic, L., Charoba, E., Benson, S.M., 2009. Core-scale experimental study of relative permeability properties of CO2 and brine in reservoir rocks. Energy Procedia 1 (1), 3515–3522.

Pham, T.D., Kobayashi, M., Adachi, Y., 2015. Adsorption of anionic surfactant sodium dodecyl sulfate onto alpha alumina with small surface area. Colloid Polym. Sci. 293 (1), 217–227.

Pingo-Almada, M.B., et al., 2013. Experimental investigation on the effects of very low salinity on Middle Eastern sandstone corefloods. In: Presented at the SPE European Formation Damage.

Pollack, N.R., Enick, R.M., Mangone, D.J., Morsi, B.I., 1988. Effect of an aqueous phase on CO2/tetradecane and CO2/Maljamar-crude-oil systems. SPE Reserv. Eng. 3 (02), 533–541.

Pope, G.A., Nelson, R.C., 1978. A chemical flooding compositional simulator. Soc. Pet. Eng. J. 18 (05), 339–354.

Pourafshary, P., Moradpour, N., 2019. Hybrid EOR methods utilizing low-salinity water. In: Enhanced Oil Recovery Processes—New Technologies. IntechOpen.

Puntervold, T., Strand, S., Austad, T., 2007. Water flooding of carbonate reservoirs: effects of a model base and natural crude oil bases on chalk wettability. Energy Fuels 21 (3), 1606–1616.

Qiao, C., Johns, R., Li, L., 2016. Understanding the chemical mechanisms for low salinity waterflooding. In: 78th EAGE Conference and Exhibition 2016. vol. 2016, pp. 1–13.

Radke, C.J., Somerton, W.H., 1978. Enhanced recovery with mobility and reactive tension agents. Oil Gas J.

Ragab, A, Mansour, E.M., 2021. Enhanced Oil Recovery: Chemical Flooding. Geophysics and Ocean Waves Studies 51.

Rahimi, V., Bidarigh, M., Bahrami, P., 2017. Experimental study and performance investigation of miscible water-alternating-CO2 flooding for enhancing oil recovery in the Sarvak formation. Oil Gas Sci. Technol. 72 (6), 35.

Ramanathan, R., Shehata, A.M., Nasr-El-Din, H.A., 2015. Water alternating CO2 injection process—does modifying the salinity of injected brine improve oil recovery. In: OTC Brasil. Offshore Technology Conference.

Ramanathan, R., Shehata, A.M., Nasr-El-Din, H.A., 2016. Effect of rock aging on oil recovery during water-alternating-CO2 injection process: an interfacial tension, contact angle, coreflood, and CT scan study. In: SPE Improved Oil Recovery Conference. Society of Petroleum Engineers.

Reisberg, J., Doscher, T.M., 1956. Interfacial phenomena in crude oil-water systems. Prod. Mon. 21 (1), 43–50.

RezaeiDoust, A., Puntervold, T., Strand, S., Austad, T., 2009. Smart water as wettability modifier in carbonate and sandstone: a discussion of similarities/differences in the chemical mechanisms. Energy Fuels 23 (9), 4479–4485.

RezaeiDoust, A., Puntervold, T., Austad, T., 2011. Chemical verification of the EOR mechanism by using low saline/smart water in sandstone. Energy Fuels 25 (5), 2151–2162.

Roehl, P.O., Choquette, P.W., 2012. Carbonate Petroleum Reservoirs. Springer Science & Business Media.

Romanuka, J., Hofman, J.P., Ligthelm, D.J., Suijkerbuijk, B.M., Marcelis, A.H., Oedai, S., Brussee, N.J., van der Linde, A., Aksulu, H., Austad, T., 2012, April. Low salinity EOR in carbonates. In: SPE Improved Oil Recovery Symposium. OnePetro.

Rose, S.C., Buckwatter, J.F., Woodhall, R.J., 1989. The Design Engineering Aspects of Waterflooding.

Rosen, M.J., Kunjappu, J.T., 2012. Surfactants and Interfacial Phenomena. John Wiley & Sons.

Rossen, W.R., 1996. Foams in enhanced oil recovery. In: Foams: Theory, Measurements and Applications. vol. 57, pp. 413–464.

Sagir, M., Tan, I.M., Mushtaq, M., Talebian, S.H., 2014. FAWAG using CO2 philic surfactants for CO2 mobility control for enhanced oil recovery applications. In: SPE Saudi Arabia Section Technical Symposium and Exhibition.

Sagir, M., Mushtaq, M., Tahir, M.B., Tahir, M.S., Ullah, S., Shahzad, K., Rashid, U., 2018. CO2 foam for enhanced oil recovery (EOR) applications using low adsorption surfactant structure. Arab. J. Geosci. 11 (24), 1–13.

Sahni, V., Dean, R.M., Britton, C., Kim, D.H., Weerasooriya, U., Pope, G.A., 2010. The role of co-solvents and co-surfactants in making chemical floods robust. In: SPE Improved Oil Recovery Symposium.

Salager, J.L., Morgan, J.C., Schechter, R.S., Wade, W.H., Vasquez, E., 1979. Optimum formulation of surfactant/water/oil systems for minimum interfacial tension or phase behavior. Soc. Pet. Eng. J. 19 (02), 107–115.

Salehi, M.M., Hekmatzadeh, A., Sajjadian, V.A., Masoumi, M., 2017. Simulation of polymer flooding in one of the Iranian oil fields. Egypt. J. Pet. 26 (2), 325–330.

Samanta, A., Ojha, K., Mandal, A., 2011. Interactions between acidic crude oil and alkali and their effects on enhanced oil recovery. Energy Fuels 25 (4), 1642–1649.

Scheihing, M.H., Thompson, R.D., Seifert, D., 2002. Multiscale reservoir description models for performance prediction in the Kuparuk River Field, North Slope of Alaska. In: SPE Western Regional/AAPG Pacific Section Joint Meeting.

Seethepalli, A., Adibhatla, B., Mohanty, K.K., 2004. Physicochemical interactions during surfactant flooding of fractured carbonate reservoirs. SPE J. 9 (04), 411–418.

Shabib-Asl, A., Ayoub, M.A., Saaid, I.M., Valentim, P.P.J., 2015. Experimental investigation into effects of crude oil acid and base number on wettability alteration by using different low salinity water in sandstone rock. J. Jpn. Pet. Inst. 58 (4), 228–236.

Shaddel, S., Tabatabae-Nejad, S.A., 2015. Alkali/surfactant improved low-salinity waterflooding. Transp. Porous Media 106 (3), 621–642.

Shah, D.O., Walker, R.D., Hsieh, W.C., Shah, N.J., Dwivedi, S., Nelander, J., Pepinsky, R., Deamer, D.W., 1976. Some structural aspects of microemulsions and co-solubilized systems. In: SPE Improved Oil Recovery Symposium.

Shakeel, M., Pourafshary, P., Rehan Hashmet, M., 2020. Hybrid engineered water–polymer flooding in carbonates: a review of mechanisms and case studies. Appl. Sci. 10 (17), 6087.

ShamsiJazeyi, H., Verduzco, R., Hirasaki, G.J., 2014. Reducing adsorption of anionic surfactant for enhanced oil recovery: part I. Competitive adsorption mechanism. Colloids Surf. A Physicochem. Eng. Asp. 453, 162–167.

Shariatpanahi, S.F., Hopkins, P., Aksulu, H., Strand, S., Puntervold, T., Austad, T., 2016. Water based EOR by wettability alteration in dolomite. Energy Fuels 30 (1), 180–187.

Sharma, H., Dufour, S., Arachchilage, G.W.P.P., Weerasooriya, U., Pope, G.A., Mohanty, K., 2015. Alternative alkalis for ASP flooding in anhydrite containing oil reservoirs. Fuel 140, 407–420.

Shehata, A.M., Nasr-El-Din, H.A., 2015, April. Zeta potential measurements: Impact of salinity on sandstone minerals. In: SPE International Symposium on Oilfield Chemistry. OnePetro.

Sheng, J., 2010. Modern Chemical Enhanced Oil Recovery: Theory and Practice. Gulf Professional Publishing.

Sheng, J.J., 2013. Surfactant enhanced oil recovery in carbonate reservoirs. In: Enhanced oil recovery field case studies. Gulf Professional Publishing, pp. 281–299.

Sheng, J.J., 2014. Critical review of low-salinity waterflooding. J. Pet. Sci. Eng. 120, 216–224.

Sheng, J.J., 2015. Status of surfactant EOR technology. Petroleum 1 (2), 97–105.

Sheng, J.J., Leonhardt, B., Azri, N., 2015. Status of polymer-flooding technology. J. Can. Pet. Technol. 54 (02), 116–126.

Shi, S., Wang, Y., Wang, L., Jin, Y., Wang, T., Wang, J., 2015. Potential of spontaneous emulsification flooding for enhancing oil recovery in high-temperature and high-salinity oil reservoir. J. Dispers. Sci. Technol. 36 (5), 660–669.

Shiraki, R., Dunn, T.L., 2000. Experimental study on water–rock interactions during CO2 flooding in the Tensleep Formation, Wyoming, USA. Appl. Geochem. 15 (3), 265–279.

Simjoo, M., Rezaei, T., Andrianov, A., Zitha, P.L.J., 2013. Foam stability in the presence of oil: effect of surfactant concentration and oil type. Colloids Surf. A Physicochem. Eng. Asp. 438, 148–158. https://doi.org/10.1016/J.COLSURFA.2013.05.062.

Skauge, A., Palmgren, O., 1989. Phase behavior and solution properties of ethoxylated anionic surfactants. In: SPE International Symposium on Oilfield Chemistry.

Solairaj, S., 2011. *New method of predicting optimum surfactant structure for EOR* (Doctoral dissertation). Doctoral dissertation.

Somasundaran, P., Fuerstenau, D.W., 1966. Mechanisms of alkyl sulfonate adsorption at the alumina-water interface. J. Phys. Chem. 70 (1), 90–96.

Somasundaran, P., Krishnakumar, S., 1997. Adsorption of surfactants and polymers at the solid-liquid interface. Colloids Surf. A Physicochem. Eng. Asp. 123, 491–513.

Sorbie, K.S., 1991. Introduction to polymer flooding. In: Polymer-Improved Oil Recovery. Springer, pp. 1–5.

Sorbie, K.S., 2013. Polymer-Improved Oil Recovery. Springer Science & Business Media.

Southwick, J.G., van den Pol, E., van Rijn, C.H., van Batenburg, D.W., Boersma, D., Svec, Y., Mastan, A.A., Shahin, G., Raney, K., 2016. Ammonia as alkali for alkaline/surfactant/polymer floods. SPE J. 21 (01), 10–21.

Speight, J.G., 2016. Introduction to Enhanced Recovery Methods for Heavy Oil and Tar Sands. Gulf Professional Publishing.

Spildo, K., Johannessen, A.M., Skauge, A., 2012. Low salinity waterflood at reduced capillarity. In: SPE Improved Oil Recovery Symposium. Society of Petroleum Engineers.

Srisuriyachai, F., 2008. Evaluation of Alkali Flooding Combined With Intermittent Flow in Carbonate Reservoir.

Standnes, D.C., Austad, T., 2003. Wettability alteration in carbonates: interaction between cationic surfactant and carboxylates as a key factor in wettability alteration from oil-wet to water-wet conditions. Colloids Surf. A Physicochem. Eng. Asp. 216 (1–3), 243–259.

Stumm, W., Morgan, J.J., 2012. Aquatic Chemistry: Chemical Equilibria and Rates in Natural Waters. vol. 126 John Wiley & Sons.

Suijkerbuijk, B.M., Hofman, J.P., Ligthelm, D.J., Romanuka, J., Brussee, N., van derLinde, H.A., Marcelis, A.H., 2012. Fundamental investigations into wettability and low salinity flooding by parameter isolation. In: SPE Improved Oil Recovery Symposium.

Sun, S., Zhang, T., 2020. Reservoir Simulations: Machine Learning and Modeling. Gulf Professional Publishing.

Taber, J.J., Martin, F.D., Seright, R.S., 1997. EOR screening criteria revisited—Part 1: introduction to screening criteria and enhanced recovery field projects. SPE Reserv. Eng. 12 (03), 189–198.

Tadros, T., 2013. Encyclopedia of Colloid and Interface Science. Springer.

Tagavifar, M., Jang, S.H., Sharma, H., Wang, D., Chang, L.Y., Mohanty, K., Pope, G.A., 2018. Effect of pH on adsorption of anionic surfactants on limestone: experimental study and surface complexation modeling. Colloids Surf. A Physicochem. Eng. Asp. 538, 549–558.

Tan, K.L., Hameed, B.H., 2017. Insight into the adsorption kinetics models for the removal of contaminants from aqueous solutions. J. Taiwan Inst. Chem. Eng. 74, 25–48.

Tang, G.Q., Morrow, N.R., 1997. Salinity, temperature, oil composition, and oil recovery by waterflooding. SPE Reserv. Eng. 12 (04), 269–276.

Tang, G.-Q., Morrow, N.R., 1999. Influence of brine composition and fines migration on crude oil/brine/rock interactions and oil recovery. J. Pet. Sci. Eng. 24 (2–4), 99–111.

Tavassoli, S., Korrani, A.K.N., Pope, G.A., Sepehrnoori, K., 2016. Low salinity surfactant flooding—a multimechanistic enhanced-oil-recovery method. SPE J. 21 (03), 744–760.

Tawfik, M.S., Karpyn, Z., Johns, R., 2019. Multiscale study of chemically-tuned waterflooding in carbonate rocks using micro-computed tomography. In: IOR 2019—20th European Symposium on Improved Oil Recovery. 2019, pp. 1–23.

Tawfik, M.S., Karpyn, Z.T., Johns, R.T., 2020. Effect of oil chemistry on the performance of low salinity waterflooding: an integrated experimental approach. In: SPE Annual Technical Conference and Exhibition.

Teklu, T.W., Alameri, W., Graves, R.M., Kazemi, H., Al Sumaiti, A.M., 2016. Low-salinity water-alternating-CO2 EOR. J. Pet. Sci. Eng. 142, 101–118.

Theng, B.K.G., 2012. Polymer behaviour at clay and solid surfaces. In: Developments in Clay Science. vol. 4. Elsevier, pp. 47–75.

Türksoy, U., Bağci, S., 2000. Improved oil recovery using alkaline solutions in limestone medium. J. Pet. Sci. Eng. 26 (1–4), 105–119.

Unsal, E., ten Berge, A., Wever, D.A.Z., 2018. Low salinity polymer flooding: lower polymer retention and improved injectivity. J. Pet. Sci. Eng. 163, 671–682.

Vetter, O.J., Farone, W.A., 1987. Calcium carbonate scale in oilfield operations. In: SPE Annual Technical Conference and Exhibition.

Wang, X., Wang, X., Hu, W., Wan, Y., Cao, J., Lv, C., Wang, R., Cui, M., 2017. Supercritical CO2-involved water–rock interactions at 85°C and partial pressures of 10–20 MPa: sequestration and enhanced oil recovery. Energy Explor. Exploit. 35 (2), 237–258. https://doi.org/10.1177/0144598716687933.

Wang, J., Ayirala, S.C., AlSofi, A.M., Al-Yousef, A.A., Saudi Aramco, 2018. Smartwater synergy with surfactant polymer flooding for efficient oil mobilization in carbonates. In: SPE EOR Conference at Oil and Gas West Asia. Society of Petroleum Engineers.

Willhite, G.P., 1986. Waterflooding.

Winsor, P.A., 1954. Solvent Properties of Amphiphilic Compounds. Butterworths Scientific Publications.

Xing, D., Wei, B., McLendon, W., Enick, R., McNulty, S., Trickett, K., Mohamed, A., Cummings, S., Eastoe, J., Rogers, S., 2012. CO2-soluble, nonionic, water-soluble surfactants that stabilize CO2-in-brine foams. SPE J. 17 (04), 1172–1185.

Xu, X., Saeedi, A., Rezaee, R., Liu, K., 2015. Investigation on a novel polymer with surface activity for polymer enhanced CO2 foam flooding. In: SPE International Symposium on Oilfield Chemistry.

Yang, D., Tontiwachwuthikul, P., Gu, Y., 2005. Interfacial tensions of the crude oil+ reservoir brine+ CO2 systems at pressures up to 31 MPa and temperatures of 27 C and 58 C. J. Chem. Eng. Data 50 (4), 1242–1249.

Yu, K.M.K., Steele, A.M., Zhu, J., Fu, Q., Tsang, S.C., 2003. Synthesis of well-dispersed nanoparticles within porous solid structures using surface-tethered surfactants in supercritical CO2. J. Mater. Chem. 13 (1), 130–134.

Yu, Z., Yang, S., Liu, L., Li, S., Yang, Y., 2012. An experimental study on water-rock interaction during water flooding in formations saturated with CO2. Acta Petrol. Sin. 33 (6), 1032.

Zendehboudi, S., Shafiei, A., Bahadori, A., James, L.A., Elkamel, A., Lohi, A., 2014. Asphaltene precipitation and deposition in oil reservoirs—technical aspects, experimental and hybrid neural network predictive tools. Chem. Eng. Res. Des. 92 (5), 857–875.

Zhang, D., 2005. Surfactant-Enhanced Oil Recovery Process for a Fractured, Oil-Wet Carbonate Reservoir. PhD thesis, Rice University.

Zhang, P., Austad, T., 2006. Wettability and oil recovery from carbonates: effects of temperature and potential determining ions. Colloids Surf. A Physicochem. Eng. Asp. 279 (1–3), 179–187.

Zhang, Y., Sarma, H., 2012. Improving waterflood recovery efficiency in carbonate reservoirs through salinity variations and ionic exchanges: a promising low-cost smart-waterflood approach. In: Abu Dhabi International Petroleum Conference and Exhibition.

Zhang, Y., Xie, X., Morrow, N.R., 2007. Waterflood performance by injection of brine with different salinity for reservoir cores. In: SPE Annual Technical Conference and Exhibition.

Zhang, Y., Kogure, T., Chiyonobu, S., Lei, X., Xue, Z., 2013. Influence of heterogeneity on relative permeability for CO2/brine: CT observations and numerical modeling. Energy Procedia 37, 4647–4654.

Zhang, R., Zhang, J., Zhang, X., Dou, C., Han, R., 2014. Adsorption of Congo red from aqueous solutions using cationic surfactant modified wheat straw in batch mode: kinetic and equilibrium study. J. Taiwan Inst. Chem. Eng. 45 (5), 2578–2583.

Zhao, D., Yin, D., 2021. Effect of fluid/rock interactions on physical character of tight sandstone reservoirs during CO2 flooding. Geofluids 2021 (6), 1–14.

Zhong, W., Claverie, J.P., 2013. Probing the carbon nanotube-surfactant interaction for the preparation of composites. Carbon 51, 72–84.

Zhu, B.-Y., Gu, T., 1991. Surfactant adsorption at solid-liquid interfaces. Adv. Colloid Interface Sci. 37 (1–2), 1–32.

Zhu, Y., Tian, J., Hou, Q., Luo, Y., Fan, J., 2017. Studies on nanoparticle-stabilized foam flooding EOR for a high temperature and high salinity reservoir. In: Abu Dhabi International Petroleum Exhibition & Conference.

Ziegler, V.M., Handy, L.L., 1981. Effect of temperature on surfactant adsorption in porous media. Soc. Pet. Eng. J. 21 (02), 218–228.

Zitha, P., Felder, R., Zornes, D., Brown, K., Mohanty, K., 2011. Increasing Hydrocarbon Recovery Factors. Society of Petroleum Engineers, pp. 1–9.

Chapter 8

Nanoparticles in upstream applications

Muhammad Arif and Ravi Shankar Kumar
Department of Petroleum Engineering, Khalifa University, Abu Dhabi, United Arab Emirates

8.1 Introduction

Global energy consumption is increasing continuously, and thus, there is a growing need for new and effective methods for enhanced oil recovery (EOR). In hydrocarbon reservoirs, the varying nature of reservoir properties, operating conditions such as temperature, pressure, and salinity, and wettability changes upon interaction with injected fluids can have a significant impact on hydrocarbon recovery (Arif et al., 2020). Moreover, formation damage due to clay swelling and scale deposition (due to the mixing of two incompatible types of water) can also be detrimental to production (Taheri et al., 2011). Generally, chemicals or additives used in the oil field have preexisting limitations, and beyond that, they may degrade and show a reduction in overall efficiency in oil field operations. Thus, the development of efficient chemicals for EOR and other oil field operations and the associated technology is the need of the hour to meet these demands.

Nanoparticles (NPs), a class of advanced materials, have recently seen a remarkable interest and growth in various scientific and industrial applications such as pharmaceutical (Thomas et al., 2019), food science (Singh et al., 2017), agriculture, electrical, and electronics (Matsui, 2005), automobile (Mathew et al., 2019), surface and coating, lubrication (Jiang et al., 2017), photovoltaic applications (Altavilla and Ciliberto, 2017), heat transfer (Tiwari et al., 2013), and oil field applications (Al-Anssari et al., 2017a,b,c; Kumar et al., 2021a,b). This massive utilization is due to their inherent properties such as high surface-to-volume ratio, high surface potential, favorable electrical properties, and superior heat conductivity which generally helps NPs to regulate the required operations at the nanoscale. Several NPs such as SiO_2 (Corredor-Rojas et al., 2018), TiO_2 (Hu et al., 2016), CuO (Mahmoud et al., 2021), Al_2O_3 (Orodu et al., 2019), ZnO (Afolabi et al., 2018), and $ZnTiO_3$ (Edalatfar et al., 2021) have been investigated at the laboratory scale and demonstrated successful results for

Developments in Petroleum Science, Vol. 78. https://doi.org/10.1016/B978-0-323-99285-5.00001-6

oil field applications, viz. drilling fluid optimization, fluid loss control, preventing fine migration, stimulation fluids, and enhanced oil recovery (EOR). There are, however, certain limitations with regard to NP applications in terms of fluid behavior, physical and thermal stability, and flow properties at reservoir conditions which influence their operational efficiency. Therefore, this chapter is devoted to the understanding of NPs nature, and their intrinsic properties, and summarizes the recent research advances in the applications of NPs in drilling, EOR, stimulation (acidizing and fracturing), and formation damage.

8.2 Nanoparticles applications in drilling

Nanoparticles have been investigated for several drilling applications. These applications include, but are not limited to, reducing filtrate invasion/loss, mud cake thickness, shale swelling, etc. For these applications, NPs are added as an additive to conventional mud (water-based). The sections mentioned below describe the NP applications in drilling.

8.3 Nanoparticles in drilling fluids

In the oil field, drilling fluid is an essential component of drilling operations which is generally used to remove the rock cuttings during drilling operations, lubrication, and cooling of the drill bit (Fink, 2003). It also maintains the balance between formation pressure and hydrostatic pressure to eventually prevent well collapse and contamination of formation from hydrocarbon fluids. However, the reservoir conditions such as high pressure, high temperature, and reservoir salinity generally affect their performance by deteriorating the rheological properties, stability, and efficacy, and thus limit the overall efficiency of the drilling operations (Alcheikh and Ghosh, 2017), which can be addressed by the incorporation of novel additives such as nanomaterials in the drilling fluid system. The use of NPs in drilling fluid is one of the recent advancements in potential oil field applications where the inherent properties of NPs may counter the ongoing drilling fluid challenges. NPs in drilling fluid generally demonstrated several advantages such as reduced formation damage, controlled fluid loss, enhanced thermal properties, improved lubrication, heat transfer capabilities, and favorable rheological properties.

The most common use of NPs in drilling fluid is as a nanocomposite formulated by the inclusion of NPs in a polymeric material such as natural polymers (starch, xanthan gum, and guar gum), modified natural polymers (carboxymethylcellulose, hydroxypropyl starch, and lignosulfonate), and synthetic polymers (polyacrylamide and partially hydrolyzed polyacrylamide) (Medhi et al., 2020; Zoveidavianpoor and Samsuri, 2016). These polymers are generally used in the drilling fluid as a thickening agent which demonstrates better capabilities in terms of weight-on-bit (WOB), lubricity, fluid loss, and viscosity. However, the use of polymer has limitations at high-salinity and high-pressure high-temperature (HPHT) conditions. With the increase in reservoir depth, pressure (Fyfe et al., 1978), temperature (Criss, 2020), and salinity vary significantly which may have an adverse impact on conventional drilling fluid and the associated drilling efficiency. Furthermore, drilling through shale rocks is quite challenging. To address these issues, the use of NPs as an additive in drilling fluids has been investigated recently. Table 8.1 presents a summary of NPs used in drilling fluid applications.

It is thus clear from Table 8.1 that NPs have been tested for a wide range of drilling applications, e.g., rheology, filtration properties, fluid loss, thermal properties, and viscosity. Several researchers have investigated the use of NPs and polymer composites in drilling fluid and found a remarkable improvement in drilling performance. For instance, Sadeghalvaad and Sabbaghi (2015) synthesized a polymer-based nanocomposite of titanium dioxide (TiO_2), NPs, and polyacrylamide (PAM) as an additive for a water-based drilling fluid system called nanoenhanced drilling fluid. The addition of nanocomposites demonstrated superior performance over the conventional drilling fluid system, e.g., an improved viscosity of nanomodified drilling fluid, a decrease in thickness of mud cake, and a lower fluid loss with an increase in nanocomposite volume in the drilling fluid. A similar study by Cheraghian et al. (2018) investigated a nanocomposite of clay and silica (clay/SiO_2) NPs for water-based drilling fluid under high-temperature reservoir conditions. This proposed novel system (water-based drilling fluid and clay/SiO_2 NPs) also showed improved rheological properties and filtration control even at high temperatures. As an example, with the increase in silica NP concentration from 0.3 to 0.5 wt% in drilling fluid mud, plastic viscosity of the drilling mud increased from 20 to 23 cP while apparent viscosity increased from 28 to 28 cP (Fig. 8.1). Moreover, the use of nanocomposite showed higher thermal stability as compared to conventional water-based drilling fluid systems. Furthermore, a composite of zinc oxide (ZnO) NPs and acrylamide was synthesized by Aftab et al. (2016) which enhanced the properties and performance of water-based drilling fluid. In particular, the results showed the enhancement in rheological properties at elevated temperature conditions. Nanoparticles in drilling fluid can also work as shale inhibitors (note that such inhibitors reduce the interaction of water with clays and thus prevent clay swelling). In this context, ZnO NPs and acrylamide demonstrated reduced filtrate volume which thus leads to shale inhibition in drilling shale zones and prevent shale swelling, especially at elevated temperature.

Generally, the size of NPs varies from 1 to 100 nm which depicts a high volume-to-surface ratio. This allows them to dissipate more heat and exhibit enhanced thermal stability. The surface potential of NPs also helps to maintain physical stability. Thus, the uniform distributions of these NPs show compressive strength to enhance the drilling fluid system which results in stable rheological properties, thermal properties, filter loss control, and shale inhibition. Thus, in summary, this section confirms that the use of NPs in drilling

TABLE 8.1 Summary of NPs used in drilling fluid applications

References	Nanoparticle type	Size (nm)	NP concentration	Drilling fluids type	Study area
Sadeghalvaad and Sabbaghi (2015)	TiO_2	10–15	0.5–10 wt%	WBM-bentonite	Rheology and filtration properties
Sabbaghi et al. (2015)	TiO_2	20	0.1–0.3% volume fraction	WBM-bentonite	Thermal Properties
Cheraghian et al. (2018)	SiO_2	20	0.5 wt%	WBM-bentonite	Rheology and filtration properties
Aftab et al. (2016)	ZnO	100	0.1 ppb	WBM-barite	Rheology and fluid loss
Wang et al. (2018)	Fe_3O_4	10–20	0.05–0.5 wt%	WBM-bentonite	Rheology and thermal properties
William et al. (2014)	CuO and ZnO	50	0.1–0.5 wt%	XG and water	Thermal properties
Sabbaghi and Rezvani (2020)	Graphene	2.71	0.5 wt%	OBM	Rheology and fluid loss
Saboori et al. (2019)	CuO	4	Acrylamide/CuO: 10/1	WBM-bentonite	Rheology and thermal properties

XG—xanthan gum; OBM—oil-based mud; WBM—water-based mud.

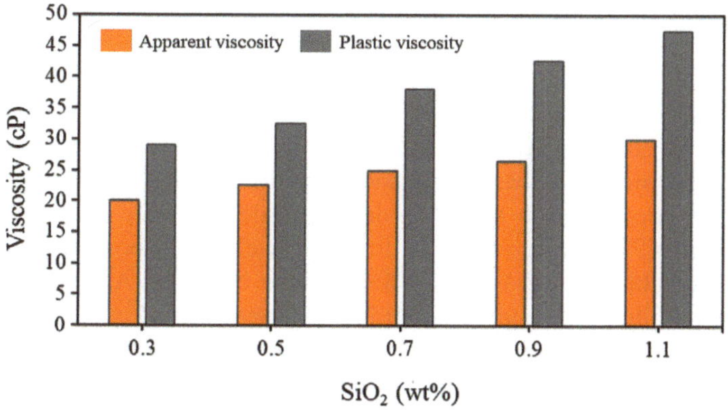

FIG. 8.1 Apparent and plastic viscosity of SiO_2 NPs added drilling fluid and their variation with NPs concentration (wt%). *Adapted from Jain, R., Mahto, V., & Sharma, V.P. (2015). Evaluation of polyacrylamide-grafted-polyethylene glycol/silica nanocomposite as potential additive in water based drilling mud for reactive shale formation. J. Nat. Gas Sci. Eng., 26, 526–537.*

fluid shows improved drilling performance and thus has the potential as an alternative fluid system (i.e., NP as an additive in drilling fluid) over the conventional drilling fluid system.

8.4 Nanoparticle in a filter cake

The filter cake formed by drilling fluid is a thin layer deposited on a permeable surface, i.e., formation wall. Generally, the filter cake coats the borehole wall and restricts drilling fluid to invade the formation, thus preventing formation damage or permeability impairment (Civan, 2016). It is expected that the filter cake should be impermeable and thin for successful drilling operations. However, under the influence of varying subsurface conditions, the thicker and permeable filter cake is formed which poses a significant threat to the wellbore stability during drilling and leads to differential pipe sticking, torque and drag, and filtration loss (DeGeare et al., 2003). Furthermore, deep well drilling offers more challenges due to the fact that their associated issues of high pressure and temperature greatly affect the filter cake condition and have a significant impact on successful drilling operations.

Thus, the identification of novel sustainable opportunities and cost-effective drilling fluids is crucial. As a result, the inclusion of NPs in the drilling fluid is an alternative that can address these challenges and is favorable for successful drilling operations (Mikhienkova et al., 2018). The inclusion of NPs in the drilling fluid does not only provide support to the required filter cake but also provides lubrication to the bit, controls filtration loss, and improves the rheological performance of conventional drilling fluid (Cheraghian, 2021). Several NPs have been investigated and reported in the literature suggesting that the use of NPs in the conventional drilling fluid can decrease the thickness of filter cake and render it impermeable which is requisite for successful drilling operation (Aftab et al., 2020; Al-Saba et al., 2018; Ismail et al., 2016; Perween et al., 2018; Rafieefar et al., 2021; Sadegh Hassani et al., 2016). The NPs-added drilling fluid is a novel technique that is designed for efficient drilling operations at a lower cost. The drilling fluid properties can be enhanced by the addition of drilling fluid at a very low concentration of NP in drilling fluid (Perween et al., 2018). In shale rocks, NPs provide inhibition to shale swelling by providing a thinner and impermeable filter cake at the walls of the borehole. A graphical representation of filter cake and filtration loss is provided in Fig. 8.2. Generally, the filter cake formed depicts pore size in micron range which leads to a high filtrate loss. However, NPs-added drilling fluid demonstrates lower filtrate loss by reducing the pore sizes of the filter cake. Fig. 8.2 also shows a typical laboratory filtration loss experimental setup.

Several NPs have been explored for their application as a drilling fluid additive (Table 8.2 presents a summary of some of the studies.). Recently, Razak Ismail et al. (2014) reported filtration loss using Al_2O_3 and TiO_2 NP additives and found filtration loss was reduced by 22% for 2 wt% of Al_2O_3, and 38% for 2 wt% TiO_2—which thus suggests better performance when TiO_2 is used as an additive. Moreover, Shakib et al. (2016) reported that 0.2 wt% of TiO_2 NPs can lead to a tenfold reduction in filtrate loss—suggesting a significant decrease. Generally, the size of NPs ranges from 1 to 100 nm whereas the size of filter cake pores generally varies in the microrange. The inclusion of these NPs in the drilling fluids fills the micron-sized pores of the thin filter cake as well as helps to reduce filtrate loss (Figs. 8.2 and 8.3).

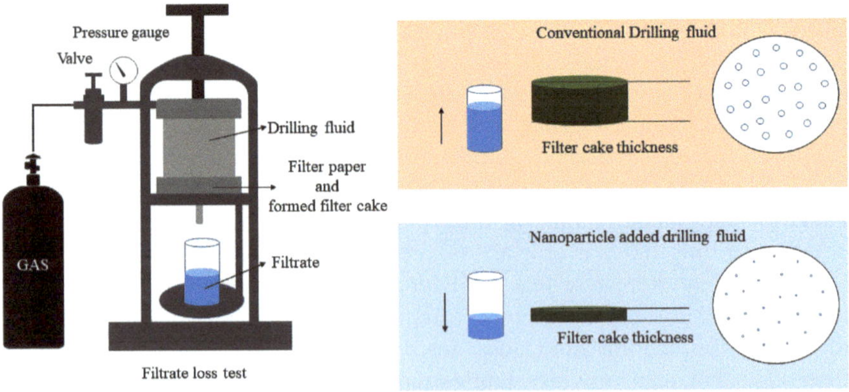

FIG. 8.2 Schematic showing filter test unit used for filtrate loss test and filter cake test; thick filter cake formed by conventional drilling fluid with high filtrate; and thin filter cake created by NPs with reduced filtrate.

TABLE 8.2 Nanoparticles in drilling fluid application

Study	Nanoparticles	NP concentration	Study area	Remark
Al-Saba et al. (2018)	TiO_2 CuO	0.5 vol% 1.5 vol%	Rheological properties, filtration loss in shale	Water adsorption reduction and viscosity improvement
Sadegh Hassani et al. (2016)	MWCNT -SiO2	0.01–2 wt%	Heat transfer, rheological properties	Improved rheological stability and heat transfer
Ismail et al. (2016)	MWCNT SiO_2	0.001–0.2 ppb	Lubricity, rheological properties, filtration characteristics	High lubricity, improved rheological properties, and reduced filtration loss
Perween et al. (2018)	$ZnTiO_3$	0.05–0.3 wt%	Filtration loss, thermal stability, rheological properties	Reduced filtration loss, high thermal stability, improved rheology
Aftab et al. (2020)	TiO_2-bentonite nanocomposite	0.8 g	Filtration loss, lubricity, viscosity	Improved lubricity, reduced shell swelling, improved viscosity

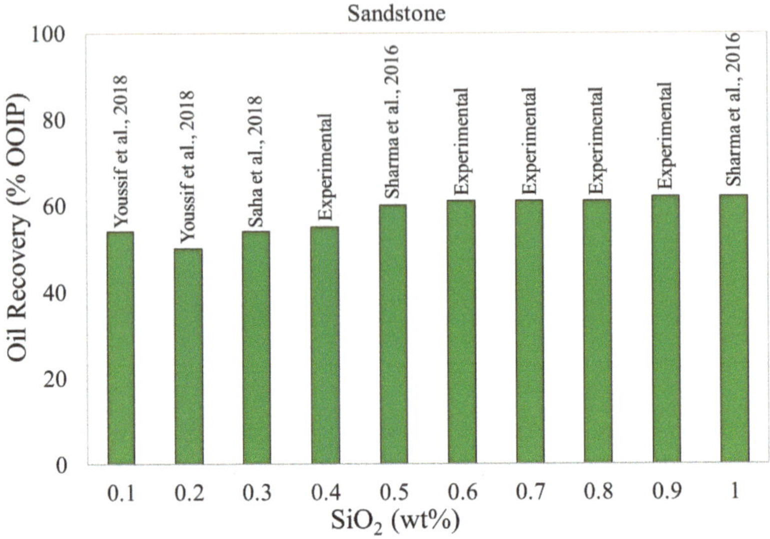

FIG. 8.3 A summary of SiO_2 NPs used for oil recovery in sandstone media (Saha et al., 2018; Sharma et al., 2016; Youssif et al., 2018).

8.5 Nanoparticles for formation damage during drilling

Drilling fluid is one of the key contributors to formation damage. This is because the invaded drilling fluids tend to block pore spaces and lead to permeability impairment—which eventually leads to a decline in hydrocarbon production. Thus, the design of drilling fluid is the major challenge that can reduce the formation damage. Several studies have been reported in the literature concerning the investigation of rheological properties, filtrate loss, mud cake thickness of water-based drilling fluid under the influence of chemical additives such as polymer and surfactant which can reduce the chances of formation damage (Ahmad et al., 2017). However, polymer and surfactant may show limitations at extremely high temperature and pressure conditions and thus not able to prevent formation damage (Mohammadi and Mahani, 2020).

Such limitation of drilling fluid thus necessitates research on materials that can curtail the challenges of formation damage during drilling. Nanoparticles (NPs) are possibly one of the materials which possess high surface area and can significantly improve the properties of drilling fluid. Several studies have been reported in the literature suggesting the use of different NPs can improve the intrinsic properties of drilling fluid; however, silica NPs are extensively explored due to their high performance in drilling operations (Agista et al., 2018; Mohammadi and Mahani, 2020). The use of silica NPs in drilling fluid produces several advantages as follows:

- Improve rheological properties in drilling fluid
- Enhance the cutting transport capability of drilling fluid

- Reduced filtration loss of drilling fluid
- Desired thickness of mud cake
- Improved lubricity

The use of NPs reduces the chances of formation damage via blocking the pores through the formation of stable mud cake which limits the deeper invasion of water-based drilling fluid in the formation (Mohammadi and Mahani, 2020). Moreover, low pH of drilling fluid exhibits less damage, whereas the severity of formation damage increases with high-pH mud. In addition, Mohammadi and Mahani (2020) reported that the required NPs concentration was higher for a high-pH environment (0.5 wt%) compared to low-pH (0.3 wt%) to counter the severity of formation damage. However, some literature data also suggest that the use of silica NPs can reduce formation damage via controlling the fine migration, changing the wettability of rock toward a strong water-wet state, and shale swelling inhibition.

Formation damage due to drilling fluid (with or even without the inclusion of NPs) is fairly investigated in literature via core-scale observations. However, the microscale (using micromodels) and pore-scale observations (using X-ray micro-CT imaging) are relatively limited. Although it is very clear that the use of silica NPs in drilling fluid is beneficial and can reduce the formation damage, yet future studies should focus on pore-scale observations to enhance confidence in NPs application in reducing formation damage.

8.6 Nanoparticle for cementing applications

In the oil field, cementing operations are performed after the completion of the drilling operation to strengthen the drilled well via making a strong bond between steel casing and formations. It also separates the casing and the reservoir fluids to prevent corrosion. Moreover, cementing is one of the challenging processes where different factors affect the cement bindings such as fluid loss, settling time, and mechanical strength. The recent studies suggest that the aforementioned issues can be addressed using NPs. Several NPs are used to improve the cement slurry quality and can be used in well-cementing operations as suggested in the literature. However, the stability of nanofluid is a key challenge in cementing applications. The stability of these NPs in cement slurry is important because the agglomerated NPs may cause loose cementing of oil well and can also propagate in the reservoir formation which led to pore blockage and other issues. Recently, Maagi et al. (2019) conducted a fluid loss test of a cement slurry after the addition of three different NPs (SiO_2, TiO_2, and Al_2O_3) and suggested that the addition of NPs in cement slurry can control the cement flow and their properties, resulting in reducing the filtration loss more significantly. In addition, the use of binary and tertiary NPs in cement slurry is more effective than the sole NPs system. Moreover, the study also suggested that excessive use of NPs in cement slurry reduces the slurry fluidity. Similarly, Baig et al. (2017) examined zeolite NPs for cement slurry and suggested that the nanozeolite can be a potential additive for cement slurry which produces early strength and makes cement bonds potentially active which reduce the chances of corrosion of casing due to its denser structure and high compressive strength.

Now, it is clear that the addition of NPs in cement slurry improves the key parameter such as density, settling time, filtration loss, and viscosity, which is beneficial for the compressive strength of the cementing operations. Also, an extensive laboratory investigation will be a major concern before the field trial by keeping the field conditions as a prime parameter.

8.7 Application of nanoparticles for enhanced oil recovery

The ongoing demand for hydrocarbon triggered the decline of oil reserves globally and turned into a major challenge for the producers concerned with the recovery of remaining (residual) oil from oil reservoirs. Thus, improving oil recovery from the current reservoirs holds the key to meeting the current and future challenges of global energy demands. The primary and secondary stages of oil recovery are only able to produce 30% of oil from the reservoir; the remaining 70% is still unrecoverable for which tertiary recovery is commonly used. The last stage of oil recovery, i.e., tertiary recovery, is also known as enhanced oil recovery (EOR), which is composed of several methodologies such as thermal, miscible, chemical, and microbial EOR methods. In miscible flooding, the injected gas mix with trapped oil improves microscopic displacement efficiency while the overall volumetric efficiency is low (Seyyedsar et al., 2017). In thermal EOR, hot fluid or steam is injected which provides heat to the reservoir which decreases oil viscosity to make the oil more mobile. However, heat may be lost during injection in deep reservoirs and limits its applicability to shallow reservoirs containing viscous oil. Moreover, chemical flooding involves the injection of polymer solution, surfactants, alcohol, and alkali. While polymer flooding provides a favorable mobility ratio ($M < 1$), surfactant flooding is targeted to reduce the interfacial tension and alter the rock wettability from oil wet to water wet (Negin et al., 2017). These chemicals are typically affected by harsh subsurface conditions, e.g., high salinity, and thus demonstrate lower efficiency at such conditions. Therefore, there is a need to implement new technology to overcome the shortcomings of all these methods. The new technology should ensure minimum adsorption of chemicals on the rock surface, a significant reduction in IFT, and ability to withstand high salinity, alter wettability, and improve properties (e.g., viscosity and density) of injected fluids. The emergence of nanotechnology is thus a part of the chemical EOR process to enhance the production from oil reservoirs, and the following sections discuss some published datasets on increase in oil recovery due to nanofluids and the associated mechanisms.

8.8 Oil recovery using nanofluids

Several researchers use a variety of nanoparticles at different concentrations for exploring applications for enhanced oil recovery (Table 8.3). Sharma et al. (2016) examined the potential of SiO_2 NPs in a base fluid of polyacrylamide and found improvement in the rheological profile of the formulated nanofluid and reduction in IFT followed by improved oil recovery in a sandstone. Figs. 8.3 and 8.4 provide a summary of SiO_2 NPs used for sandstone and carbonate media, respectively. It is clear from Fig. 8.3 that the use of silica NPs of more than 0.5 wt% possibly enhances the oil recovery for sandstone media as per the data available in literature which can be able to produce more than 60% OOIP. While the use of silica NP in carbonate rocks is less reported, still small presence of these NPs in carbonate reservoirs can improve the oil recovery and maximum oil recovery can be obtained at 0.6 wt% of silica NPs.

The key mechanisms associated with enhanced oil recovery using nanofluids include wettability alteration generally toward more water wet (Maghzi et al., 2011) and interfacial tension (IFT) reduction (Kumar et al., 2021a,b). Typically, reservoir rocks depict pores of micrometer sizes; thus, nanosized particles of nanofluid can easily access reservoir pores without choking pore throats and reduce the interfacial tension and change the wettability of rock from oil wet to water wet (Al-Anssari et al., 2017a,b,c). The change in wettability can be defined by two mechanisms: (a) by adsorbing on rock surface and (b) by getting adsorbed to the liquid surface. Generally, the first mechanism is found to be more effective. Moreover, the change in rock wettability from oil wet to water wet is also attributed to the NPs wedge shape structure as reported by Wasan et al. (2011).

Kumar et al. (2020a,b) conducted nanofluid flooding experiments using silica–titania nanocomposites formulated in a base fluid of HPAM and recorded significant improvements in rheological properties and oil recovery at high-temperature conditions. It is widely agreed (and shown at laboratory scale) that the inclusion of NPs and polymers as a composite nanofluid is favorable for IFT reduction and wettability alteration toward water wet and can enhance the oil production from the reservoir (though other theories do exist where mixed-wet rocks are reported to yield better oil recovery). In addition, it is recommended to use the combination of NPs, surfactant, and polymers (NSP) as EOR agents, and as a result, ultra-low IFT and strongly water-wet conditions can be achieved which are responsible for higher oil recovery (Pal et al., 2020; Saxena et al., 2019). The use of NPs in the oil recovery process is a novel approach where nanosized particles generally change the reservoir key parameters at the nanoscale level and ultralow IFT and strong water-wet conditions can be achieved. The following subsections describe studies of wettability alteration and IFT reduction using NPs.

TABLE 8.3 Summary of different NPs used for oil recovery applications

Study	Porous media	Base fluid	Surfactant (wt%)	NPs (wt%)	Study area	Cumulative oil recovery
Saha et al. (2018)	Sandstone	Xanthan gum	–	SiO_2 (0.1–0.3)	Wettability, oil recovery	54%
Maghzi et al. (2011)	Glass model	PAM	–	SiO_2 (0.1)	Wettability, oil recovery	55%
Sharma et al. (2016)	Sandstone	PAM	Anionic	SiO_2 (0.5–2.0)	IFT, wettability, oil recovery	60%
Hu et al. (2016)	Sandstone	Brine	Anionic, cationic, nonionic	TiO_2 (5–500) ppm	Wettability, oil recovery	41%
Yousefvand and Jafari (2015)	Sandstone	HPAM	Anionic	TiO_2 (2.3)	Mobility control, oil recovery	52%
Ehtesabi et al. (2014)	Sandstone	DI water	–	TiO_2 (0.01–0.05)	Wettability, IFT, oil recovery	55%
Kumar et al. (2021a,b)	Sandstone	PAM	Anionic, nonionic	SiO_2 (0.5) + TiO_2 (0.05; 0.1)	Wettabilty, IFT, oil recovery, NP retention	>70%

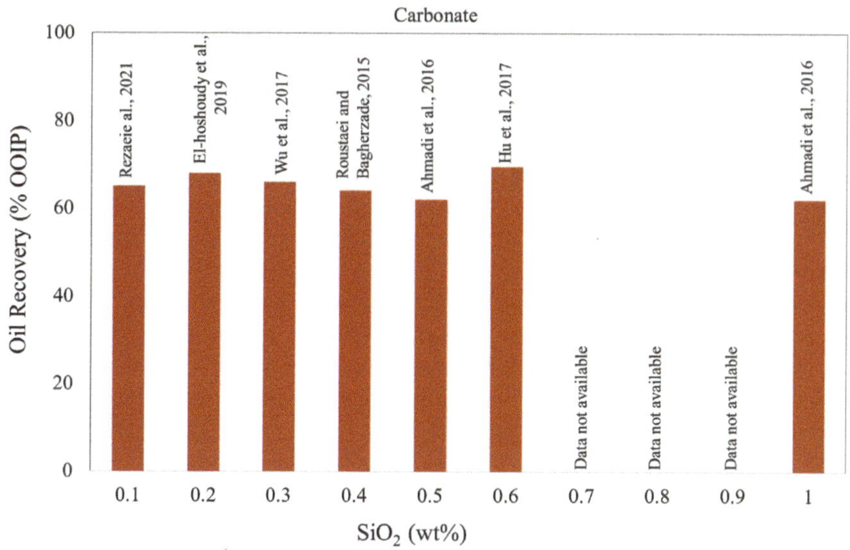

FIG. 8.4 A summary of SiO_2 NPs used for oil recovery in carbonate media (El-Hoshoudy et al., 2019; Hu et al., 2017; Roustaei and Bagherzadeh, 2015; Wu et al., 2017; Rezaei et al., 2021; Ahmadi et al., 2016).

8.9 Wettability alteration using nanoparticles

Nanofluids have been used as chemical enhanced oil recovery (EOR) agents in sandstones and carbonates. The main mechanism behind the incremental oil recovery using nanofluids is found to be wettability alteration (Al-Anssari et al., 2018a,b). Several studies have been conducted to investigate the effect of using nanofluids on wettability alteration in both sandstones and carbonates.

Saha et al. (2018) studied the effect of three different silica nanofluid concentrations on Berea sandstone. The use of 0.1, 0.3, and 0.5 wt% concentrations resulted in contact angle reduction from 86.2° (without nanofluid) to 28° (at 0.1 wt%), 25° (at 0.3 wt%), and 21°, (at 0.5 wt%), respectively. This shows that the increase in nanofluid concentration leads to a lower contact angle which reflects a wettability shift toward a more water-wet state. Al-Anssari et al. (2017a,b,c) studied the effect of silica nanofluids on Iceland spar pure calcite. The results showed the use of 0.2 wt% concentration of silica nanofluids reduced the contact angle from 121° to 70°. Additionally, Aghajanzadeh et al. (2019) investigated the effect of silica nanofluids on wettability alteration using contact angle measurements on both limestone and dolomite rocks. For the case of limestone rocks, the use of two nanofluid concentrations of 0.05 and 0.15 wt% led to the reduction of contact angle from 167° to 117.2° and 74.5°, respectively. For the case of dolomite, the contact angle was reduced to 110° and 68° using the 0.05 and 0.15 wt% concentrations, respectively. Moreover, Ali et al. (2019) used three different concentrations of 0.05, 0.1, and 0.2 wt% of polymer-coated nanocomposite (NC) on carbonate pellets. The contact angle was reduced from 132.6° to 86°, 69.6°, and 57°, respectively. Furthermore, Tohidi et al. (2021) used silica nanofluids of 0.03 wt% concentration on carbonates and observed a significant reduction in contact angle from 180° to 67.5°.

From the literature above, it can be concluded that wettability alteration toward a more water-wet state is achieved at a higher concentration of nanofluids. However, it should be noted that there is an optimal nanofluid concentration after which the reduction in contact angle measurements and wettability alteration will not be significant. The crucial factors that affect the nanofluid optimal concentration are brine salinity, surface charge, chemical composition, and the effect of the nanofluid treatment on surface roughness (Al-Anssari et al., 2015, 2017a,b,c).

8.10 Change in interfacial tension using nanoparticles

Nanofluids generally tend to reduce the IFT at the oil/brine interface. Hendraningrat et al. (2013) measured the IFT of crude oil/brine/nanofluid system using a spinning drop tensiometer at ambient conditions. The IFT of crude oil/brine (3 wt% NaCl) was ~19.2 mN/m which was reduced to nearly half (10 mN/m) after adding 0.01 wt% silica nanofluid. However, with a further increase in concentration to 0.05 wt %, a relatively small decline in IFT was noticed (IFT = 7.9 mN/m). While an increase in nanofluid concentration leads to a reduction in IFT, an optimal nanofluid concentration exists for achieving the desired reduction in IFT beyond which little or no further reduction is found.

Fan and Striolo (2012) performed a mechanistic evaluation of the IFT of the decane/water system upon exposure to nanofluids. The dissipative particle dynamic (DPD) simulations were used to estimate IFTs, and the proposed algorithm (named as micro-Wilhelmy plate (MWP) algorithm) takes into consideration the NP–NP interactions. The IFT reduction was found to be a function of surface area per NP. With a decrease in surface area per NP, a more significant reduction was observed in decane/water IFT. This is because the corresponding NP density at the interface increases with the decrease in surface area per NP. Moreover, NPs with greater desorption energy tend to depict a greater reduction in IFT (Fan and Striolo, 2012)—suggesting better efficiency.

Furthermore, Zargartalebi et al. (2014) measured the IFT of the oil/brine system using fumed silica (AEROSIL) and surfactant composite. At original conditions (i.e., without NP or surfactant), oil/brine IFT measured 16 mN/m which was reduced to ~4 mN/m after adding 2000 ppm sodium dodecyl sulfate (SDS) surfactant—suggesting a notable decline. Subsequently, upon the addition of 1000 ppm of fumed silica, IFT further reduced to ~2.5 mN/m—albeit the reduction was relatively lower compared to surfactant alone.

Recently, Al-Anssari et al. (2018a,b) reported IFT of CO_2/brine/nanofluid systems at high-pressure and high-temperature conditions with implications for CO_2 geosequestration. The observed IFT increased with temperature but decreased with increasing pressure. As an example, CO_2/brine IFT was 40 mN/m for brine of 5 wt% $CaCl_2$ without nanoparticle addition. However, with the addition of 0.1 wt% silica nanoparticles, IFT was reduced to 36 mN/m. Interestingly, the trends of IFT change were found to be a function of the hydrophilicity of nanoparticles. Thus, hydrophobic silica demonstrated a greater decrease in CO_2/brine IFT compared to that hydrophilic silica. This is because hydrophilic silica exhibits a tighter packing of NPs in the bulk water phase compared to hydrophobic silica particles which tend to move closer to the CO_2/water interface and thus result in greater IFT reduction.

Thus, in summary, the general trend is that IFT of oil/brine or CO_2/brine systems tends to decrease with the addition of NPs. However, the rate of decrease and associated methods of measurement are often contradictory and need further investigation. Moreover, the composition of crude oil and brine in question can also influence the reported behavior.

8.11 Factors affecting nanofluid stability

The synthesis of stable nanofluids is one of the key parameters and prerequisites for long-term usage in the oil field. Therefore, a stable nanofluid is a prerequisite for nanofluid application in enhanced oil recovery and can be characterized by the uniform dispersion of nanoparticles in base fluid (Abdullah et al., 2018) such as water, oil, glycol, and polymeric solutions. Several NPs have been explored for oil recovery applications such as SiO_2, TiO_2, CuO, Al_2O_3, ZnO, Fe_3O_4, and graphene. However, silica NPs (SiO_2) are extensively explored due to their abundance in the subsurface. In addition, the nanofluid prepared with these NPs may tend to agglomerate and show sedimentation with the course of time and surrounding conditions. Thus, understanding stability issues is a key parameter and can be categorized as follows:

a). Kinetic stability: Particles remain in suspension due to Brownian motion
b). Dispersion stability: Particles remain in suspension without agglomeration
c). Thermodynamic stability: Electrostatic repulsion and steric repulsion
d). Chemical stability: Higher viscosity of base fluid which may hold the NPs in suspension

8.12 Agglomeration caused by storage time

Agglomeration of dispersed particles may be observed with time and the surrounding conditions (Kumar and Sharma, 2018). Nanofluids may lose their thermophysical property due to agglomeration and cluster formation which is an important aspect of their industrial applications. Generally, the NPs remain in suspension due to the inter-particle repulsive forces acting on the adjacent particles. These repulsive forces reduce with time, and net forces acting on the suspended NPs show attraction with the neighboring particle and form clusters due to the agglomeration. As a result, agglomerated NPs become bulky and show sedimentation within the nanofluid with time which may block the pore throat and reduce the oil recovery. However, challenges to the physical stability caused by agglomeration can be addressed by introducing different additives to the sole nanofluids. Kumar and Sharma (2018) prepared silica nanofluid of 0.5 wt% in the base fluid of polymers and found that the stability can be improved by introducing another similar charged NPs (TiO_2) in the silica nanofluid.

8.13 Agglomeration caused by salinity

The nanofluid stability can be greatly affected by the surrounding conditions where it is exposed to harsh conditions such as highly saline reservoirs (23,000–100,000 ppm or even greater) (Sheng, 2011). Generally, salt contains monovalent and divalent ions which tend to screen out the surface potential of the NPs. The screening out of surface charges is responsible for the weakening of repulsive forces acting on NPs in a nanofluid. Thus, the NPs show heavy sedimentation due to the agglomeration induced by the saline media. However, the instability caused by the saline media can be also addressed by improving the surface charge of NPs by the incorporation of surfactant in the nanofluid. Recently, silica nanofluid was stabilized by incorporating an anionic surfactant, and results showed improvement in surface potential which helped NPs to remain in suspension for a longer duration (Al-Anssari et al., 2017a,b,c). Kumar et al. (2020a,b) also investigated the stability of silica nanofluid prepared in the base solution of polyacrylamide by the addition of anionic surfactant, and significant improvement in stability and rheological properties in saline media was observed. Furthermore, oil recovery was also found to increase. The addition of surfactant generally forms micelles in the nanofluid, and these micelles enhance the surface charge of NPs via charge interaction and curtail the effect of salinity.

The mechanism which induces stability improvement can be summarized as follows:

- Smaller-sized NPs of higher surface potential are recommended for the preparation of nanofluid which increases the Brownian motion and hence decreases agglomeration
- The higher viscosity of the base fluid may reduce the chances of agglomeration and sedimentation; thus, it can improve stability
- Binary nanofluid systems may increase the stability by increasing the overall repulsive forces in the nanofluid
- Stability can be also improved by increasing the surface charge of the NPs in the nanofluid by the inclusion of surfactants. Generally, surfactants form micelles, and micelles get laid on NPs and enhance the surface potential or zeta potential (Al-Anssari et al., 2017a,b,c)

8.14 Agglomeration caused by high temperature

The thermal stability of a nanofluid is also one of the important challenges triggered due to the agglomeration of NPs. This is particularly true in high-temperature conditions when the effective kinetic energy increases and the random motion of NPs increase too. The increase in random motion of the NPs increases the collision rate between the adjacent NPs and reduces the nanofluid stability (Liu et al., 2013). Moreover, the change in viscosity of the base fluid with an increase in temperature also induces additional detrimental effects on the nanofluid agglomeration. Kumar et al. (2020a,b) investigated the effect of temperature on silica nanofluid prepared in HPAM. The increase in temperature exhibited chain scission to the polymer chain and deteriorated the base fluid viscosity, and further, the uniform distribution of NPs disturbed which produce agglomeration of the NPs. However, the agglomeration of NPs in a nanofluid can be curtailed by incorporating other NPs reported in the literature (Kumar and Sharma, 2018; Kumar et al., 2020a,b; Nasrollahzadeh et al., 2019).

8.15 Applications of nanofluids in fracturing and stimulation

There is a growing interest in using nanofluids in fracturing and stimulation applications during the last decade or so. The advantage of using nanofluids for such applications is that they can be removed easily from the fracturing fluid during the flow-back process through filtration and centrifugal separation which results in an effective cleanup of the fracturing fluid (Mao et al., 2022). In this section, the applications of nanofluids as additives in polymeric fracturing fluids, viscoelastic surfactant (VES) fracturing fluids, and foam-based fracturing fluids are briefly discussed.

8.16 Additive in polymeric fracturing fluids

The use of nanoparticles as additives for enhancing polymeric fracturing fluids has been evaluated by a few studies. Typically, favorable results are observed when titanium and zirconium nanoparticles are utilized as additives. Hurnaus and Plank (2015a) found that the titanium oxide nanoparticles are responsible for the cross-linking of hydroxypropyl guar (HPG) in fracturing fluids. The potential mechanism of the titanium nanoparticles' interaction and cross-linking with the HPG is the formation of hydrogen bonds between the OH groups present in the titanium nanoparticles' surface and the galactomannan chain of HPG. The mechanism of zirconium cross-linking on HPG has also been studied (Hurnaus and Plank, 2015b). Note that cross-linking did not occur when the zirconium nanoparticles were not present which indicates that the presence of the nanoparticles is responsible for the cross-linking effect (Hurnaus and Plank, 2015b).

The use of nanomaterials to enhance the rheological properties of the borate-cross-linked guar gels has also been studied (Alharbi et al., 2017). Nanomaterials induce the interactions within the guar polymer chains, which leads to an increase in fracturing fluid viscosity, reduces the borate–guar bonding, and minimizes fluid leak-off at downhole conditions. The cross-linking resulting from multiple interactions between the cis-hydroxyl groups on the guar polymer and the nanoparticles is proposed as the main mechanism of the improved fluid performance in presence of nanoparticles.

It should be noted that the cross-linking activities of nanoparticles can be influenced by other parameters such as pH and nanoparticle type and size (Hurnaus and Plank, 2015a). Also, despite the promising results, the cross-linking mechanisms of polymer-based gels with nano-cross-linker are still uncertain. Therefore, future studies on this area should be conducted since different forces could affect the performance of nanoparticles in the polymeric suspension (Chauhan et al., 2017).

8.17 Additive in viscoelastic surfactant fracturing fluids

Nanoparticles are also used as additives to improve the rheological properties and performance of viscoelastic surfactant (VES) fracturing fluids (Huang and Crews, 2008). Nanoparticles can adjust the fluid properties through the modifications of the nanoparticles' surfaces. That is because nanoparticles have high surface force such as the van der Waals which can result in pseudo-cross-linking of nanoparticles and surfactant micelles (Huang et al., 2010). Chauhan et al. (2017) investigated the effect of silica nanoparticles on the rheological properties of VES-based gels formed from SDS-ALS mixed system. Their results indicated that the nanofluid-induced VES fluids demonstrated high viscosity at higher temperature and shear rate and were thermally stable. Additionally, Hanafy et al. (2018) reported that a 50% enhancement in the apparent viscosity of anionic surfactant VES fluid is observed when silica nanoparticles were used. They also added that the performance of nanofluid-enhanced VES fracturing fluids is influenced by the VES concentration, operating temperature, and nanoparticles shape.

8.18 Additive in foam-based fracturing fluids

Nanofluids have also been used as additives in foam-based fracturing fluids. The rheological properties and foam stability can be enhanced when a network of particle aggregate is formed (Alyousef et al., 2017; Yekeen et al., 2018). Adsorption and aggregation of nanoparticles at the gas–liquid interface of foam are considered the main mechanisms of foam stability improvement in the presence of the nanoparticles (Fameau and Salonen, 2014; Kumar and Mandal, 2017; Yekeen et al., 2018).

Nanofluid-enhanced foam-based fracturing fluids can be polymer-based, worm-like micelles (WLMs), or polyelectrolyte complex nanoparticles (PECNPs). For nanoparticles-enhanced polymer-based foam, Emrani et al. (2017) investigated the effect of silica nanoparticles on the rheological properties and the mobility control performance of AOS surfactant-stabilized foam fracturing fluids and guar gum at high temperature and salinity. The results suggested that the surfactant AOS improves the MRF by 300% as compared to the NaCl solution. Importantly, the addition of SiO_2 NPs and guar gum with CO_2 foams provides better foam stability as well as mobility reduction factor (MRF). In addition, it was observed that the stability remained constant at a high temperature of 250 °F while the MRF improved to ~28 (dimensionless) as compared to that of 77 °F. The observed change in mobility of foams is observed due to a change in viscosity of the specimen foam which generally improves with an addition of polymer and NPs in CO_2 foam (Alyousef et al., 2017; Ibrahim and Nasr-El-Din, 2018; Kumar et al., 2021a,b; Yekeen et al., 2018).

8.19 Challenges and outlook

This study presented a brief overview to emphasize the use of NPs in several oil field applications. It is clear that nanofluids have demonstrated notable potential for several key subsurface applications, e.g., as an additive in drilling fluids, EOR, and stimulation fluids as evidenced by numerous laboratory investigations. NPs have the potential to enhance the property of injected fluid such as viscosity and density, can also alter reservoir rock and fluid properties such as rock wettability, reduction in IFT, reduction in oil viscosity, reduction of mobility ratio, and formation damage, and thus can improve the sweep efficiency of injected fluid. While the current data indicate the effectiveness of nanofluids in oil field applications and consensus on the mechanistic understanding of the mechanisms pertinent to NPs superiority, still several issues remain unaddressed, which are identified as follows:

- Systematic investigations are required to determine the stability of NPs combined with a range of EOR fluids such as surfactants, polymers, and low-salinity brine and its performance evaluation for oil recovery
- Also, the effect of two or more combinations of NPs and their retention study may extend their usages in oil field, which are very few in literature and still in progress as for real oil field application
- While the NP-assisted EOR has been largely investigated at core scale, the micro- to pore-scale evaluations have received relatively less attention—which future studies should address
- Wettability alteration and IFT dataset for a broad range of brine and crude oil compositions are still limited, and thus, further laboratory investigations are required

In summary, despite the notable benefits of NPs to enhance hydrocarbon production and contribute to the global energy needs, several challenges need to be evaluated before the use of nanoparticles in real oil field applications to reduce the uncertainties and scientific knowledge gaps for field-scale implementation. Stable dispersion of nanofluid which requires no agglomeration in the fluid system is one of the major challenges. The type of nanofluid and the associated cost under pertinent reservoir conditions are another challenge. Typically, the studies on NPs are predominantly based on laboratory investigation while the field trials are limited. Kanj et al. (2011) reported the first field trial for reservoir nanoagents (carbon-based fluorescent nanoparticles) called Arab-A-Dots in the Ghawar oilfield, Saudi Arabia. The method used in this field was Huff and Puff mode, and high cumulative recovery of 86% was recorded. Tables 8.4 and 8.5 present a succinct summary of the recent advances and the applicability of NPs in the hydrocarbon industry which includes a series of laboratory and field testing—suggesting the extensive outlook of nanoparticles application in the upstream industry.

TABLE 8.4 Summary of NPs used in different oil field applications such as drilling, cementing, well stimulation, and enhanced oil recovery

Laboratory test				
Study	*NPs*	*Study parameter*	*Applications*	*Remark*
Wang et al. (2019)	SiO_2	Fluid loss and gas hydrate formations	Drilling	Reduced fluid loss as compared to pure water
McDonald (2012)		Shale stability and drilling cost/time control		Reduced filter cake thickness
Hoelscher et al. (2012)		Fluid loss and rheology		Reduced fluid loss and improved rheology in shale formations
Perween et al. (2018)	$ZnTiO_3$	Fluid loss, rheology, and thermal stability		Reduced filtrate loss volume, improved rheological and thermal properties of the mud
Bég et al. (2018)	SiO_2+TiO_2	Lubricity, rheology, and mechanical properties		Improved stability, lubricity, and mechanical properties of the drilling fluid
Ponmani et al. (2014)	CuO & ZnO	Stability and efficiency		Improved thermal stability and efficiency of nanodrilling fluid compared to microdrilling fluid
Jalal et al. (2012)	SiO_2	Mechanical strength of cement	Cementing	Improved mechanical strength of the cement using NPs
Pang et al. (2014)		Cement settling time and mechanical strength		Accelerated settling time of cement slurry and improved mechanical strength of cement using NPs
Vipulanandan et al. (2018)	Al_2O_3	Electrical and strength		Improved electrical resistivity and comprehensive strength of cement using NPs
Vipulanandan et al. (2015)	Fe_2O_3	Sensing and strength		Improved sensing properties and comprehensive strength of cement using NPs
Jafariesfad et al. (2016)	MgO	Settling time and chemical shrinkage		Accelerated settling time and reduced chemical shrinkage of cement slurry

Continued

TABLE 8.4 Summary of NPs used in different oil field applications such as drilling, cementing, well stimulation, and enhanced oil recovery—Cont'd

Laboratory test				
Study	*NPs*	*Study parameter*	*Applications*	*Remark*
Gurluk et al. (2013)	MgO & ZnO	Rheological properties	Well stimulation	Improved rheological properties of fracturing fluid
Singh et al. (2018)	SiO_2	Permeability improvement and acid encapsulation		Improved permeability fractures and formation of microencapsulated acid
Fakoya and Shah (2018)		Filtration characteristic and rheology		Improved filtration characteristic and rheological properties for hydraulic fracturing applications
Li et al. (2019)		Adsorption and permeability impairment		Reduced adsorption capacity and permeability impairment
Saha et al. (2018)	SiO_2	Wettability, IFT, rheology, and oil recovery	EOR	Wettability alteration to water wet from oil wet, reduced IFT between oil and EOR fluid, improved rheology and oil recovery
Kumar et al. (2020a,b)		Effect of saline environment on rheology and oil recovery		Improved rheological properties and oil recovery in saline environment using surfactant
Al-Anssari et al. (2018a,b)		Stability of nanofluid and wettability of calcite		Improved stability and wettability alteration to water wet using NPs and surfactant at reservoir conditions
Al-Anssari et al. (2017a,b,c)		Wettability of calcite/CO_2/brine system		Wettability alteration of calcite/CO_2/brine system using NPs and suggestions for enhanced CO_2 storage potential
Ibrahim and Nasr-El-Din (2018)		Stability of foams and oil recovery		Improved foam stability, sweep efficiency, and oil recovery
Kumar et al. (2021a)	SiO_2+TiO_2	Stability, rheology, EOR, and retention		Improved stability, rheology, and oil recovery, and reduced NP retention in porous media
Kumar et al. (2021b)		Foam stability, foam rheology, thermal stability		Improved stability, rheology, and thermal stability recommended for EOR
Hogeweg et al. (2018)	Al_2O_3 & TiO_2	Oil viscosity and oil recovery		Reduced oil viscosity and improved oil recovery
Ehtesabi et al. (2015)	TiO_2	Transport and retention of NPs, sweep efficiency, and oil recovery		Transport of NPs in porous media and NPs retention study, increment of sweep efficiency, and oil recovery

TABLE 8.5 Summary of NPs used in different oil field applications as a field trail for enhanced oil recovery

Field trial				
Study	*NPs*	*Study parameter*	*Applications*	*Remark*
Franco et al. (2017)	Al_2O_3 and SiO_2	Formation damage remediation and inhibition in Columbia	Well performance/oil and gas recovery	(a) The oil production rate was increased by 300 bbl/day; injection time was of 8 months (b) Oil production rate was increased by 134 bbl/day; gas production rate was increased by 1MMSCF/day
Kanj et al. (2011)	Carbon-based fluorescent nanoparticle (Arab-A-Dots)	Stability analysis in harsh reservoir conditions and oil recovery at Ghawar field, Saudi Arabia	Oil recovery	(a) High stability was suggested for the NPs and also recommended for oil recovery (b) High recovery of 86% was recorded with this field trial
Zabala et al. (2016)	–	Mobility control of crude oil in Columbia	Oil recovery	Increment in oil recovery was observed
Barroso et al. (2018)	–	Shale stability using WBM where NPs are added as stabilizing agent in Brazil	Drilling	(a) Shale swelling inhibition and wellbore stability were greatly improved (b) Same drilling fluid was reused which reduced the 15% operational cost

Acknowledgement

The authors would like to acknowledge Khalifa University for providing support under the project FSU-2020-18.

References

Abdullah, M., Malik, S.R., Iqbal, M.H., Sajid, M.M., Shad, N.A., Hussain, S.Z., Razzaq, W., Javed, Y., 2018. Sedimentation and stabilization of nano-fluids with dispersant. Colloids Surf. A Physicochem. Eng. Asp. 554, 86–92.

Afolabi, R.O., Paseda, P., Hunjenukon, S., Oyeniyi, E.A., Zhang, Z., 2018. Model prediction of the impact of zinc oxide nanoparticles on the fluid loss of water-based drilling mud. Cogent. Engineering 5 (1). Article 1514575.

Aftab, A., Ismail, A.R., Khokhar, S., Ibupoto, Z.H., 2016. Novel zinc oxide nanoparticles deposited acrylamide composite used for enhancing the performance of water-based drilling fluids at elevated temperature conditions. J. Petrol. Sci. Eng. 146, 1142–1157.

Aftab, A., Ali, M., Arif, M., Panhwar, S., Saady, N.M.C., Al-Khdheeawi, E.A., Mahmoud, O., Ismail, A.R., Keshavarz, A., Iglauer, S., 2020. Influence of tailor-made TiO2/API bentonite nanocomposite on drilling mud performance: towards enhanced drilling operations. Appl. Clay Sci. 199, 105862.

Aghajanzadeh, M.R., Ahmadi, P., Sharifi, M., Riazi, M., 2019. Wettability modification of oil-wet carbonate reservoirs using silica-based nanofluid: an experimental approach. J. Petrol. Sci. Eng. 178, 700–710.

Agista, M., Guo, K., Yu, Z., 2018. A state-of-the-art review of nanoparticles application in petroleum with a focus on enhanced oil recovery. Appl. Sci. 8 (6), 871.

Ahmad, H.M., Kamal, M.S., Murtaza, M., Al-Harthi, M.A., 2017. Improving the drilling fluid properties using nanoparticles and water-soluble polymers. In: SPE-188140-MS: SPE Kingdom of Saudi Arabia Annual Technical Symposium and Exhibition, 24 April 2017, Dammam, Saudi Arabia.

Ahmadi, M.A., Ahmad, Z., Phung, L.T.K., Kashiwao, T., Bahadori, A., 2016. Evaluation of the ability of the hydrophobic nanoparticles of SiO2 in the EOR process through carbonate rock samples. Pet. Sci. Technol. 34 (11−12), 1048–1054.

Al-Anssari, S., Barifcani, A., Wang, S., Maxim, L., Iglauer, S., 2015. Wettability alteration of oil-wet carbonate by silica nanofluid. J. Colloid Interface Sci. 461, 435–442.

Al-Anssari, S., Arif, M., Wang, S., Barifcani, A., Iglauer, S., 2017a. Stabilising nanofluids in saline environments. J. Colloid Interface Sci. 508, 222–229.

Al-Anssari, S., Nwidee, L.N., Arif, M., Wang, S., Barifani, A., Lebedev, M., Iglauer, S., 2017b. Wettability Alteration of Carbonate Rocks Via Nanoparticle-Anionic Surfactant Flooding at Reservoirs Conditions. In: *SPE-189203-MS*: SPE Symposium: Production Enhancement and Cost Optimisation, 07–08 November 2015, Kuala Lumpur, Malaysia.

Al-Anssari, S., Arif, M., Wang, S., Barifcani, A., Lebedev, M., Iglauer, S., 2017c. Wettability of nano-treated calcite/CO2/brine systems: implication for enhanced CO2 storage potential. Int. J. Greenh. Gas Control. 66, 97–105.

Al-Anssari, S., Barifcani, A., Keshavarz, A., Iglauer, S., 2018a. Impact of nanoparticles on the CO2-brine interfacial tension at high pressure and temperature. J. Colloid Interface Sci. 532, 136–142.

Al-Anssari, S., Arif, M., Wang, S., Barifani, A., Lebedev, M., Iglauer, S., 2018b. Wettability of nanofluid-modified oil-wet calcite at reservoir conditions. Fuel 211, 405–414.

Al-Saba, M.T., Al Fadhli, A., Marafi, A., Hussain, A., Bander, F., Al Dushaishi, M.F., 2018. Application of nanoparticles in improving rheological properties of water based drilling fluids. In: SPE-192239-MS: SPE Kingdom of Saudi Arabia Annual Technical Symposium and Exhibition, 23 April 2018, Dammam, Saudi Arabia.

Alcheikh, I.M., Ghosh, B., 2017. A comprehensive review on the advancement of non-damaging drilling fluids. Int. J. Petrochem. Res. 1 (1), 61–72.

Alharbi, A., Liang, F., Al-Muntasheri, G.A., Li, L., 2017. In: Nanomaterials-Enhanced High Pressure Tolerance of Borate-Crosslinked Guar Gels [Paper Presentation]. SPE-188817-MS, Abu Dhabi International Petroleum Exhibition & Conference (UAE), Abu Dhabi. Society of Petroleum Engineers.

Ali, J.A., Kolo, K., Manshad, A.K., Stephen, K.D., 2019. Potential application of low-salinity polymeric-nanofluid in carbonate oil reservoirs: IFT reduction, wettability alteration, rheology, and emulsification characteristics. J. Mol. Liq. 284, 735–747.

Altavilla, C., Ciliberto, E., 2017. Inorganic Nanoparticles: Synthesis, Applications, and Perspectives. CRC Press.

Alyousef, Z., Almobarky, M., Schechter, D., 2017. Enhancing the stability of foam by the use of nanoparticles. Energy Fuel 31, 10620–10627.

Arif, M., Abu-Khamsin, S.A., Zhang, Y., Iglauer, S., 2020. Experimental investigation of carbonate wettability as a function of mineralogical and thermo-physical conditions. Fuel 264, 116846.

Baig, M.T., Rahman, M.K., Al-Majed, A., 2017. Application of Nanotechnology in Oil Well Cementing. In: SPE-187543-MS: SPE Kuwait Oil and Gas Show and Conference, 15 October 2017, Kuwait city, Kuwait.

Barroso, A.L., Marcelino, C.P., Leal, A.B., Odum, D.M., Lucena, C., Masculo, M., Castro, F., 2018. New Generation Nano Technology Drilling Fluids Application Associated to Geomechanic Best Practices: Field Trial Record in Bahia - Brazil. In: SPE-28731-MS: Annual Offshore Technology Conference, 30 April 2018, Houstan, Texas, USA.

Bég, O.A., Espinoza, D.E.S., Kadir, A., Shamshuddin, M., Sohail, A., 2018. Experimental study of improved rheology and lubricity of drilling fluids enhanced with nano-particles. Appl. Nanosci. 8 (5), 1069–1090.

Chauhan, G., Ojha, K., Baruah, A., 2017. Effects of nanoparticles and surfactant charge groups on the properties of VES gel. Braz. J. Chem. Eng. 34, 241–251.

Cheraghian, G., 2021. Nanoparticles in drilling fluid: a review of the state-of-the-art. J. Mater. Res. Technol. 13, 737–753.

Cheraghian, G., Wu, Q., Mostofi, M., Li, M.C., Afrand, M., Sangwai, J.S., 2018. Effect of a novel clay/silica nanocomposite on water-based drilling fluids: improvements in rheological and filtration properties. Colloids Surf. A Physicochem. Eng. Asp. 555, 339–350.

Civan, F., 2016. Cake filtration: mechanism, parameters and modeling. In: Reservoir Formation Damage, pp. 295–341.

Corredor-Rojas, L.M., Hemmati-Sarapardeh, A., Husein, M.M., Dong, M., Maini, B.B., 2018. Rheological behavior of surface modified silica nanoparticles dispersed in partially hydrolyzed polyacrylamide and xanthan gum solutions: experimental measurements, mechanistic understanding, and model development. Energy Fuel 32 (10), 10628–10638.

Criss, R.E., 2020. Thermal models of the continental lithosphere. In: Heat Transport and Energetics of the Earth and Rocky Planets, pp. 151–174.

DeGeare, J., Haughton, D., McGurk, M., 2003. Pipe sticking. The Guide to Oilwell Fishing Operations. Gulf Professional Publishing.

Edalatfar, M., Yazdani, F., Salehi, M.B., 2021. Synthesis and identification of ZnTiO3 nanoparticles as a rheology modifier additive in water-based drilling mud. J. Petrol. Sci. Eng. 201, 108415.

Ehtesabi, H., Ahadian, M.M., Taghikhani, V., Ghazanfari, M.H., 2014. Enhanced heavy oil recovery in sandstone cores using TiO2 nanofluids. Energy Fuel 28 (1), 423–430.

Ehtesabi, H., Mahdi Ahadian, M., Taghikhani, V., 2015. Enhanced heavy oil recovery using TiO2 nanoparticles: investigation of deposition during transport in core plug. Energy Fuel 29 (1), 1–8.

El-Hoshoudy, A.N., Gomaa, S., Hassan, A., Attia, A.M., 2019. Effects of alkaline/polymer/nanofluids on oil recovery at harsh reservoir conditions. Pet. Coal 61 (6), 1552–1567.

Emrani, A.S., Ibrahim, A.F., Nasr-El-Din, H.A., 2017. Mobility control using nanoparticle stabilized CO2 foam as a hydraulic fracturing fluid. In: SPE-185863-MS: SPE Europe Featured at 79th EAGE Conference and Exhibition, 12–15 June, 2017, Paris, France.

Fakoya, M.F., Shah, S.N., 2018. Effect of silica nanoparticles on the rheological properties and filtration performance of surfactant-based and polymeric fracturing fluids and their blends. SPE Drill. Complet. 33 (2), 100–114.

Fameau, A.-L., Salonen, A., 2014. Effect of particles and aggregated structures on the foam stability and aging. C. R. Phys. 15, 748–760.

Fan, H., Striolo, A., 2012. Nanoparticle effects on the water-oil interfacial tension. Phys. Rev. E 86 (5), 051610.

Fink, J.K., 2003. Drilling muds. Oil Field Chemicals. Gulf Professional Publishing.

Franco, C.A., Zabala, R., Cortés, F.B., 2017. Nanotechnology applied to the enhancement of oil and gas productivity and recovery of Colombian fields. J. Petrol. Sci. Eng. 157, 39–55.

Fyfe, W.S., Price, N.J., Thompson, A.B., 1978. Fluids in the earth's Crust: Their Significance in Metamorphic, Tectonic, and Chemical Transport Processes. Elsevier.

Gurluk, M.R., Nasr-El-Din, H.A., Crews, J.B., 2013. Enhancing the performance of viscoelastic surfactant fluids using nanoparticles. In: SPE-164900-MS: 75th European Association of Geoscientists and Engineers Conference and Exhibition, 10 June 2013, London, United Kingdom.

Hanafy, A., Najem, F., Nasr-El-Din, H.A., 2018. Impact of nanoparticles shape on the VES performance for high temperature applications. In: SPE-190099-MS, SPE Western Regional Meeting, 22–26 April, 2018, Garden Grove, California, USA.

Hendraningrat, L., Li, S., Torsæter, O., 2013. A coreflood investigation of nanofluid enhanced oil recovery. J. Petrol. Sci. Eng. 111, 128–138.

Hoelscher, K.P., De Stefano, G., Riley, M., Young, S., 2012. Application of nanotechnology in drilling fluids. In: SPE-157031-MS: SPE International Oilfield Nanotechnology Conference, 12 June 2012, Noordwijk, The Netherlands.

Hogeweg, A.S., Hincapie, R.E., Foedisch, H., Ganzer, L., 2018. Evaluation of Aluminium Oxide and Titanium Dioxide Nanoparticles for EOR Applications. In: SPE-190872-MS: SPE Europec Featured at 80th EAGE Conference and Exhibition 11 June 2018, Copenhagen, Denmark.

Hu, Z., Azmi, S.M., Raza, G., Glover, P.W.J., Wen, D., 2016. Nanoparticle-assisted water-flooding in Berea sandstones. Energy Fuel 30 (4), 2791–2804.

Hu, Z., Haruna, M., Gao, H., Nourafkan, E., Wen, D., 2017. Rheological properties of partially hydrolyzed polyacrylamide seeded by nanoparticles. Ind. Eng. Chem. Res. 56 (12), 3456–3463.

Huang, T., Crews, J.B., 2008. Nanotechnology applications in viscoelastic surfactant stimulation fluids. SPE Prod. Oper. 23 (04), 512–517.

Huang, T., Crews, J.B., Agrawal, G., 2010. Nanoparticle pseudocrosslinked micellar fluids: optimal solution for fluid-loss control with internal breaking. In: SPE 128067, SPE International Symposium and Exhibition on Formation Damage Control, 10–12 February, 2010, Lafayette, Louisiana, USA.

Hurnaus, T., Plank, J., 2015a. Behavior of titania nanoparticles in cross-linking hydroxypropyl guar used in hydraulic fracturing fluids for oil recovery. Energy Fuel 29, 3601–3608.

Hurnaus, T., Plank, J., 2015b. Crosslinking of guar and HPG based fracturing fluids using ZrO2 nanoparticles. In: SPE-173778-MS, SPE International Symposium on Oilfield Chemistry, the Woodlands, Texas, 13–15 April. 2015.

Ibrahim, A.F., Nasr-El-Din, H., 2018. An Experimental Study for the Using of Nanoparticle/VES Stabilized CO2 Foam to Improve the Sweep Efficiency in EOR Applications. In: SPE-191534-MS: SPE Annual Technical Conference and Exhibition, 24 September 2018, Dallas, Texas, USA.

Ismail, A.R., Aftab, A., Ibupoto, Z.H., Zolkifile, N., 2016. The novel approach for the enhancement of rheological properties of water-based drilling fluids by using multi-walled carbon nanotube, nanosilica and glass beads. J. Petrol. Sci. Eng. 139, 264–275.

Jafariesfad, N., Gong, Y., Geiker, M.R., Skalle, P., 2016. Nano-Sized MgO with Engineered Expansive Property for Oil Well Cement Systems. In: SPE-180038-MS: SPE Bergen One Day Seminar, 20 April 2016, Grieghallen, Bergen, Norway.

Jalal, M., Mansouri, E., Sharifipour, M., Pouladkhan, A.R., 2012. Mechanical, rheological, durability and microstructural properties of high performance self-compacting concrete containing SiO2 micro and nanoparticles. Mater. Des. 34, 389–400.

Jiang, C.-C., Cao, Y.-K., Xiao, G.-Y., Zhu, R.-F., Lu, Y.-P., 2017. A review on the application of inorganic nanoparticles in chemical surface coatings on metallic substrates. RSC Adv. 7 (13), 7531–7539.

Kanj, M.Y., Rashid, M.H., Giannelis, E.P., 2011. Industry first field trial of reservoir Nanoagents. In: SPE-142592-MS: Middle East Oil and Gas Show and Conference, 25 September 2011.

Kumar, S., Mandal, A., 2017. Investigation on stabilization of CO2 foam by ionic and nonionic surfactants in presence of different additives for application in enhanced oil recovery. Appl. Surf. Sci. 420, 9–20.

Kumar, R.S., Sharma, T., 2018. Stability and rheological properties of nanofluids stabilized by SiO2 nanoparticles and SiO2-TiO2 nanocomposites for oilfield applications. Colloids Surf. A Physicochem. Eng. Asp. 539, 171–183.

Kumar, R.S., Chaturvedi, K.R., Iglauer, S., Trivedi, J., Sharma, T., 2020a. Impact of anionic surfactant on stability, viscoelastic moduli, and oil recovery of silica nanofluid in saline environment. J. Petrol. Sci. Eng. 195, 107634.

Kumar, R.S., Narukulla, R., Sharma, T., 2020b. Comparative effectiveness of thermal stability and rheological properties of nanofluid of SiO2-TiO2 nanocomposites for oil field applications. Ind. Eng. Chem. Res. 59 (35), 15768–15783.

Kumar, R.S., Al-Arbi Ganat, T., Sharma, T., 2021a. Performance evaluation of silica nanofluid for sand column transport with simultaneous wettability alteration: an approach to environmental issue. J. Clean. Prod. 303, 127047.

Kumar, R.S., Sinha, A.S.K., Sharma, H., Sharma, T., 2021b. High performance carbon dioxide foams of nanocomposites of binary colloids for effective carbon utilization in enhanced oil recovery applications. J. Mol. Liq. 343, 117659.

Li, Y., Guo, J., Wang, S., Yang, R., Lu, Q., 2019. Reducing Hydroxypropyl Guar Gum Adsorption on Rock by Silica Nanoparticles for Tight Reservoir Damage Remediation. In: SPE- IPTC-19561-MS: International Petroleum Technology Conference, 22 March 2019, Beijing, China.

Liu, Y., Sun, C., Bolin, T., Wu, T., Liu, Y., Sternberg, M., Sun, S., Lin, X.-M., 2013. Kinetic pathway of palladium nanoparticle sulfidation process at high temperatures. Nano Lett. 13 (10), 4893–4901.

Maagi, M.T., Lupyana, S.D., Gu, J., 2019. Effect of Nano-SiO2, Nano-TiO2 and Nano-Al2O3 addition on fluid loss in oil-well cement slurry. Int. J. Concr. Struct. Mater. 13 (1), 1–6.

Maghzi, A., Mohebbi, A., Kharrat, R., Ghazanfari, M.H., 2011. Pore-scale monitoring of wettability alteration by silica nanoparticles during polymer flooding to heavy oil in a five-spot glass micromodel. Transp. Porous Media 87 (3), 653–664.

Mahmoud, O., Mady, A., Dahab, A.S., Aftab, A., 2021. Al_2O_3 and CuO nanoparticles as promising additives to improve the properties of KCl-polymer mud: an experimental investigation. The. Can. J. Chem. Eng.

Mao, Z., Cheng, L., Liu, D., Li, T., Zhao, J., Yang, Q., 2022. Nanomaterials and technology applications for hydraulic fracturing of unconventional oil and gas reservoirs: a state-of-the-art review of recent advances and perspectives. ACS Omega 7, 29543–29570.

Mathew, J., Joy, J., George, S.C., 2019. Potential applications of nanotechnology in transportation: A review. J. King Saud Univ. Sci. 31 (4), 586–594.

Matsui, I., 2005. Nanoparticles for electronic device applications: A brief review. J. Chem. Eng. Japan 38 (8), 535–546.

McDonald, M., 2012. A Novel Potassium Silicate for Use in Drilling Fluids Targeting Unconventional Hydrocarbons. In: Spe-162180-MS: SPE Canadian Unconventional Resources Conference, 30 October 2012, Calgary, Alberta, Canada.

Medhi, S., Chowdhury, S., Gupta, D.K., Mazumdar, A., 2020. An investigation on the effects of silica and copper oxide nanoparticles on rheological and fluid loss property of drilling fluids. J. Pet. Explor. Prod. Technol. 10 (1), 91–101.

Mikhienkova, E.I., Minakov, A.V., Zhigarev, V.A., Matveev, A.V., 2018. The effect of nanoparticles additives on filtration properties of drilling muds with microparticles. J. Phys. 12127.

Mohammadi, M., Mahani, H., 2020. Insights into the pore-scale mechanisms of formation damage induced by drilling fluid and its control by silica nanoparticles. Energy Fuel 34 (6), 6904–6919.

Nasrollahzadeh, M., Issaabadi, Z., Sajadi, S.M., 2019. Green synthesis of cu/Al_2O_3 nanoparticles as efficient and recyclable catalyst for reduction of 2,4-dinitrophenylhydrazine, methylene blue and Congo red. Compos. Part B Eng. 166, 112–119.

Negin, C., Ali, S., Xie, Q., 2017. Most common surfactants employed in chemical enhanced oil recovery. Petroleum 3 (2), 197–211.

Orodu, K.B., Afolabi, R.O., Oluwasijuwomi, T.D., Orodu, O.D., 2019. Effect of aluminum oxide nanoparticles on the rheology and stability of a biopolymer for enhanced oil recovery. J. Mol. Liq. 288, 110864.

Pal, N., Verma, A., Ojha, K., Mandal, A., 2020. Nanoparticle-modified gemini surfactant foams as efficient displacing fluids for enhanced oil recovery. J. Mol. Liq. 310, 113193.

Pang, X., Boul, P.J., Jimenez, W.C., 2014. Nanosilicas as accelerators in Oilwell cementing at low temperatures. SPE Drill. Complet. 29 (01), 98–105.

Perween, S., Beg, M., Shankar, R., Sharma, S., Ranjan, A., 2018. Effect of zinc titanate nanoparticles on rheological and filtration properties of water based drilling fluids. J. Petrol. Sci. Eng. 170, 844–857.

Ponmani, S., William, J.K.M., Samuel, R., Nagarajan, R., Sangwai, J.S., 2014. Formation and characterization of thermal and electrical properties of CuO and ZnO nanofluids in xanthan gum. Colloids Surf. A Physicochem. Eng. Asp. 443, 37–43.

Rafieefar, A., Sharif, F., Hashemi, A., Bazargan, A.M., 2021. Rheological behavior and filtration of water-based drilling fluids containing graphene oxide: experimental measurement, mechanistic understanding, and modeling. ACS Omega 6 (44), 29905–29920.

Razak Ismail, A., Seong, T.C., Buang, N.A., Rosli, W., Sulaiman, W., 2014. Improve performance of water-based drilling fluids using nanoparticles. In: 5th Sriwijaya International Seminar on Energy and Environmental Science and Technology, 10–11 September 2014, Palembang, Indonesia.

Rezaei, A., Khodabakhshi, A., Esmaeili, A., Razavifar, M., 2021. Effects of initial wettability and different surfactant-silica nanoparticles flooding scenarios on oil-recovery from carbonate rocks. Petroleum. In press.

Roustaei, A., Bagherzadeh, H., 2015. Experimental investigation of SiO2 nanoparticles on enhanced oil recovery of carbonate reservoirs. J. Pet. Explor. Prod. Technol. 5 (1), 27–33.

Sabbaghi, S., Rezvani, A., 2020. Experimental investigation of rheological & filtration properties and thermal conductivity of water-based drilling fluid enhanced. Powder Technol. 368, 323–341.

Sabbaghi, S., Saboori, R., Barahoei, M., 2015. Enhanced heat transfer of drilling fluid with Titania Nanofluid. Res. Rev.: J. Eng. Technol. 4 (3), 12–17.

Saboori, R., Sabbaghi, S., Kalantariasl, A., 2019. Improvement of rheological, filtration and thermal conductivity of bentonite drilling fluid using copper oxide/polyacrylamide nanocomposite. Powder Technol. 353, 257–266.

Sadegh Hassani, S., Amrollahi, A., Rashidi, A., Soleymani, M., Rayatdoost, S., 2016. The effect of nanoparticles on the heat transfer properties of drilling fluids. J. Petrol. Sci. Eng. 146, 183–190.

Sadeghalvaad, M., Sabbaghi, S., 2015. The effect of the TiO2/polyacrylamide nanocomposite on water-based drilling fluid properties. Powder Technol. 272, 113–119.

Saha, R., Uppaluri, R.V.S., Tiwari, P., 2018. Silica nanoparticle assisted polymer flooding of heavy crude oil: emulsification, rheology, and wettability alteration characteristics. Ind. Eng. Chem. Res. 57 (18), 6364–6376.

Saxena, N., Kumar, A., Mandal, A., 2019. Adsorption analysis of natural anionic surfactant for enhanced oil recovery: the role of mineralogy, salinity, alkalinity and nanoparticles. J. Petrol. Sci. Eng. 173, 1264–1283.

Seyyedsar, S.M., Farzaneh, S.A., Sohrabi, M., 2017. Investigation of low-density CO2 injection for enhanced oil recovery. Ind. Eng. Chem. Res. 56 (18), 5443–5454.

Shakib, J.T., Kanani, V., Pourafshary, P., 2016. Nano-clays as additives for controlling filtration properties of water-bentonite suspensions. J. Petrol. Sci. Eng. 138, 257–264.

Sharma, T., Iglauer, S., Sangwai, J.S., 2016. Silica nanofluids in an oilfield polymer polyacrylamide: interfacial properties, wettability alteration, and applications for chemical enhanced oil recovery. Ind. Eng. Chem. Res. 55 (48), 12387–12397.

Sheng, J.J., 2011. Surfactant Flooding. In: Modern Chemical Enhanced Oil Recovery. Gulf Professional Publishing.

Singh, T., Shukla, S., Kumar, P., Wahla, V., Bajpai, V.K., 2017. Application of nanotechnology in food science: perception and overview. Front. Microbiol. 8 (1501).

Singh, R., Tong, S., Panthi, K., Mohanty, K.K., 2018. Nanoparticle-Encapsulated Acids for Stimulation of Calcite-Rich Shales. In: SPE-URTEC-2897114-MS: SPE/AAPG/SEG Unconventional Resources Technology Conference, 23 July 2018, Houston, Texas, USA.

Taheri, A., Zahedzadeh, M., Masoudi, R., Ataei, A., Roayaei, E., 2011. Simulation and experimental studies of mineral scale formation effects on performance of Sirri-C oil field under water injection. Iran. J. Chem. Chem. Eng. 30 (3), 9–24.

Thomas, S., Grohens, Y., Pottathara, Y.B., 2019. Industrial Applications of Nanomaterials. Elsevier.

Tiwari, A.K., Ghosh, P., Sarkar, J., 2013. Heat transfer and pressure drop characteristics of CeO2/ water nanofluid in plate heat exchanger. Appl. Therm. Eng. 57 (1–2), 24–32.

Tohidi, Z., Jafari, A., Omidkhah, M., 2021. Electrolyte nanofluid performance on the oil detachment from an oil-wetted carbonate surface: water channel formation using molecular dynamics simulation. J. Petrol. Sci. Eng. 196, 108006.

Vipulanandan, C., Krishnamoorti, R., Mohammed, A., Boncan, V., Narvaez, G., Hughes, B., Head, B., Pappas, J.M., 2015. Iron Nanoparticle Modified Smart Cement for Real Time Monitoring of Ultra Deepwater Oil Well Cementing Applications. In: SPE-OTC-25842-MS: Annual Offshore Technology Conference, 04 May 2015, Houston, Texas, USA.

Vipulanandan, C., Mohammed, A., Ganpatye, A.S., 2018. Smart Cement Performance Enhancement with NanoAl2O3 for Real Time Monitoring Applications Using Vipulanandan Models. In: SPE-OTC-28880-MS: Annual Offshore Technology Conference, 30 April 2018, Houston, Texas, USA.

Wang, Z., Wu, Y., Luo, P., Tian, Y., Lin, Y., Guo, Q., 2018. Poly (sodium p-styrene sulfonate) modified Fe3O4 nanoparticles as effective additives in water-based drilling fluids. J. Petrol. Sci. Eng. 165, 786–797.

Wang, R., Liu, T., Ning, F., Ou, W., Zhang, L., Wang, Z., Peng, L., Sun, J., Liu, Z., Li, T., Sun, H., Jiang, G., 2019. Effect of hydrophilic silica nanoparticles on hydrate formation: insight from the experimental study. J. Energy Chem. 30, 90–100.

Wasan, D., Nikolov, A., Kondiparty, K., 2011. The wetting and spreading of nanofluids on solids: role of the structural disjoining pressure. Curr. Opin. Colloid Interface Sci. 16 (4), 344–349.

William, J.K.M., Ponmani, S., Samuel, R., Nagarajan, R., Sangwai, J.S., 2014. Effect of CuO and ZnO nanofluids in xanthan gum on thermal, electrical and high pressure rheology of water-based drilling fluids. J. Petrol. Sci. Eng. 117, 15–27.

Wu, Y., Chen, W., Dai, C., Huang, Y., Li, H., Zhao, M., He, L., et al., 2017. Reducing surfactant adsorption on rock by silica nanoparticles for enhanced oil recovery. J. Petrol. Sci. Eng. 153, 283–287.

Yekeen, N., Manan, M.A., Idris, A.K., Padmanabhan, E., Junin, R., Samin, A.M., Oguamah, I., 2018. A comprehensive review of experimental studies of nanoparticles-stabilized foam for enhanced oil recovery. J. Petrol. Sci. Eng. 164, 43–74.

Yousefvand, H., Jafari, A., 2015. Enhanced oil recovery using polymer/nanosilica. Procedia Mater. Sci. 11, 565–570.

Youssif, M.I., El-Maghraby, R.M., Saleh, S.M., Elgibaly, A., 2018. Silica nanofluid flooding for enhanced oil recovery in sandstone rocks. Egypt. J. Pet. 27 (1), 105–110.

Zabala, R., Franco, C.A., Cortés, F.B., 2016. Application of Nanofluids for Improving Oil Mobility in Heavy Oil and Extra-Heavy Oil: A Field Test. In: SPE-179677-MS: DOE Improved Oil Recovery Symposium Proceedings, 11 April 2016, Tulsa, Oklahoma, USA.

Zargartalebi, M., Barati, N., Kharrat, R., 2014. Influences of hydrophilic and hydrophobic silica nanoparticles on anionic surfactant properties: interfacial and adsorption behaviors. J. Petrol. Sci. Eng. 119, 36–43.

Zoveidavianpoor, M., Samsuri, A., 2016. The use of nano-sized tapioca starch as a natural water-soluble polymer for filtration control in water-based drilling muds. J. Nat. Gas Sci. Eng. 34, 832–840.

Chapter 9

Molecular simulations in upstream applications

Elkhansa Elbashier[a], Giuliano Carchini[a], Mohammed Saad[a,b], Ibnelwaleed A. Hussein[a,b], Golibjon R. Berdiyorov[c], and Santiago Aparicio[d]

[a]*Gas Processing Center, College of Engineering, Qatar University, Doha, Qatar*
[b]*Department of Chemical Engineering, College of Engineering, Qatar University, Doha, Qatar*
[c]*Qatar Environment and Energy Research Institute, Hamad Bin Khalifa University, Doha, Qatar*
[d]*Department of Chemistry, University of Burgos, Burgos, Spain*

9.1 Introduction

Computer simulations are considered to be a good tool to supplement experimental studies on the behavior of different gas molecules and particles on porous materials and surfaces. They can provide nanoscopic details about the properties of small gas molecules on different surfaces and interfaces which are not accessible experimentally. Seaton et al. were the first to use molecular simulations to evaluate the distribution of the porous media toward the estimation of ultimate recovery (Seaton et al., 1989). Other studies were done in petroleum engineering using molecular simulation methods to investigate the fluid–fluid and fluid–solid interactions at nano- and microscales. This chapter introduces the most common simulation methods, how they are applied to petroleum engineering, and what type of parameters can each method provide. Moreover, the simulation work is applied to various problems in the oil and gas research, including enhanced oil recovery (EOR), enhanced gas recovery (EGR), the estimation of ultimate gas recovery (EUR), and the scale inhibition and removal which are described in the following sections of the chapter.

9.2 Computational chemistry and molecular simulations

This section starts with details about the theoretical concepts behind the molecular simulation methods. The section is divided into two parts, considering the two main and most useful approaches for upstream characterization: (i) quantum chemistry and (ii) molecular dynamics simulations.

Developments in Petroleum Science, Vol. 78. https://doi.org/10.1016/B978-0-323-99285-5.00011-9

9.2.1 Quantum chemistry

Quantum chemistry (QC) is the most accurate approach in computational materials science with electrons and atomic nuclei being the building blocks for any materials description. On the contrary, the force-field-based molecular dynamic methods use atoms, pseudo-atoms, or even amino acids as building blocks, and the interactions between these species are obtained by fitting the experimental findings or the results from the first-principles calculations. Quantum chemistry applies a quantum mechanics (QM) approach (i.e., the solution of the Schrödinger equation) to find energies, charges, reaction paths, transition states, and solvation energies of the considered model system. It is, however, not a simple approach, being extremely computationally demanding, and requiring that the user has some knowledge about its inner workings, to interpret the results. Different options will yield quite different results, some more correct than others. Computers might do the calculations, but these must be done considering the suitable theoretical approaches for each type of problem, and humans must interpret the results. Thus, knowledge is a key to QC methods, as well as many other computational fields in physics, chemistry, and engineering.

9.2.1.1 *Schrödinger equation*

In quantum mechanics, any given system is described by a complex wave-function. However, the wave-function cannot be directly related to any measurable physical quantity. The wave-function can be used to find the probability of a given particle inside a certain volume as follows:

$$P = \iiint \Psi^*(x, y, z, t)\Psi(x, y, z, t)\,dxdydz \tag{9.1}$$

where P is the probability and Ψ^* (x, y, z, t) is the complex conjugate of Ψ (x, y, z, t). The wave-function is obtained by solving the Schrödinger equation, which was introduced by Erwin Schrödinger in 1962. In its time-independent form, the Schrödinger equation can be written in the following form:

$$\widehat{H}\Psi = E\Psi \tag{9.2}$$

where Ψ is the eigenstate, E is the corresponding lowest energy eigenvalue, and $\widehat{H}$ is the electron Hamiltonian which includes different information about the atom, such as the kinetic and potential energies and the energies of the electron–nuclei interactions as well.

9.2.1.2 *Introduction to density functional theory*

Density functional theory (DFT) is considered one of the most common methods for theoretical simulations of electronic structures. The basics of the DFT relay on finding the ground state of electrons in a system for the many-body time-independent Schrodinger equation:

$$\Psi(\{\mathrm{r}\}) = \Psi(r_i, r_2, \ldots, r_n)$$

$$\hat{H}\Psi = \left[\hat{T} + \hat{V} + \hat{U}\right]\Psi = \left[\sum_{i=1}^{N}\left(-\frac{h}{2m_i}\nabla_i^2\right) + \sum_{i=1}^{N} V(r_i) + \sum_{i<1}^{N} U\left(r_i, r_j\right)\right]\Psi = E\Psi \tag{9.3}$$

where $\hat{V}$ and $\hat{T}$ are the potential and kinetic energies, respectively, and $\hat{U}$ is the electro–electron interaction energy which is described by the many-particle equation. This equation depends on the coordination of each electron in the system (total of n electrons), and the solution of the energy function depends on all the possible wave-functions with the number M^n. Thus, the complexity and computational cost of solving the energy equation increase exponentially with the number of electrons. Hence, the concept of the DFT method is to reduce the computational cost by defining a relation between the single-particle density $\rho(r)$ that relies only on the spatial coordination and the wave-function $\Psi(\{\mathrm{r}\})$. Thus, DFT converts the many-particle equation (Eq. 9.3), to a single particle equation without the term $\hat{U}$.

In the Kohn-Sham theory, the DFT energy is given by the following equation:

$$E_{DFT}[\rho] = T_S[\rho] + E_{ne}[\rho] + J[\rho] + E_{xc}[\rho] \tag{9.4}$$

where the subscript S in the kinetic energy term indicates that the energy is calculated from a Slater determinant (determinant of a matrix of spin orbitals). The $E_{xc}[\rho]$ term (exchange-correlation contributions) quality determines the quality of the DFT results. The value of $E_{xc}[\rho]$ was approximated by various researchers, leading to different so-called functionals, and expected values were suggested, such as the Becke 3 parameter Lee-Yang-Parr (B3LYP) hybrid functional that is usually used in literature, and it is given by the following equation:

$$E_{XC}^{B3LYP} = (1-a)E_X^{LSDA} + aE_X^{exact} + b\Delta E_X^{B88} + (1-c)E_C^{LSDA} + cE_C^{LYP} \tag{9.5}$$

where the three parameters, (a $\cong$0.2, b $\cong$0.7, and c $\cong$0.8) are fitted to experimental data, E_c is electron correlation energy, E_X is the exchange contribution energy, *LSDA* stands for local spin density approximation, and the E_X^{exact} is the exact exchange energy for the Kohn-Sham orbitals. However, the exact exchange energy could not be determined from the given equation since the Kohn-Sham orbitals are not equal to the Hartree–Fock orbitals. Recall that the Kohn-Sham orbitals are for noninteracting systems, E_X^{B88} is a correction to the exchange energy from the LSDA method proposed by Becke, and E_C^{LSDA} is the electron correlation energy contribution from the LSDA method.

The various available methods exist for a greater range of applications of DFT, including the transition metal complex applications. Some other applications where DFT has shown remarkable success are the study of chemical potential and electron activities allowing the estimation of the chemical properties of a system. Thus, applying DFT to understand the interaction of gas molecules with different surfaces is highly effective as it provides information about adsorption energy, binding affinity, charge analysis, trapping potential, and other valuable information.

9.2.1.3 Basis set

The distribution of electrons around an atom in an atomic orbital is described by a one-electron function. In the context used in this section, the atomic orbitals have a different name, and they are defined as basis functions when describing a basis set. To apply the quantum theory to molecules, molecular orbital linear combination of atomic orbitals (MO-LCAO) is used; that is, by taking linear combinations of atomic orbitals, or basis functions, it is possible to construct molecular orbitals. A molecular orbital, in the MO-LCAO theory, is thus a mathematical combination of atomic orbitals. There have been proposed many basis functions. Two of the most used are the Slater-type orbital (STO) and Gaussian-type orbital (GTO).

However, both of these basis functions have their limitations. The Slater-type orbital requires much more computational power to solve the given problem than the Gaussian-type orbitals. On the other hand, the Gaussian-type orbitals are inferior to the Slater-type orbitals in terms of accuracy. The Gaussian-type orbitals have a zero slope at the nucleus position, while the Slater-type orbitals have a cusp. The discontinuous behavior of the Slater-type orbitals at the nucleus is desired, and as such the Gaussian-type orbitals fail to reproduce this behavior. Another problem with the Gaussian-type orbitals is that they fall off too quickly far away from the nucleus, due to the r^2 dependence in the exponential function. Despite their differences, both Slater-type orbitals and Gaussian-type orbitals can be used to construct basis sets. However, due to the deficiencies in the GTOs, additional GTOs have to be used to achieve the same accuracy as STO. Usually, three GTOs are needed to represent one STO. Still, it is computationally faster to use multiple GTOs per STO.

The minimal basis sets, like STO, lack the flexibility to represent the valance orbitals in different molecular environments; hence, multiple combined basis sets are needed to describe one orbital. On the other hand, using a combination of split-valance basis sets provides different spatial extents for the valance orbital, giving the electron density more flexibility, and allowing it to adjust to different molecular environments. There are various types of split-basis sets; the most common basis sets are the Pople basis sets described generally as X-YZg. The X represents the number of combined Gaussian functions used to describe a single basis set, while for the valence orbitals, split-valance double basis sets are used, each has a specific number of combined Gaussian functions, Y and Z. More basis sets could be used, triplet or quadratic (X-YZWg or X-YZWVg). Examples of the split-valence basis sets are 3-21 G and 6-311 G, which are used in different studies (Günsel et al., 2019; Pandya et al., 2019; Tüzün and Saripinar, 2020).

To further increase the accuracy of the basis sets, polarization functions can be added as well. Polarization functions are higher angular momentum functions. For a hydrogen atom, this would mean polarizing the s-orbitals by adding a p-function to the basis set. Similarly, the p-orbitals of heavier atoms could be polarized by adding a d-function. Furthermore, when dealing with systems that have diffused electronic distributions, e.g., excited states and anions, the addition of a set of diffuse functions could also be considered. The diffused set functions can extend to a large distance away from the basis function center or the nucleus as they have small exponents in the exponential functions. Other basis sets are available, each with different accuracies and limitations, such as the plane-wave basis sets described separately in the next section as they are most popular recently.

9.2.1.4 Plane-waves and pseudo-potentials

The atomic state structure of a solid-state material, e.g., a rock, is described in the form of unit cells of atoms. All the atoms inside the unit cell are described by the same wave-function provided by the Schrödinger equation; thus, it is difficult to use the local basis set to describe all the electrons available, as it will form an infinite number of eigenvalues that will yield zero. Hence, plane-waves are used to describe a system as a unit cell and apply periodic boundary conditions to form a large system in totally making them more practically efficient for solids than basis sets. Although, typically, the plane-waves and the wave-function describe the whole system, including the core and valence electrons, one should consider that some electrons will have high-energy interaction, such as the core electrons close to the nuclei, while the most essential electrons to consider are the valence electrons. Hence, the plane-waves of the core electrons are mostly excluded to reduce the number of plane-waves and computational effort. Additionally, to limit the number of plane-waves used, the cutoff value for the number of plane-waves used should be specified. Also, one should consider that having a minimal cutoff value means neglecting many plane-waves yielding a reduced results' accuracy. Thus, a balance between the accuracy and the computational cost should be considered when selecting the energy cutoff value.

Another critical point is that the plane-waves could describe both the core and valence electrons; however, this is not practical due to the high amount of plane-waves needed, as mentioned before. Thus, pseudo-potential is used to describe the core electrons, keeping only the valance electrons described by pseudo-wave-functions with fewer nodes. This combination of pseudo-potential and plane-waves gives an assumption that the nuclei and core electrons are almost frozen, keeping the valance electron, most involved in chemical interactions, in consideration. The combination of plane-waves and pseudo-potential allows for the study of large systems with a specified unit cell, which opened the door for applying DFT methods in various studies, including surface analysis, adsorption, and oil and gas investigations.

9.2.1.5 Introduction to molecular dynamic methods

Although the methods based on static quantum chemistry, such as DFT, provide promising results for the analysis of small systems, it is limited in most cases to systems containing up to a few hundred atoms. Additionally, static quantum methods do not give enough information about the dynamic behaviors of the systems. Nevertheless, quantum methods were developed to be simulated with time using the extended DFT method, the Ab-initio molecular dynamics (AIMD). In this method, the electronic states of the atoms and molecules are calculated by quantum chemistry, such as DFT, along with applying Newton's equations for dynamic motions. However, the consideration of the electronic states increased the efforts required for calculations. Thus, the AIMD method is limited to smaller systems and shorter times. Nevertheless, the AIMD method is widely applied in petroleum applications, such as surface and adsorption analysis, scale inhibition/removal, desolvation of rocks, and polymers on the surface.

As mentioned, AIMD is not suitable to simulate larger systems and long simulation time; thus, molecular dynamic methods are based on Newton's equations, and they succeed in analyzing the physical movements of atoms and molecules. Molecular dynamic methods allow the atoms/molecules to move dynamically with time, and their trajectories are analyzed numerically and statistically to calculate transport and thermodynamic properties such as diffusion coefficient. In these methods, the parameters of the atomic interactions derived from DFT can be used to develop a classical force field that governs the behavior of the atoms of the systems. Unfortunately, the development of the force field is complex, and they are not easily transferable, meaning that each force field works only on a specific set of systems. Moreover, reactions are neglected unless we consider reactive force fields, which are more complicated. Systems with thousands of atoms/molecules, such as applications in biosystems and large nanostructure, clusters, aggregation, and absorption surfaces, can be approached considering classical molecular dynamic methods. The power of the classical molecular dynamics method could be applied to larger systems (thousand or hundreds of thousands of atoms), and it can provide knowledge about the transport and density of gases adsorbed on surfaces. Thus, oil and gas applications could be analyzed using the classical molecular dynamics method with force fields. The following subsections will describe more basic information about potential energy in molecular dynamics, force fields, ensembles, thermostats, and barostats.

9.2.1.6 Basics: Potential energy of the system

Molecular dynamics simulations use classical mechanics assumptions to describe potential energy (U) as the summation of all possible interactions between the atoms in a system, ignoring the quantum effects. The interaction terms include the following:

(a) Bonded terms (U_{bonded}), which are the intramolecular interactions, and
(b) Nonbonding terms ($U_{nonbonded}$), which are the interactions between any atoms that are not connected directly.

So, the total potential energy of the system could be written as follows:

$$U(r^N) = U_{bonded}(r^N) + U_{nonbonded}(r^N) \quad (9.6)$$

where r^N are the total bodies in the system. Fig. 9.1A illustrates the general expression of the intramolecular potential force of a single molecule, which includes the energies of bond stretching, bond bending, and torsion:

$$U_{bonded}(r^N) = \frac{1}{2}\sum_{bond} k_b (r_{ij} - r_0)^2 + \frac{1}{2}\sum_{bending} k_\theta (\theta_{ijk} - \theta_0)^2 + \frac{1}{2}\sum_{torsion}\sum_{n} k_\phi^n \left(1 + \cos\left(n\phi_{ijkl} - \phi_0\right)\right) \quad (9.7)$$

where the bond term expresses the vibration potential between two adjacent atoms in the same plane separated by the equilibrium bond distance *r* between points *i* and *j*. The second term, bending, expresses the potential energy between two bonds, including three consecutive atoms of the molecules, at points *i*, *j*, and *k*. The bonding and bending terms are usually

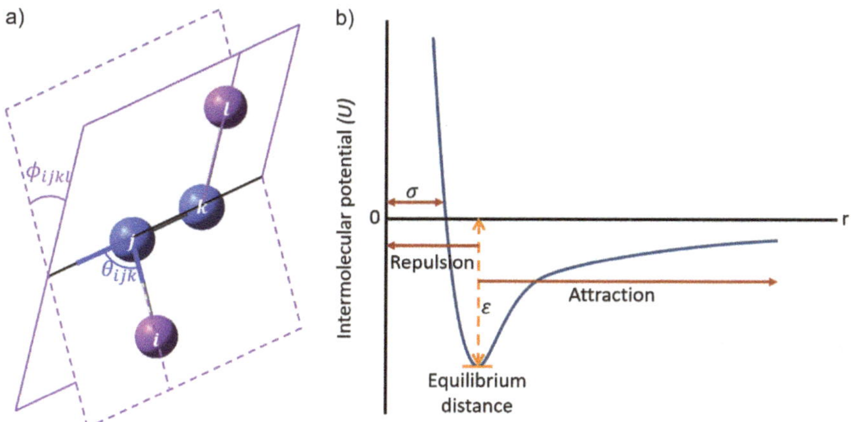

FIG. 9.1 (A) Scheme representing bonding terms of the intramolecular interactions of a molecule. (B) Lennard-Jones potential.

presented as harmonic and quadratic forms, respectively. The last term refers to the torsion potential, which describes the dihedral angle between three bonds connecting the points i, j, k, and l. In complex molecules and systems, an additional combination between the three terms is required, such as bond–bond energy, or bond–torsion energy.

The nonbonded interactions, $U_{nonbonded}(r^N)$, are described using interatomic interaction potentials, such as the widely considered Lennard-Jones "6-12" type potentials as shown in Fig. 9.1B and expressed in the following formula:

$$U_{nonbonded}\left(r^N\right) = 4\varepsilon_{ij}\left[\left(\frac{\sigma_{ij}}{r_{ij}}\right)^{12} - \left(\frac{\sigma_{ij}}{r_{ij}}\right)^{6}\right] \quad (9.8)$$

where σ_{ij} is the distance between points i and j, where $U_{nonbonded}=0$, and ε_{ij} is the depth of the potential well which corresponds to the strength of the attraction between the two atoms.

Additionally, the distance at the maximum depth of the potential well represents the equilibrium distance at which the attraction and repulsion forces between the two atoms are zero. The two atoms tend to stay at this equilibrium distance till an external force is excreted on them. When the two atoms are closer than this distance, the potential energy goes to positive values reflecting the high repulsion force between the atoms, as their orbitals will start to interfere. Thus, distance σ_{ij} is known as the Van der Waals radius, and it corresponds to the minimum distance between two atoms. The inverse of the 12th and 6th power terms in the Lennard-Jones equation corresponds to the repulsive and attraction forces, respectively. Other modifications to the two terms and their powers are done to improve the representation of the interactions between the two atom electron clouds.

To include the effect of other intermolecular interactions, such as neat electrostatic interactions, other terms are added, such as Coulomb potentials that describe the electrostatic charges as follows:

$$U_{Coulomb}\left(r_{ij}\right) = \frac{q_1 q_2}{4\pi\varepsilon_0 r_{ij}} \quad (9.9)$$

where ε_0 is vacuum permittivity, q_1 and q_2 are atomic charges. The nonbonded interactions are calculated in MD simulations as a many-body problem, which requires extensive computational cost. Hence, to reduce the calculation efforts, the pairwise summation is used to calculate all nonbonded interactions.

The bonded and nonbonded terms are described by differential equations and parameters, where each of them considers specific sets of potentials and parameters mentioned previously. The selection of the force fields depends on the characteristics of the studied systems. Oil and gas applications use mostly long-chain molecules, which are represented generally by two types of force fields: explicit and united atom force fields. The explicit force field expresses each atom in the system explicitly, while the united atom force field groups the atoms to form a pseudo-atom, e.g., a —CH_3 group considering four atoms, i.e., four sites, is replaced by a single site representing the whole methyl group, thus reducing computational cost but losing the quality of systems description. Various types of force fields along with their key modifications are mentioned with each oil and petroleum application according to the system requirements.

9.2.1.7 Ensemble

In the molecular dynamic analysis, the thermodynamic parameters of the system are defined by a set of conditions known as ensembles that represent the parameters on a larger realistic macroscopic scale. The selection of the ensemble depends on the target of the simulation; thus, various ensembles are provided in Table 9.1.

The dynamic simulation mainly works as a microcanonical ensemble, according to Newtonian MD with total conserved energy, however to address and simulate the temperatures and pressures, canonical or isobaric/isothermal ensembles, and modifications to the Newtonian scheme. Various types of thermostats and barostats were developed to simulate the temperature and pressure in a system, respectively. The most straightforward thermostat is based on applying velocity rescaling while adjusting the system's volume for barostat could simulate pressure. However, more sophisticated thermostats and barostats were developed to simulate different ensembles correctly. Samples of thermostats are Nose-Hoover, Anderson, and Langevin thermostats.

Nose-Hoover thermostat applies degrees of freedom related to momenta and time in the system, forming a thermal reservoir, while the Anderson thermostat re-randomizes the momenta of the system to simulate the temperature. On the other hand, barostats are all based on Anderson barostat, which applies a degree of freedom to the system related to the volume parameter. Extended barostats include the Nose-Hoover, the Parrinello-Rahman, and the Martyna-Tuckerman-Tobias-Klein (MTTK). Nose-Hoover barostats were the first to apply Anderson for molecular simulation; however, they were limited to large systems only. The development of Parrinello-Rahman allows for the shape of simulation and includes external stress, which gives it the ability to study the influence of stress on the structural deformations of a rock. Finally, the MTTK is an extension to the Nose-Hoover barostat and thermostat, forming a time-reversible thermostat that works for small systems as well. The Parrinello-Rahman and the MTTK barostats can be combined with the Nose-Hoover thermostat.

TABLE 9.1 Various available ensembles with their conditions.

Ensemble	Name	Molecular no. (N)	Volume (V)	Temperature (T)	Pressure (P)	Chemical potential	Energy (E)
Canonical	NVT	Fixed	Fixed	Fixed	–	–	–
Microcanonical	NVE	Fixed	Fixed	–	–	–	Fixed
Isobaric-isothermal	NPT	Fixed	–	Fixed	Fixed	–	–
Grand canonical	μVT	–	Fixed	Fixed	–	Fixed	–

9.3 Practical details

9.3.1 Building a model

The first step in molecular simulations is to build a model system with a suitable representation of the experimental setup while trying to address most of the real environmental conditions. A suitable theoretical approach is also required to suit the created model and get results close to the practical approach.

Fig. 9.2 shows steps toward creating a representative model system of a nanopore structure:

a. The first step is to determine the type of rock formation that you need to analyze and determine its constitution (*e.g.*, the shale formations were modeled in literature by inorganic minerals such as clay, quartz, and calcite (Le et al., 2015; Wang et al., 2016; Zhai et al., 2014)).
b. The bulk of the real sample, such as pyrite (corrosion inhibition/removal) (Jalab et al., 2021; Onawole et al., 2019a, 2020a), quartz, calcite, or other carbonaceous samples (Fang et al., 2020), is imported from crystallography data.
c. The bulk is cleaved to create a surface, slit nanopore (Fig. 9.2), or cylindrical nanopore representing the desired system.
d. The environmental conditions such as vacuum, solvent, gas molecules, temperature, or pressure are applied to represent the real reservoir situation.
e. Finally, a suitable theoretical approach is defined to describe the interactions between atoms and molecules.

9.3.2 Periodic boundary conditions

One of the main objectives of molecular simulation is to create a bridge between the microscopic cell, *i.e.*, the model, and the macroscopic system, *i.e.*, the real sample, allowing for an accurate estimation of the macroscopic properties by applying the simulation to a unit cell. The computationally feasible cell size, as well as the amount of the material considered for simulations (i.e., a few nanometers in size for the most powerful molecular simulation method), cannot exceed thousands of molecules. Therefore,

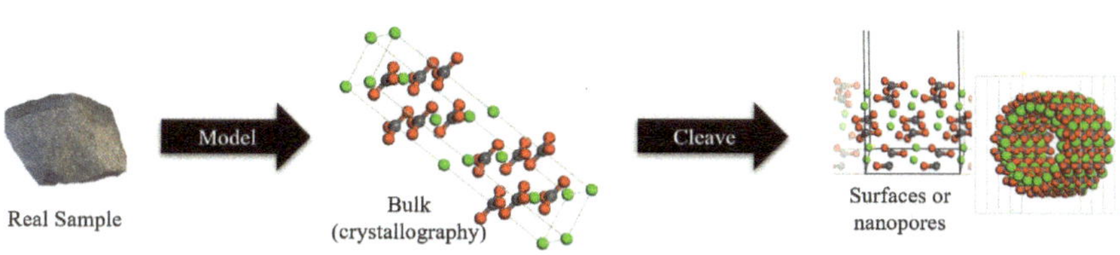

FIG. 9.2 Modeling of calcium carbonate system as an example of the carbonaceous reservoir.

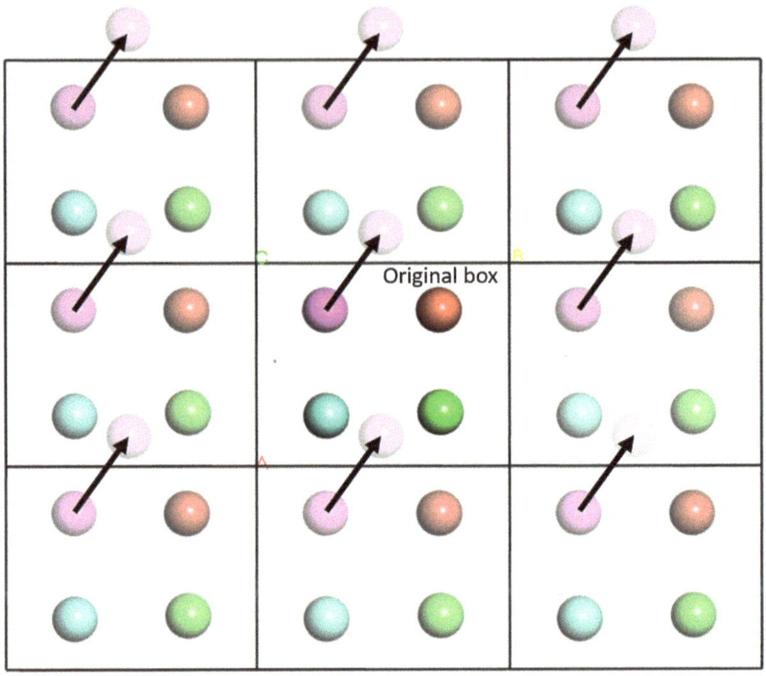

FIG. 9.3 Representation of the periodic boundary conditions in molecular modeling.

periodic boundary conditions are applied to a single cell to represent larger systems, by surrounding the unit cell with its duplicated images as shown in Fig. 9.3. In this way, spurious effects on the simulated cell limits are also eliminated to mimic a continuous material. Based on this setting, an image of an atom will enter the box on one side if it leaves the original box from the other side, according to the minimal image convention as demonstrated in the figure.

9.3.3 Continuum solvation models

The solvation is a key parameter to allow the molecular simulation model to reach the level of experimental results. There are two methods for applying the solvents as shown in Fig. 9.4:

a. Implicit model: a method to represent solvent as a continuous medium.
b. Explicit model: addition of individual "explicit" solvent molecules to represent the solvent.

Classical approaches, like classical MD or Monte Carlo simulations, have traditionally been modeling liquids and solvation-free energies. These methods

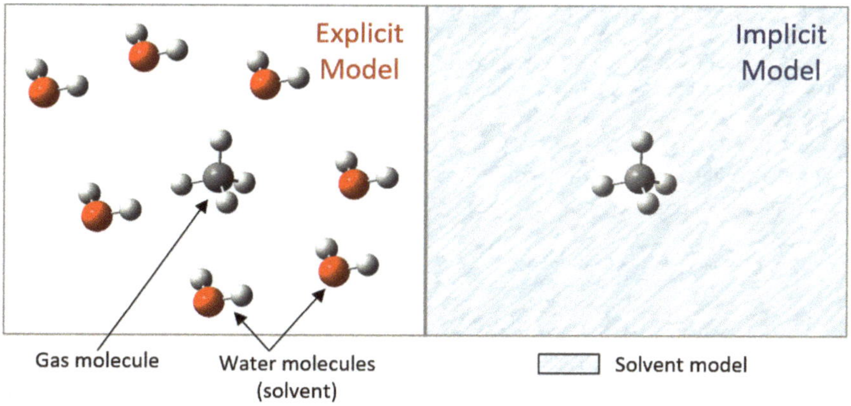

FIG. 9.4 Explicit (left panel) and implicit (i.e., continuum) (right panel) models of salvation.

use solvent molecules to model the liquid explicitly, and they are capable computationally of running a large number of molecules, including the system with the solvent. In addition, the explicit model is used in quantum chemistry methods, such as DFT, and in Ab-initio MD as well by adding a few solvent molecules close to the molecule/surface of interest. However, depending on the solvent, the number of solvent molecules needed can be substantial and result in corresponding computational efforts. This is time-consuming and requires extensive computational power. Continuum solvation models replace the explicit solvent molecules with an implicit solvent in the form of a continuum. The solute is then modeled in a solvent reaction field as a cavity (with partial atomic charges assigned to the atoms) in a dielectric. In some studies, both solvation methods are applied, so that an implicit model is applied, while some molecules are simulated explicitly around the molecule/surface of interest to interact directly with them.

9.3.4 Application of molecular simulation in oil and gas engineering

The current section deals with the actual application of the methods in the oil and petroleum investigations. The next sections describe different problems, including the enhanced oil and gas recovery, estimation of ultimate recovery, and scale removal. For each problem, both DFT and MD were used to get information of interest, such as adsorption, diffusion coefficient, binding energies, and others. In addition, the DFT method was used to estimate some interaction parameters and physical properties related to the molecules on a surface. At the same time, molecular dynamic simulation is a method that describes the physical properties of atoms related to their movement, predicting information about the dynamic changes in the systems.

9.3.5 Density functional theory

The parameters that could be calculated from the DFT method include the following:

a. Surface energy (γ), which reflects the stability and reactivity of the surfaces toward explaining which has more affinity to interact with different molecules.

$$\gamma = \frac{(E_{SURF} - nE_{bulk})}{A_{tot}} \tag{9.10}$$

b. The one-molecule adsorption energy (E_{ads}) reflects the interaction affinity of a particle to a surface/nanopore. Binding energy and interaction energies are calculated in some works by the same idea presented in the following equation of E_{ads}:

$$E_{ads} = \frac{E_{SURF+NX} - E_{SURF} - (N \times E_X)}{N} \tag{9.11}$$

c. where E_{surf} is the energy of the optimized surface, A is the area of the surface, E_{bulk} is the energy of the optimized bulk, n is the number of bulk in the surface slap, X corresponds to the type of gas, N is the number of gas molecules on the surface, $E_{surf+NX}$ is the energy of the surface with the molecule, and E_x is the energy of the optimized gas molecule.

d. The adsorption capacity of a specific molecule in a surface or a nanopore structure. The maximum capacity is observed when the E_{ads} is zero, which means that the surface/nanopore does not have more affinity to interact and adsorb the particle.

$$\mathbf{E_c} = \frac{\mathbf{E_{ads}}}{\mathbf{NA}} \tag{9.12}$$

e. Energy barrier (ΔE) is a method for finding saddle points and minimum energy paths between known reactants and products by calculating the nudged elastic band (NEB) between the different states of a process. This method can be used to quantify the mechanism of oil/gas–rock interactions.

f. Charge analysis by either electron difference densities (EDDs) or electron localization functions (ELF).

9.3.6 Molecular dynamics

The analyses done using molecular simulations include the following:

a. Density profile and radial distribution function (RDF) describe how density varies as a function of distance from a reference particle. This parameter is used to describe the adsorption affinity, amount of gas or molecules adsorbed, or the layering behavior of the particles on a specific surface. The RDF could also describe the selectivity of a solid to the different particles studied.

b. The diffusion coefficient (D) is the main parameter to describe the transport behavior of a particle on a surface. There are different equations to calculate the D; however, the general idea is the same. Collel *et al.* used the following equation (Collell et al., 2014):

$$D = \lim_{t \to \infty} \frac{\left\langle [r_i(t) - r_i(0)]^2 \right\rangle}{6t} \tag{9.13}$$

where D is the diffusion coefficient, t is the time, $r_i(t)$ is the position of the molecule i at time t, and $r_i(0)$ is the initial position of the molecule i.

c. The interfacial tensions (IFTs) of different systems are used to investigate the influence of chemicals on the IFT of the oil/water system. Fig. 9.5 shows how the wettability angle can differ for various droplets. The IFTs were calculated from the expression of pressure tensors as follows:

$$\gamma = \frac{1}{2}\left\langle L_x\left(P_{xx} - \frac{P_{yy} + P_{zz}}{2}\right)\right\rangle \tag{9.14}$$

where γ is the interfacial tension, L_x is the length of the box in the x-direction, P_{xx}, P_{yy}, and P_{zz} are the pressure tensor components in x, y, and z directions, respectively.

d. Cumulative net ionic charge Q(r). It is used in some studies to analyze the influence of ions and charges on an interface:

$$Q(r) = \int_0^r [C_+(r) - C_-(r)]dr \tag{9.15}$$

where $C_+(r)$ or $C_-(r)$ is the positive or negative charge distribution, respectively, defined by $C_{+,\ -}(r) = C_{+,\ -}(r) l^3 / NR^3$. R is the maximum radius for the system, defined with the origin located at the center of mass of the dodecane droplet, N is the number of configurations in the simulations, and l is the number of layers.

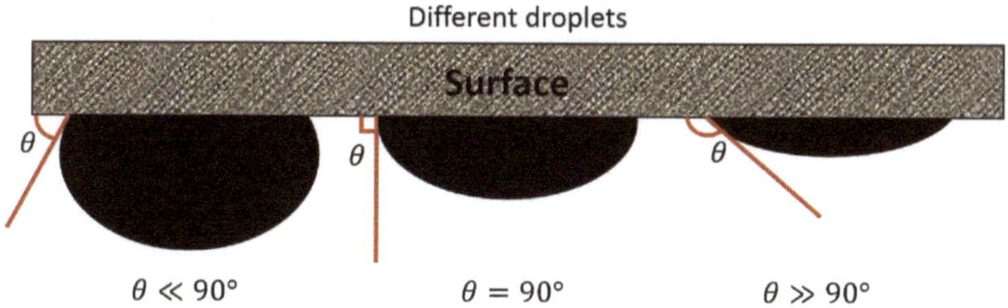

FIG. 9.5 Different wettability angles.

9.4 Enhanced oil recovery

9.4.1 Introduction to the process

The processes of enhanced oil recovery (EOR) include heating, chemical and gas floods, and steam combustion. The chemical and gas floods are identified by the material and fluids injected for recovery. Gas floods include flue gas, hydrocarbons, and carbon dioxide, while chemical floods include alkalis, surfactants, and polymers which are usually used as a mixture. Molecular simulations are used in the EOR to investigate the interaction and adsorption of the oil molecules in different minerals and the change of those interactions and interfacial tension in the presence of material used for the recovery. Experimental methods were used to investigate the behavior of crude oil on the surface of different minerals. For instance, the adsorption of crude oil on a mineral covered with water film was studied by Fathi et al. (2011) who concluded that the polarity of the oil components affects the adsorption of the crude oil. A similar conclusion on the effect of polarity was provided by Rezaei Gomari and Hamouda (2006), which lead to further studies on the polarity effect (Farooq et al., 2011; Neirinck et al., 2010; Puntervold et al., 2007). Although various experimental analyses were done, further experimental studies are challenging; in particular, it is still unknown the cause behind the change in the oil adsorption behavior with polarity and the added chemicals. To reveal the cause of the alteration, nano- and microscopic scale analysis is required, and here is where the molecular simulation is introduced to the EOR analysis. Molecular simulation is used to analyze the properties, dynamics, and detailed information related to the EOR process. The next section describes how molecular simulation was used lately in EOR investigations.

9.4.2 Applications of molecular simulation

As mentioned in the previous sections, various methods are used in the EOR process, including chemicals to alter the oil–surface interactions. Molecular simulations can provide a molecular-level insight into the interactions between components, which is generally difficult to obtain from experiments. The types of interactions include the dynamics of the oil flow and replacement by used chemicals through density profile, radial distribution function, and the alteration in the interfacial tension (IFT) wettability, which results in oil droplets detachment.

Jha et al. (2021) used molecular simulations to investigate the CO_2 EOR mechanism and the CO_2 geosequestration. In their study, they analyzed the mechanism of fluid–fluid and rock–fluid interaction between different ionic species of a low-salinity surfactant nanofluid, with $scCO_2/CO_2$ and CO_2–decane mixture on a limestone surface. Their model included water mixed with salts, the cationic surfactant (hexadecyltrimethylammonium bromide or CTAB) with ZrO_2 nanoparticles surfactant, and CO_2 molecules with conditions of 20 MPa pressure and 343 K temperature. The model used is shown in Fig. 9.6.A, and the distribution in terms of density along the z-direction

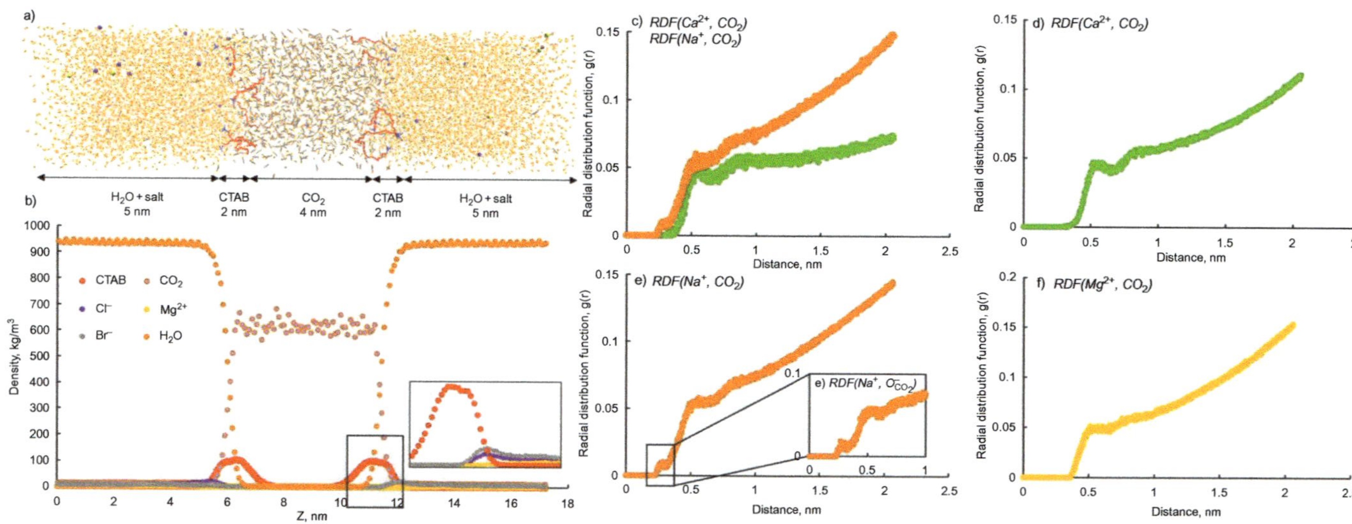

FIG. 9.6 Figure (A) shows a model system for molecular dynamics simulations and (B) shows the density profile of the species in the system. (C–F) Radial distribution functions of CO_2 and different ions in the salt mixture (Jha et al., 2021).

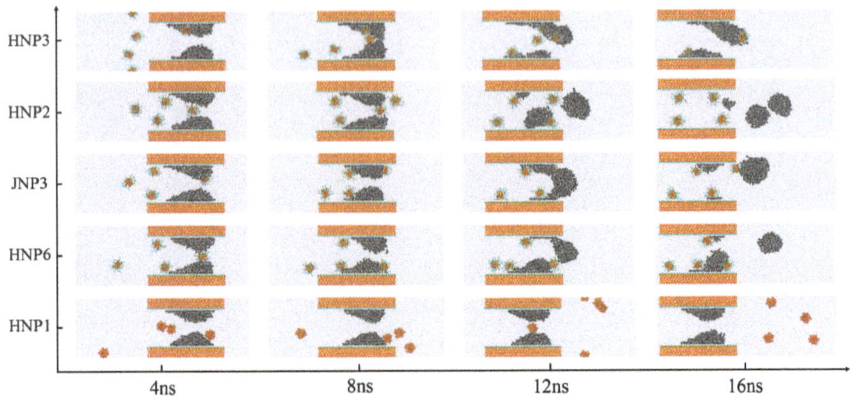

FIG. 9.7 The displacement of oil molecules by different types of chemicals. The figure shows the change in oil wettability on the methylated quartz nanochannel during the nanofluid injection process (Li, 2020).

is plotted in Fig. 9.6.B. Then, the result of the radial distribution function for ions in the system (Fig. 9.6.C–F) demonstrated how the monovalent cations are distributed evenly while the divalent cations are concentrated in bulk as they create peak distributions. These results were later used to explain why divalent ions had a greater effect on the IFT than monovalent cations.

In another study by Li (2020), the dynamic behavior of oil droplets in a slit nanopore of silica was investigated with the addition of various surfactants (Fig. 9.7) and how the IFT changes. They used molecular dynamics to visualize the behavior of the droplets' displacement. As seen in the figure, the oil was attached to the surface at the beginning of the simulation (4n); then, the surfactant decreased the oil–surface interaction. In their work, they calculated the IFT and interaction energy alteration for each surfactant, agreeing with the visualized detachment.

9.5 Enhanced gas recovery

9.5.1 Introduction to the process

The competitive adsorption of both CH_4 and CO_2 on the surface of nanoporous materials, such as carbonaceous nanopores, that occur during the process of EGR has been widely investigated using molecular simulation methods. As demonstrated in Fig. 9.8, the main objective is to shed some light on the competitive adsorption and the displacement mechanism of CH_4 and CO_2 on a surface.

Multiple studies have used DFT, and MD methods to study the diffusion properties of the gas in shale formations and carbonated reservoirs. Different models of surfaces were considered to describe the structure of the porous media inside the shale and carbonate formations.

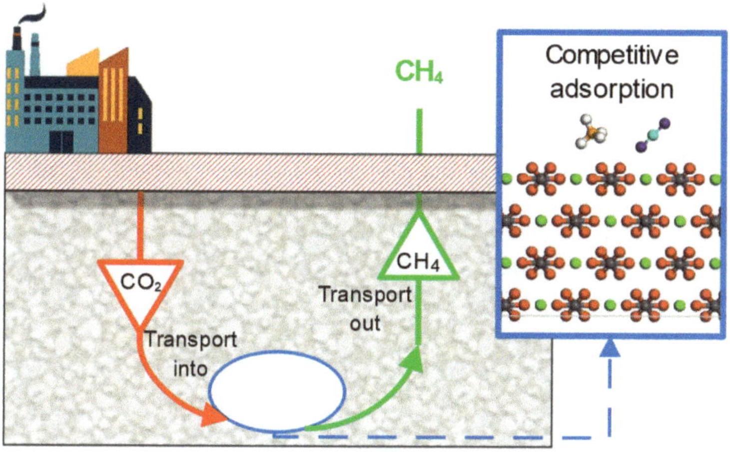

FIG. 9.8 Representation of EGR process by CO_2 insertion showing the part where molecular simulation is employed.

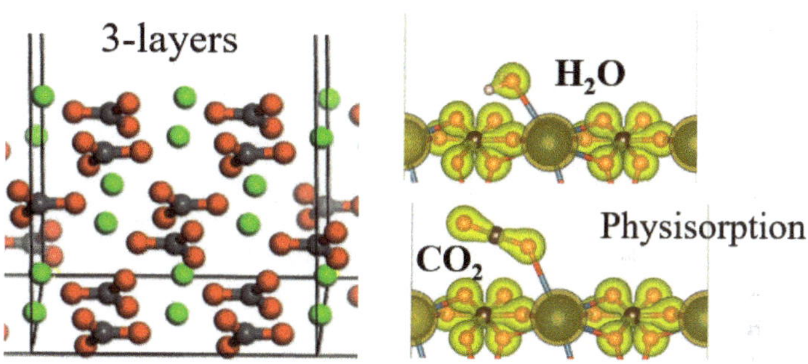

FIG. 9.9 Model of adsorption and charge distribution of CO_2 and H_2O in calcite surface (Carchini et al., 2020).

9.5.2 Adsorption preferability between CH_4 and CO_2 by DFT

In a previous study done by our simulation group, we applied the DFT calculation to study the interaction of CH_4, CO_2, N_2, H_2S, and H_2O on the most stable surface of calcite, which is calcite (104) (Carchini et al., 2020). In this work, a slab with three layers was used to represent the reservoir wall (Fig. 9.9). The study revealed that the adsorption of all the gases is weak, with CO_2, and H_2O having the strongest interaction. This observation suggests that the presence of CO_2 could affect the gas recovery process, as they are more interacting with the surface, thus leading to methane displacement from the nanopores and increasing methane production. However, the molecules of H_2O showed higher adsorption affinity than all other molecules at adsorption sites that are not available to CO_2, which may create problems in the enhanced gas recovery process by CO_2.

Additionally, Luo et al. (2017) studied the adsorption of CH_4, C_2H_6, CO_2, and N_2 molecules on calcite surfaces considering (100) morphology. However, they used a surface model with only two layers, as demonstrated in Fig. 9.10. Therefore, the model they created is considered a weak representation of the real bulk system. Additionally, they only considered calcite (100) morphology which is not the most stable morphology of calcite (De Leeuw and Parker, 1998). Additionally, Zhang et al. have used DFT calculations to study the adsorption of CH_4 and CO_2 in different polymorphs of calcium carbonates, including calcite (104), aragonite (011)-Ca, and vaterite (010)-CO_3 as shown in Fig. 9.10.

9.5.3 The molecular dynamic of gas adsorption and dynamic in porous media

In 2017, Sun et al. (2017a) studied the adsorption properties of CH_4 and CO_2 on kerogen slit nanopores to represent shale formations as shown in the molecular structure represented in Fig. 9.11. They considered four different temperatures (298, 323, 343, and 373 K) and pressure (10–20,000 KPa). They were able to get information about the density profile of the two gases on the surface along with the selectivity, displacement rate of each gas, and nature of adsorption.

Later in the same year, the same group did a similar investigation with the same conditions but considered different models of the slit nanopore (Sun et al., 2017b). Three slit nanopores were created as follows: graphene-silica, calcite-graphene, and calcite-silica (Figs. 9.9-9.12A–C). The adsorption of CH_4 was more favored on the graphene surface, while CO_2 was much favored on the calcite surface. Additionally, they studied the effect of nanopore size on the adsorption of the gases and concluded that the adsorption of CO_2 on calcite represented by mass loading is greater than CH_4 by almost double the amount due to the significant interaction between CO_2 and calcite surface. Furthermore, they were able to calculate the diffusion coefficient of the two gases (Figs. 9.9-9.12D and E), which showed that CH_4 has greater diffusion than CO_2, which is again due to the high interaction of CO_2 to the surface, limiting it is compared to CH_4. The results of the loading mass and diffusion coefficient reveal the ability of CO_2 to detach CH_4 and replace it.

The effect of hydration on the diffusion and adsorption of methane was further studied by Bui et al. (2017) and Mohammed and Gadikota (2018). The adsorption and transportation of methane in 1 nm slit nanopore of silica, magnesium oxide, alumina, muscovite, and calcite were investigated, and a description of the relationship between density profile and diffusion coefficient was provided. They found that on almost all the surfaces, methane is diffusing isotopically; however, for calcite, the diffusion is anisotropic (Figs. 9.9-9.13). The same results were provided by Mahmoud et al. who considered a 2 nm slit nanopore of calcite, graphene, and montmorillonite surface

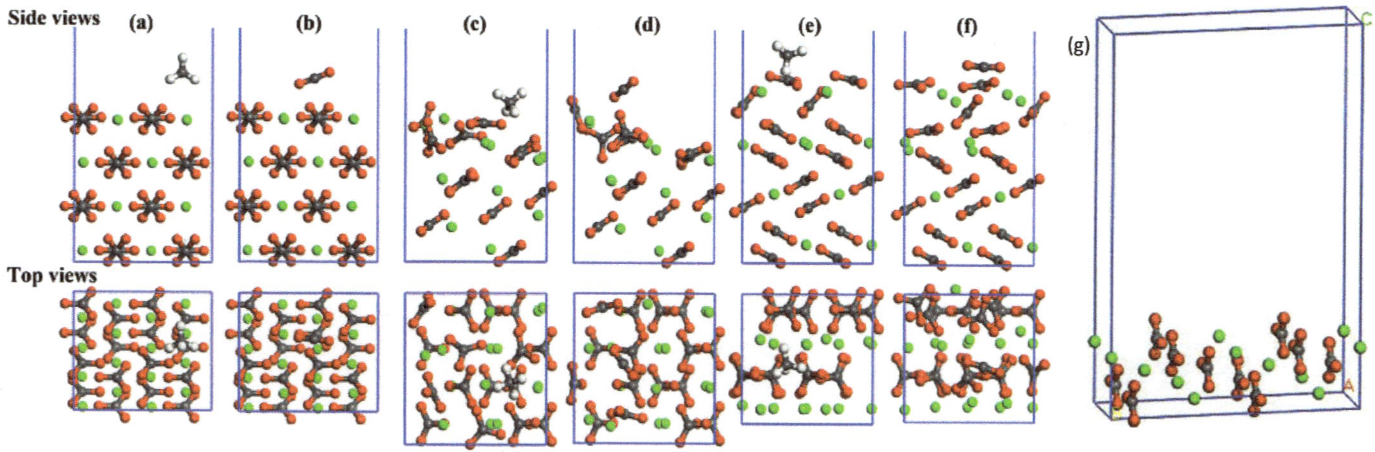

FIG. 9.10 The models of the carbonated surfaces were done by (A–F) Zhang et al. (2020) and (G) Luo et al. (2017). Figures (A), (C), and (E) and Figures (B), (D), and (F) show the adsorption of CH_4 and CO_2 on calcite (104), aragonite (011)-Ca, and vaterite (010)-CO_3 surfaces, respectively. Figure (G) shows a side view of the one-layer model of calcite (100) surface.

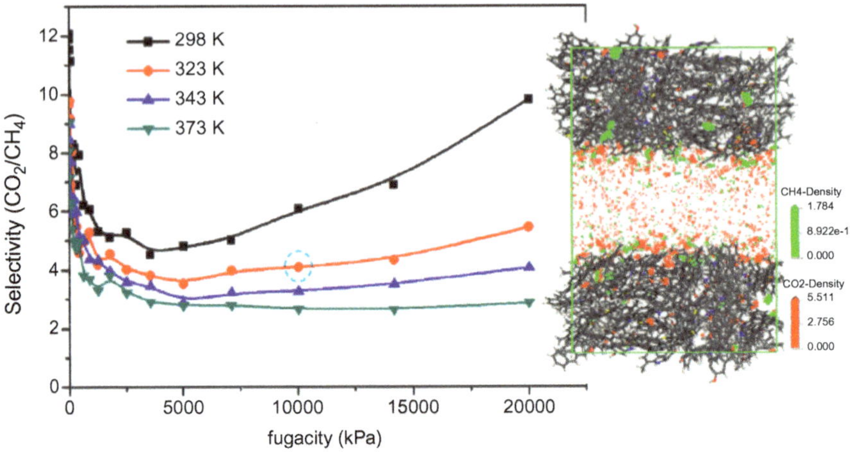

FIG. 9.11 The selectivity of CO_2/CH_4 on different applied pressure at four temperatures inside kerogen slit nanopore (Sun et al., 2017a).

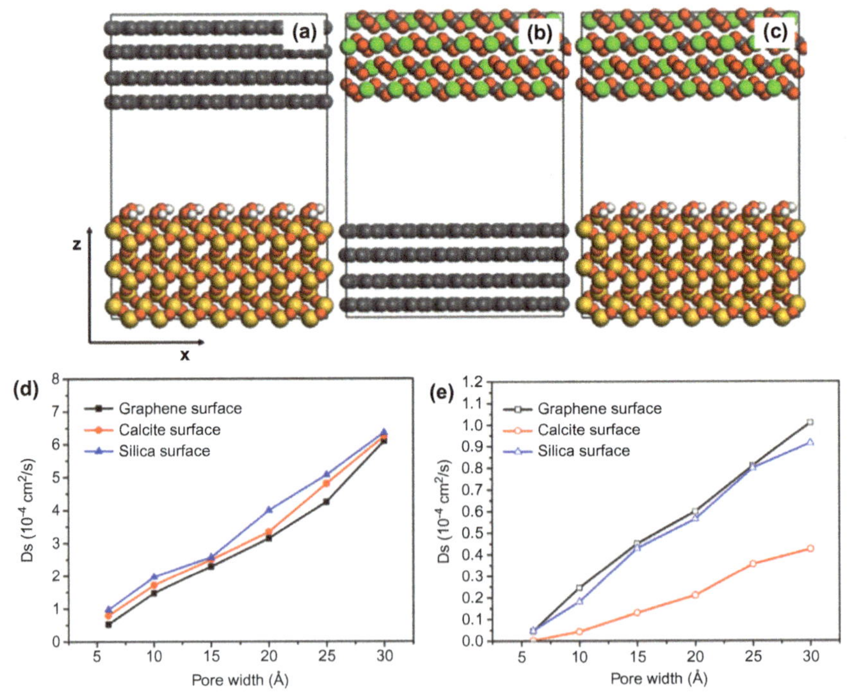

FIG. 9.12 The structures of (A) graphene-silica, (B) calcite-graphene, and (C) calcite-silica slit nanopores. The change of diffusion coefficient of (D) CH_4 and (E) CO_2 on change with the width of graphene, calcite, and silica slit nanopores (Sun et al., 2017b).

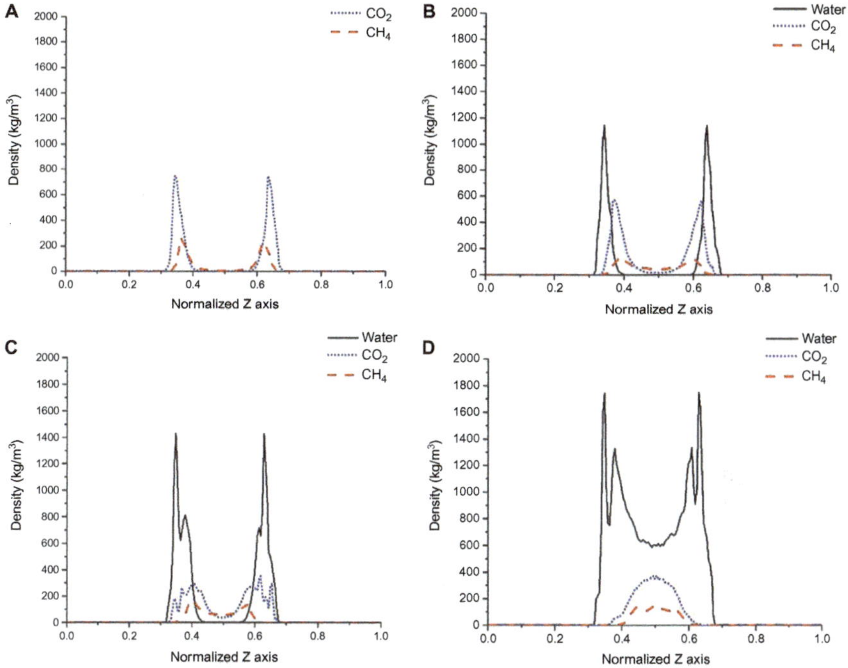

FIG. 9.13 Density profile of the adsorption of CH_4 and CO_2 on calcite slit nanopore in the presence of (A) no water, (B) 150 water molecules, (C) 500 water molecules, and (D) 1296 water molecules ($1\ g/cm^3$). The graphs reveal the anisotropic diffusion of the two gases in the presence of hydration (Mohammed and Gadikota, 2018).

with a water content of $1\ g/cm^3$. They also found that the diffusion of CH_4 is anisotropic in considered surfaces except for graphene. The explanation of the isotropic diffusion of CH_4 in the graphene surface is again explained by the strong interaction between the gas molecules and the surface (Sun et al., 2017b; Wang et al., 2018). On the other hand, the anisotropic diffusion of CH_4 on calcite is due to the distribution of ions on the surface of calcite compared to other surfaces, which affects the diffusion of water and methane (Figs. 9.9-9.13).

9.6 Estimation of ultimate recovery

9.6.1 Introduction to the process

The main point in the accurate estimation of ultimate recovery (EUR) of gas is to consider both the gas molecules that could be presently placed in the free spaces or inside the porous structure of the rock and their transport inside the rock, which is affected by the rock's pore system (Dewers et al., 2012; Heath et al., 2011). The study of gas interaction with the porous media can give an evaluation of the EUR as it affects the process of gas recovery (Stueck et al., 2016). The macroscopic properties of the fluid significantly control the recovery at the macroscopic level (Guo et al., 2020; Huang et al., 2021). There is a deviation from normal fluid behavior and transport in the tight gas reservoirs as the distance in the porous media is on the nanoscopic scale similar to the distance between the molecules of the gas trapped in the rocks (Salama et al., 2017). This change in the fluid properties is due to disruption in the thermodynamic equilibrium of the fluid inside the nanoscopic porous media. Hence, evaluating the recovery of the gas at the molecular level by computational chemistry is the best approach to get information on these details (Nelson, 2009). Different types of molecular simulations, such as DFT and MD, were used to study the gas transport properties and interactions inside the tight gas carbonates described in the next section. Molecular simulations could describe a model of the gas adsorption and transport inside the porous media. This utility includes the analysis of gas–surface interaction and the dynamic of gas inside the nanopores to estimate the affinity of the gas to various types of surfaces and the ability of the gas to be transported outside the tight porous media.

9.6.2 Analysis of gas adsorption by DFT

DFT simulations have been used widely for the study of various types of carbonaceous materials, including coal (Yang et al., 2010), activated carbon (Cheng et al., 2010; Rios et al., 2009), carbon nanotubes (Hallock and Kahng, 2004), and polymers (Carrott et al., 2010). DFT methods could provide an estimation of the electronic properties and interaction energies of the gas molecules on a surface. Few studies have used DFT methods to investigate tight gas reservoirs. In literature, the tight gas reservoir was modeled as various types of surfaces considering their different morphologies, such as the calcite structure of calcium carbonate reservoirs. A study done by our group studied the adsorption energies of CH_4 molecules at the nine different morphologies of calcite (Figs. 9.9-9.14) (Onawole et al., 2020b). In the study, the energy of each surface is also calculated to find the most stable surface and the surface with the highest adsorption affinity (highest adsorption energy). It was found that calcite (110) has a greater affinity to the surface compared to other surfaces, which results in adsorption of up to 24 molecules compared to only four molecules on the calcite (104) surface from the calculations of the capacity test by the DFT method.

In another study done by our group, we investigated the adsorption of CH_4 on the surface of calcite in the presence of point vacancies defects (Ca and CO_3- vacancies). The energy barrier analysis from the NEB calculations (Figs. 9.9-9.15) was used to explain the possibility of CH_4 migration or trapping by each vacancy and to estimate the energy required to release the methane molecule from the trapping potential. The result of the NEB

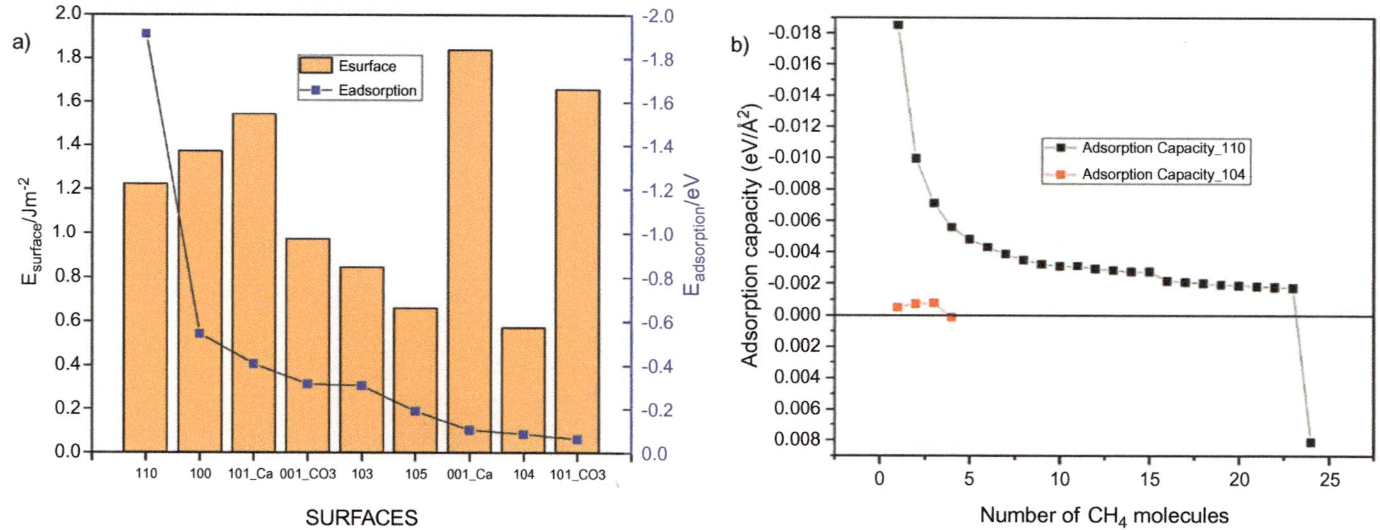

FIG. 9.14 (A) Adsorption energies of calcite at different surfaces with the corresponding surface energy of each. (B) The capacity of the gas at the most stable surface and surface that showed the highest affinity to the gas (Onawole et al., 2020b).

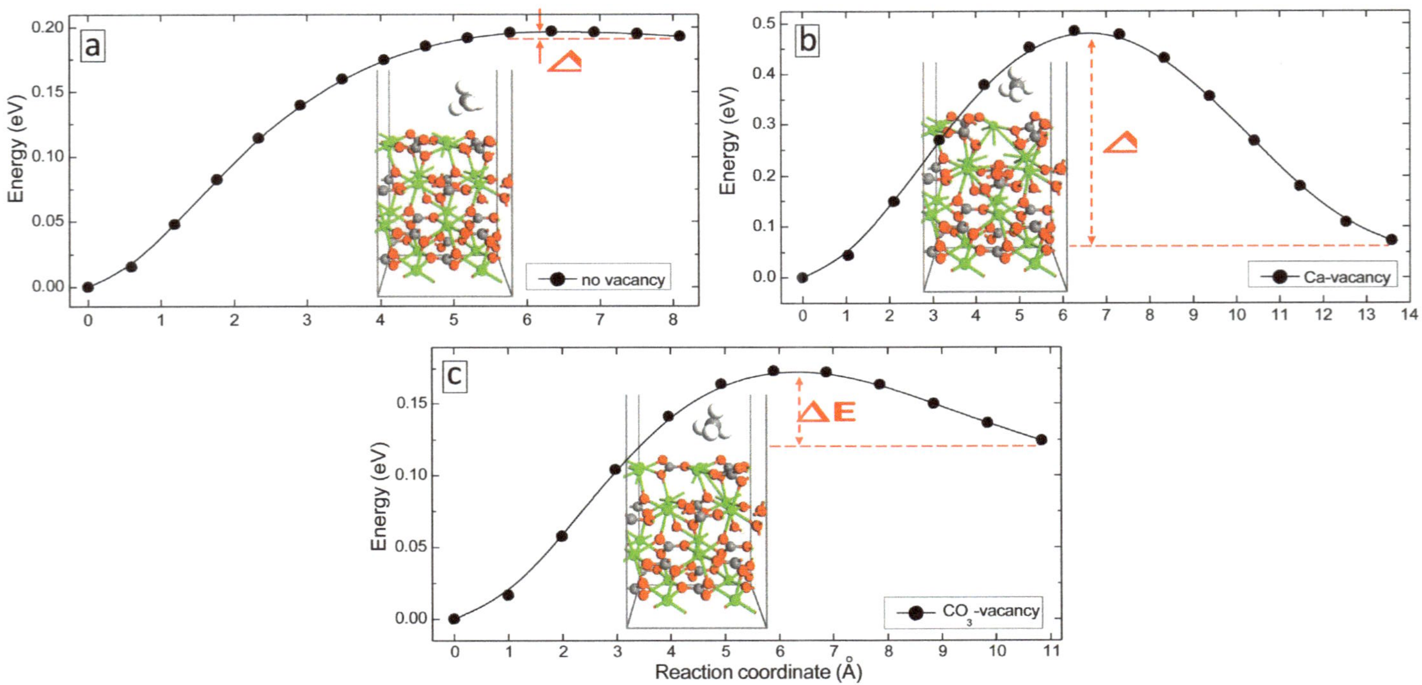

FIG. 9.15 The results of the energy barrier of the CH_4 adsorption process on vacancy free surface (A), Ca-vacancy (B), andCO_3- vacancy (C) defected surfaces are calculated by the NEB method (Berdiyorov et al., 2021).

calculations showed trapping potential in the order of Ca- > CO_3- vacancies > vacancy-free surfaces. The highest trapping of Ca-vacancy corresponds to the high adsorption affinity to CH_4 compared to the two other systems. This result concludes that the presence of point vacancies on the surface increases the trapping of CH_4, which may lead to lower diffusion.

Furthermore, natural gas consists mainly of methane along with other higher hydrocarbons, and inorganic constitutes such as carbon dioxide, hydrogen sulfide, nitrogen, and water molecules. It is crucial to consider the inorganic constitutes of the natural gas as some of them have greater interactions with rock surfaces altering the estimated natural gas properties in the case of pure hydrocarbons. Additionally, the rocks of the reservoirs experience variation in their physical properties, such as strain, caused by pressure and temperature variation and gas transport through the porous media. These variations affect the adsorption of the gases in the rock, hence, the estimation of the amount of gas in the reservoir. A previous study was done by our group aimed to analyze the adsorption of the gases present in the natural gas in the calcite surface under a range of strained surfaces (Elbashier et al., 2021). The result showed that large molecules such as large hydrocarbons (C_2H_6) and the molecules with high affinity to the surface (CO_2) are affected by strain. The stretching of the surface increased the surface affinity to CO_2 and C_2H_6 surfaces. As shown in Figs. 9.9-9.16, the charge analysis by EDD calculations showed more significant overlapping between the CO_2 and stretched surface compared to the compressed surface revealing the higher transfer of charges between the surface and the O_{atom} of CO_2.

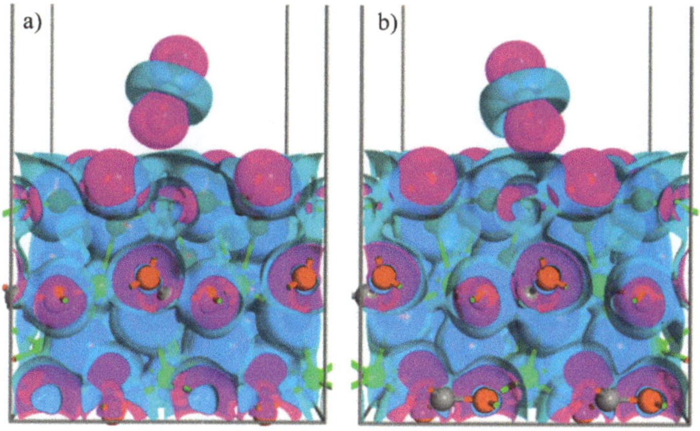

FIG. 9.16 The EDD result of CO_2 molecule adsorbed on the surfaces of calcite (104) under compression of −3% (A) and stretching of 3% (B). The isosurface plots (isovalues −0.03 1/Å3 (cyan color) and 0.03 1/Å3 (magenta color) (Elbashier et al., 2021).

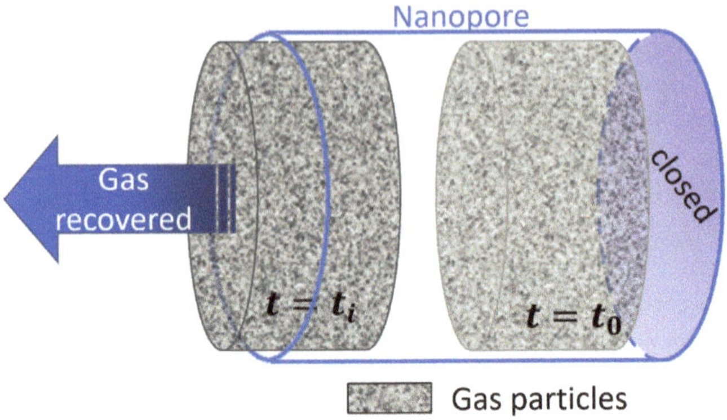

FIG. 9.17 Estimation of ultimate recovery by allowing gas flow with pressure drop using molecular dynamics method.

9.6.3 Dynamic of gas flow using molecular dynamic

An approach toward the EUR using molecular simulation is to model a porous system with a specific diameter and allow the gas to move dynamically with the pressure gradient, which could be done using the classical MD method as shown in Figs. 9.9-9.17. The gas is initially located at the end of a closed nanopore; afterward, the gas is subject to a pressure gradient from the initial state at $t = t_0$ to outside the nanopore at $t = t_i$. The amount of gas recovered outside the nanopore and the time needed for this transportation is measured using the radial distribution function and diffusion coefficient to describe the transport behavior. The same approach was used in the analysis of EOR and EGR; however, the main focus here is to assess the amount of gas recovered as a function of pressure and temperature per time and size of the nanopore.

9.7 Scale removal and inhibition

9.7.1 Introduction to the process

In oil fields, a high cost of about 1.4 billion USD annually is lost due to the problems arising from inorganic scale buildup. The presence of hydrogen sulfide (H_2S) in the gas well simulates the formation and deposition of the scale, such as iron sulfide scales, when it reacts with the ferrous ions. One major inorganic scale that blocks gas/oil wells is the iron sulfides scale. The conventional methods for scale removal, such as HCl addition, create environmental problems such as the release of toxic gases, along with the possibility of increased unfavorable corrosion of the gas/oil wells by the acid. Thus, other materials have been tested for the removal of scale, and other new environmentally friendly efficient materials have also been developed. Understanding the electronic interaction between the iron and iron sulfide molecules with the scale removal/inhibitor is crucial. Therefore, molecular simulations investigate the electronic structure and reactions' mechanisms and identify potential materials for desolvation and inhibition. Quantum chemistry methods, like DFT, provide helpful information about the scale inhibition/removal mechanism, such as binding energy, charge analysis, and molecular orbitals, that reflect the stability, reactivity, and strength of the scale removal/inhibition agent. Moreover, AIMD is used to simulate the mechanism of the scale desolvation with time, which allows to visually observe the distortion of the scale surface by the active agent. The following subsection explains the application of molecular simulation in scale removal in more detail, providing examples of studies from the literature.

9.7.2 Analysis of scale removal by DFT and AIMD methods

The nature of ferrous allows it to present in different forms of electronic spins, which affect the number of coordination, the final optimized structures, and the mechanism of its reactions. Hence, the electronic behavior of Fe is complicated due to the multiple possible electronic states, which make it difficult to estimate the correct structure and Fe complexes (Buijs et al., 2018). Here comes molecular simulation, which has excellent potential in explaining atoms and molecules' electronic structures and behaviors.

In previous research by our team, Onawole et al. investigated the interactions of different chelating agents with ferrous Fe^{2+} and ferric ions Fe^{3+}, such as EDTA, HEDTA, and DTPA (Onawole et al., 2019b). The paper compared the binding and coordination number between the different agents, for instance, DTPA containing nitrogen atoms that allowed for more coordination numbers. Furthermore, the results showed that increasing the carboxylic groups in the agents increases the chelating agent's electronegativity. The paper summarizes all the binding energies of the compounds with the two ions and further explains the nature of interaction given by the Mulliken charges and the frontier orbital analysis (molecular orbitals; HOMO and LUMO). Figs. 9.9-9.18 shows the structure of the Fe^{3+}-HEDTA complex, with the electrostatic potential (MEP) map that represents the density of electrons around the structure. Moreover, Figs. 9.9-9.18 shows the HOMO and LUMO orbitals. The results showed that the HOMO is delocalization around the carboxylic groups, indicating that they are the parts where electron transfer occurs; on the other hand, the LUMO is delocalized away from the carboxylic group. The same conclusion was observed from the MEP maps and charge analysis. The MEP map shows the charge localization around the carboxylic group, while the charge analysis clearly shows the change in the charge before and after coordinating with Ferric ions around the carboxylic groups.

Similar analyses could be done on other molecules to check if they have the potential to work as scale inhibition/removal agents. Additionally, designing

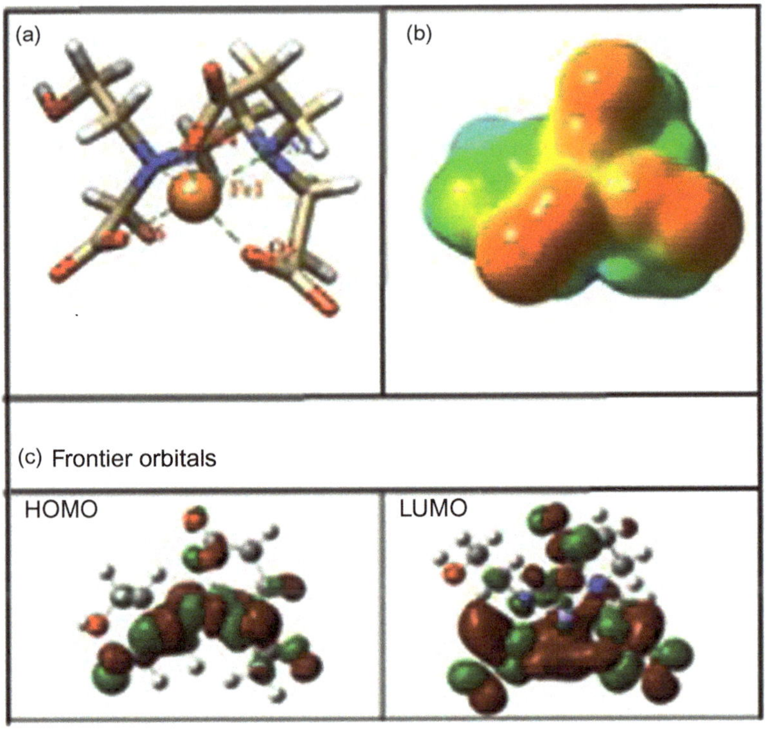

FIG. 9.18 The results of the DFT calculations for the analysis of the Fe^{3+}-HEDTA complex show the (A) optimized structure, (B) the electrostatic potential (MEP) map, and (C) the frontier orbitals (HOMO and LUMO) (Onawole et al., 2019b).

new agents is also possible, where the same type of investigations are done to study the different possible options to modify a specific agent before synthesizing it in the lab (Onawole et al., 2021). Furthermore, the effect of pH on the binding could also be addressed by applying different solvents to the system representing different solvents or adding protons to the agent representing its state at different pHs (Onawole et al., 2019a).

Periodic DFT could also be used to study the interaction of the active agents with the surface of the scale to study the adsorption and binding energy. The method could give an idea about the nature of the interaction, e.g., the formation of chemical bonds between the agents and the scale surface. Similarly, the formation of an inhibition layer on the top of an iron surface, representing the gas/oil well, could indicate the agent's potential to work as an inhibitor (Abdulazeez et al., 2019; Zeino et al., 2021).

A previous work by our group implemented the AIMD method to analyze the desolvation of the scale surface by a removal agent (Onawole et al., 2020a). The AIMD shows the interaction process as steps with time, revealing the mechanism of desolvation as shown in Figs. 9.9-9.19. In this work, the desolvation of pyrite surface (FeS) is by a chelating agent in the presence of

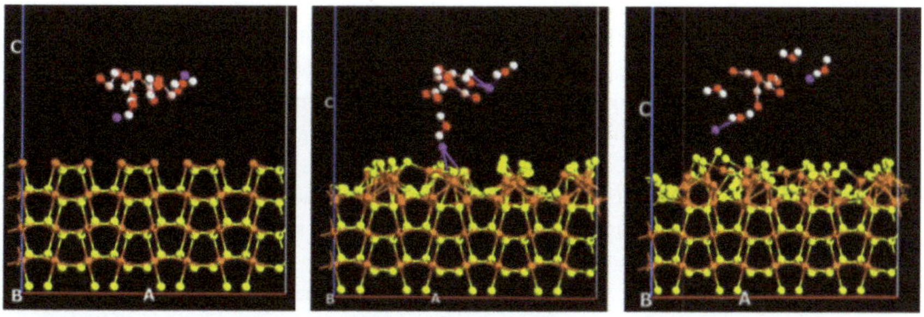

FIG. 9.19 The mechanism of pyrite surface desolvation by the agent and potassium ions calculated by the AIMD method (Onawole et al., 2020a).

potassium ions. The results of the AIMD revealed the role of K^+ ions in simulating the desolvation process by creating two bonds with the sulfur on the top layer of pyrite, causing a distortion of the surface that allows the chelating agent to chelate the ferrous/ferric ions. The same approaches, DFT, periodic DFT, and AIMD, could be employed to study the desolvation of other rocks.

9.8 Summary

An accurate estimation of the chemical and physical properties in the gas/oil reservoirs requires understanding the transport properties and the molecular interaction between the species at the nanoscale level. Therefore, molecular simulations and modeling have become important tools for investigating the petroleum parameters and pore scale. These include the estimation of the reservoirs, the long-term production, reservoir recovery, and other problems, like scale formation. Due to the confined structure, possible adsorption effects, transport, and phase behavior of the particles, the molecular interaction between the surfaces and the gas/oil particles is significant. This chapter describes how petroleum engineering problems are addressed by molecular simulations. First, an introduction to the theoretical concepts of the most used methods in molecular simulations is presented. After that, the chapter describes how these methods are used to investigate various petroleum applications as they are very useful in calculating phase interaction parameters and transport and adsorption properties. Lastly, a few applications of molecular simulation to the analysis of petroleum studies are provided, including enhanced oil recovery (EOR), enhanced gas recovery (EGR), the estimation of ultimate gas recovery (EUR), and the scale inhibition and removal.

Acknowledgments

The authors would like to acknowledge the support of the Qatar National Research Fund (a member of the Qatar Foundation) through Grant # NPRP11S-1228-170138. Furthermore, the authors acknowledge the funds provided by Qatar University provided through the Internal Grant # QUCP-CENG-2021-03.

References

Abdulazeez, I., Zeino, A., Kee, C.W., Al-Saadi, A.A., Khaled, M., Wong, M.W., Al-Sunaidi, A.A., 2019. Mechanistic studies of the influence of halogen substituents on the corrosion inhibitive efficiency of selected imidazole molecules: a synergistic computational and experimental approach. Appl. Surf. Sci. 471, 494–505. https://doi.org/10.1016/j.apsusc.2018.12.028.

Berdiyorov, G.R., Elbashier, E., Carchini, G., Hussein, I.A., Sakhaee-Pour, A., 2021. The effect of vacancy defects on the adsorption of methane on calcite 104 surface. J. Mater. Res. Technol. 14, 3051–3058. https://doi.org/10.1016/j.jmrt.2021.08.091.

Bui, T., Phan, A., Cole, D.R., Striolo, A., 2017. Transport mechanism of guest methane in water-filled nanopores. J. Phys. Chem. C 121, 15675–15686. https://doi.org/10.1021/acs.jpcc.7b02713.

Buijs, W., Hussein, I.A., Mahmoud, M., Onawole, A.T., Saad, M.A., Berdiyorov, G.R., 2018. Molecular modeling study toward development of H2S-free removal of iron sulfide scale from oil and gas wells. Ind. Eng. Chem. Res. 57, 10095–10104. https://doi.org/10.1021/acs.iecr.8b01928/suppl_file/ie8b01928_liveslides.mp4.

Carchini, G., Hussein, I., Al-Marri, M.J., Shawabkeh, R., Mahmoud, M., Aparicio, S., 2020. A theoretical study of gas adsorption on calcite for CO_2 enhanced natural gas recovery. Appl. Surf. Sci. 504, 144575. https://doi.org/10.1016/j.apsusc.2019.144575.

Carrott, P.J.M.J.M., Marques, L.M.M., Carrott, M.M.L.M.L.R., 2010. Characterisation of the porosity of polymer and carbon aerogels containing Fe, Ni or Cu prepared from 2,4-dihydroxybenzoic acid by n-nonane pre-adsorption and density functional theory. Microporous Mesoporous Mater. 131, 75–81. https://doi.org/10.1016/j.micromeso.2009.12.005.

Cheng, H.N.N., Wartelle, L.H., Klasson, K.T., Edwards, J.C., 2010. Solid-state NMR and ESR studies of activated carbons produced from pecan shells. Carbon 48, 2455–2469. https://doi.org/10.1016/j.carbon.2010.03.016 M4 - Citavi.

Collell, J., Ungerer, P., Galliero, G., Yiannourakou, M., Montel, F., Pujol, M., 2014. Molecular simulation of bulk organic matter in type II shales in the middle of the oil formation window. Energy Fuels 28, 7457–7466. https://doi.org/10.1021/ef5021632.

De Leeuw, N.H., Parker, S.C., 1998. Surface structure and morphology of calcium carbonate polymorphs calcite, aragonite, and vaterite: an atomistic approach. J. Phys. Chem. B 102, 2914–2922. https://doi.org/10.1021/jp973210f.

Dewers, T.A., Heath, J., Ewy, R., Duranti, L., 2012. Three-dimensional pore networks and transport properties of a shale gas formation determined from focused ion beam serial imaging. Int. J. Oil Gas Coal Technol. 5, 229. https://doi.org/10.1504/ijogct.2012.046322.

Elbashier, E., Hussein, I., Carchini, G., Sakhaee Pour, A., Berdiyorov, G.R., 2021. Effect of strain on gas adsorption in tight gas carbonates: a DFT study. Comput. Mater. Sci. 188, 110186. https://doi.org/10.1016/j.commatsci.2020.110186.

Fang, T., Zhang, Y., Ding, B., Yan, Y., Zhang, J., 2020. Static and dynamic behavior of CO_2 enhanced oil recovery in nanoslits: effects of mineral type and oil components. Int. J. Heat Mass Transf. 153, 119583. https://doi.org/10.1016/j.ijheatmasstransfer.2020.119583.

Farooq, U., Asif, N., Tweheyo, M.T., Sjöblom, J., Øye, G., 2011. Effect of low-saline aqueous solutions and pH on the desorption of crude oil fractions from silica surfaces. Energy Fuels 25, 2058–2064. https://doi.org/10.1021/EF1013538.

Fathi, S.J., Austad, T., Strand, S., 2011. Effect of water-extractable carboxylic acids in crude oil on wettability in carbonates. Energy Fuels 25, 2587–2592. https://doi.org/10.1021/EF200302D.

Günsel, A., Bilgiçli, A.T., Tüzün, B., Pişkin, H., Atmaca, G.Y., Erdoğmuş, A., Yarasir, M.N., 2019. Synthesis of tetra-substituted phthalocyanines bearing 2-(ethyl(m-tolyl)amino)ethanol: computational and photophysicochemical studies. J. Photochem. Photobiol. A Chem. 373, 77–86. https://doi.org/10.1016/J.JPHOTOCHEM.2018.12.038.

Guo, H., Nazari, N., Esmaeilzadeh, S., Kovscek, A.R., 2020. A critical review of the role of thin liquid films for modified salinity brine recovery processes. Curr. Opin. Colloid Interface Sci. 50, 101393. https://doi.org/10.1016/j.cocis.2020.101393.

Hallock, R.B., Kahng, Y.H., 2004. Adsorption of helium and other gases to carbon nanotubes and nanotube bundles. J. Low Temp. Phys. 134, 21–30. https://doi.org/10.1023/B:JOLT.0000012529.50135.c0.

Heath, J.E., Dewers, T.A., McPherson, B.J.O.L., Petrusak, R., Chidsey, T.C., Rinehart, A.J., Mozley, P.S., 2011. Pore networks in continental and marine mudstones: characteristics and controls on sealing behavior. Geosphere 7, 429–454. https://doi.org/10.1130/GES00619.1.

Huang, L., Zhou, W., Xu, H., Wang, L., Zou, J., Zhou, Q., 2021. Dynamic fluid states in organic-inorganic nanocomposite: implications for shale gas recovery and CO_2 sequestration. Chem. Eng. J. 411, 128423. https://doi.org/10.1016/j.cej.2021.128423.

Jalab, R., Saad, M.A., Hussein, I.A., Onawole, A.T., 2021. Calcite scale inhibition using environmental-friendly amino acid inhibitors: DFT investigation. ACS Omega 6, 32120–32132. https://doi.org/10.1021/acsomega.1c04888.

Jha, N.K., Ivanova, A., Lebedev, M., Barifcani, A., Cheremisin, A., Iglauer, S., Sangwai, J.S., Sarmadivaleh, M., 2021. Interaction of low salinity surfactant nanofluids with carbonate surfaces and molecular level dynamics at fluid-fluid interface at $ScCO_2$ loading. J. Colloid Interface Sci. 586, 315–325. https://doi.org/10.1016/J.JCIS.2020.10.095.

Le, T., Striolo, A., Cole, D.R., 2015. CO_2-C_4H_{10} mixtures simulated in silica slit pores: relation between structure and dynamics. J. Phys. Chem. C 119, 15274–15284. https://doi.org/10.1021/ACS.JPCC.5B03160/ASSET/IMAGES/ACS.JPCC.5B03160.SOCIAL.JPEG_V03.

Li, C., 2020. Molecular Simulation Study of Enhanced Oil Recovery Methods in Tight Formation. University of North Dakota.

Luo, Q., Pan, Y., Guo, P., Wang, Z., Wei, N., Sun, P., Liu, Y., 2017. First-principles calculation of adsorption of shale gas on $CaCO_3$ (100) surfaces. J. Appl. Biomater. Funct. Mater. 15, S45–S51. https://doi.org/10.5301/jabfm.5000352.

Mohammed, S., Gadikota, G., 2018. The effect of hydration on the structure and transport properties of confined carbon dioxide and methane in calcite nanopores. Front. Energy Res. 6. https://doi.org/10.3389/fenrg.2018.00086.

Neirinck, B., Soccol, D., Fransaer, J., Biest, O.V.d., Vleugels, J., 2010. Influence of short chain organic acids and bases on the wetting properties and surface energy of submicrometer ceramic powders. J. Colloid Interface Sci. 348, 654–660. https://doi.org/10.1016/J.JCIS.2010.05.009.

Nelson, P.H., 2009. Pore-throat sizes in sandstones, tight sandstones, and shales. Am. Assoc. Pet. Geol. Bull. 93, 329–340. https://doi.org/10.1306/10240808059.

Onawole, A.T., Hussein, I.A., Saad, M.A., Mahmoud, M., Ahmed, M.E.M., Nimir, H.I., 2019a. Effect of pH on acidic and basic chelating agents used in the removal of iron sulfide scales: a computational study. J. Pet. Sci. Eng. 178, 649–654. https://doi.org/10.1016/J.PETROL.2019.03.075.

Onawole, A.T., Hussein, I.A., Sultan, A., Abdel-Azeim, S., Mahmoud, M., Saad, M.A., 2019b. Molecular and electronic structure elucidation of Fe^{2+}/Fe^{3+} complexed chelators used in iron sulphide scale removal in oil and gas wells. Can. J. Chem. Eng. 97, 2021–2027. https://doi.org/10.1002/CJCE.23463.

Onawole, A.T., Hussein, I.A., Ahmed, M.E.M., Saad, M.A., Aparicio, S., 2020a. Ab initio molecular dynamics of the dissolution of oilfield pyrite scale using borax. J. Mol. Liq. 302, 112500. https://doi.org/10.1016/J.MOLLIQ.2020.112500.

Onawole, A.T., Hussein, I.A., Carchini, G., Sakhaee-Pour, A., Berdiyorov, G.R., 2020b. Effect of surface morphology on methane interaction with calcite: a DFT study. RSC Adv. 10, 16669–16674. https://doi.org/10.1039/D0RA02471F.

Onawole, A.T., Hussein, I.A., Nimir, H.I., Ahmed, M.E.M., Saad, M.A., 2021. Molecular design of novel chemicals for iron sulfide scale removal. J. Chem. 2021. https://doi.org/10.1155/2021/7698762.

Pandya, S.B., Patel, U.H., Chaudhary, K.P., Socha, B.N., Patel, N.J., Bhatt, B.S., 2019. DNA interaction, cytotoxicity and molecular structure of cobalt complex of 4-amino-N-(6-chloropyridazin-3-yl)benzene sulfonamide in the presence of secondary ligand pyridine. Appl. Organomet. Chem. 33, e5235. https://doi.org/10.1002/AOC.5235.

Puntervold, T., Strand, S., Austad, T., 2007. Water flooding of carbonate reservoirs: effects of a model base and natural crude oil bases on chalk wettability. Energy Fuels 21, 1606–1616. https://doi.org/10.1021/EF060624B.

Rezaei Gomari, K.A., Hamouda, A.A., 2006. Effect of fatty acids, water composition and pH on the wettability alteration of calcite surface. J. Pet. Sci. Eng. 50, 140–150. https://doi.org/10.1016/J.PETROL.2005.10.007.

Rios, R.B., Silva, F.W.M., Torres, A.E.B., Azevedo, D.C.S., Cavalcante, C.L., 2009. Adsorption of methane in activated carbons obtained from coconut shells using H_3PO_4 chemical activation. Adsorption 15, 271–277. https://doi.org/10.1007/s10450-009-9174-9 M4 - Citavi.

Salama, A., El Amin, M.F., Kumar, K., Sun, S., 2017. Flow and transport in tight and shale formations: a review. Geofluids. https://doi.org/10.1155/2017/4251209.

Seaton, N.A.A., Walton, J.P.R.B., Quirke, N., 1989. A new analysis method for the determination of the pore size distribution of porous carbons from nitrogen adsorption measurements. Carbon 27, 853–861. https://doi.org/10.1016/0008-6223(89)90035-3 M4 - Citavi.

Stueck, H., Houseknecht, D., Franke, D., Gautier, D., Bahr, A., Ladage, S., 2016. Shale-gas assessment: comparison of gas-in-place versus performance-based approaches. Nat. Resour. Res. 25, 315–329. https://doi.org/10.1007/s11053-015-9283-y.

Sun, H., Zhao, H., Qi, N., Li, Y., 2017a. Molecular insights into the enhanced shale gas recovery by carbon dioxide in kerogen slit nanopores. J. Phys. Chem. C 121, 10233–10241. https://doi.org/10.1021/acs.jpcc.7b02618.

Sun, H., Zhao, H., Qi, N., Li, Y., 2017b. Effects of surface composition on the microbehaviors of CH_4 and CO_2 in slit-nanopores: a simulation exploration. ACS Omega 2, 7600–7608. https://doi.org/10.1021/acsomega.7b01185.

Tüzün, B., Saripinar, E., 2020. Molecular docking and 4D-QSAR model of methanone derivatives by electron conformational-genetic algorithm method. J. Iran. Chem. Soc. 17, 985–1000. https://doi.org/10.1007/S13738-019-01835-8/FIGURES/9.

Wang, S., Feng, Q., Javadpour, F., Yang, Y.B., 2016. Breakdown of fast mass transport of methane through calcite nanopores. J. Phys. Chem. C 120, 14260–14269. https://doi.org/10.1021/ACS.JPCC.6B05511/SUPPL_FILE/JP6B05511_SI_001.PDF.

Wang, S., Feng, Q., Zha, M., Javadpour, F., Hu, Q., 2018. Supercritical methane diffusion in shale nanopores: effects of pressure, mineral types, and moisture content. Energy Fuels 32, 169–180. https://doi.org/10.1021/acs.energyfuels.7b02892.

Yang, K., Lu, X., Lin, Y., Neimark, A.V., 2010. Deformation of coal induced by methane adsorption at geological conditions. Energy Fuels 24, 5955–5964. https://doi.org/10.1021/ef100769x.

Zeino, A., Abdulazeez, I., Khaled, M., Jawish, M.W., Obot, I.B., Alhooshani, K., 2021. Mechanistic study of polyepoxy succinic acid (PESA) as green corrosion inhibitor on carbon steel in aerated NaCl Solution. Mater. Today Commun. 29, 102848. https://doi.org/10.1016/j.mtcomm.2021.102848.

Zhai, Z., Wang, X., Jin, X., Sun, L., Li, J., Cao, D., 2014. Adsorption and diffusion of shale gas reservoirs in modeled clay minerals at different geological depths. Energy Fuels 28, 7467–7473. https://doi.org/10.1021/EF5023434.

Zhang, M., Li, J., Zhao, J., Cui, Y., Luo, X., 2020. Comparison of CH_4 and CO_2 adsorptions onto calcite(10.4), aragonite(011)Ca, and vaterite(010)CO_3 surfaces: an MD and DFT investigation. ACS Omega 5, 11369–11377. https://doi.org/10.1021/acsomega.0c00345.

Chapter 10

Environmental impacts and mitigation measures of offshore oil and gas activities

Mohammad Albeldawi
Gas Processing Center, College of Engineering, Qatar University, Doha, Qatar

10.1 Introduction

Many of the environmental problems arising from the oil and gas industries and their related services have contributed to the degradation of habitats, biodiversity, water quality, air quality, and climate change. Environmental impacts include ambient air quality, marine ecology, biodiversity, and soil and groundwater contamination. However, these industries have also brought opportunities for improving the living standards of producing countries and enhanced the world's industrial economy. Therefore a balanced approach is necessary between maintaining environmental quality, controlling climate change, and hydrocarbon production. The rapid increase of oil and gas industry exploration, extraction, and plant operations has increased concerns over environmental issues, where the challenge to cope with stringent environmental regulations is mounted and pressure from local and international communities is increasing towards the need to reduce the impact of such activities at all levels. Further oil and gas expansion and developments will depend on investors' ability to develop sustainable, cleaner, more innovative, and efficient equipment, optimized unit operations, and processes that can generate fewer emissions, minimal liquid discharges, and optimum waste generation to comply with environmental regulations and standards, as well as meeting the corporate social responsibility principles. Oil and gas industry production and operations necessitate the control and management of emissions, discharges, and disposal strategies for avoiding, eliminating, or minimizing impacts and related consequences. It is equally crucial to develop an integrated emergency response and rescue plan, prepare appropriate containment facilities, and maximize reuse of treated industrial water and waste generated or minimize release into the environment.

Developments in Petroleum Science, Vol. 78. https://doi.org/10.1016/B978-0-323-99285-5.00002-8

This chapter addresses the offshore oil and gas facility development with emphasis on the central processing platform components, and the impact of offshore exploration and production on the environment. Moreover, it identifies key related regulations and legislation and highlights safety and risk management approaches, relevant sustainability impact identification and definitions, impact mitigation mechanisms, and monitoring requirements.

10.2 Offshore facilities—Central processing platform (CPP)

The upstream sector involves looking for potential underwater or underground crude oil and natural gas fields, drilling exploratory wells, and subsequently drilling and operating the wells to recover the hydrocarbons to the surface (Wikipedia, 2019). The offshore platforms are designed to conduct the exploration and production operations as well as process the extracted oil and gas. The central processing platform (CPP) is one of the main parts of an offshore platform, where the produced hydrocarbon is pumped from the wellhead through the riser for processing (drilling formulas.com, 2017). The essential function of the CPP is the separation of the produced fluid into oil, gas, and water (Fig. 10.1). The purpose of the offshore process facilities is to receive and combine hydrocarbon gas and condensate from the various production well streams and dehydrate water content to an acceptable level before transfer to the onshore processing facilities. Offshore water removal of the hydrocarbon fluids is necessary to minimize the corrosion potential and avoid

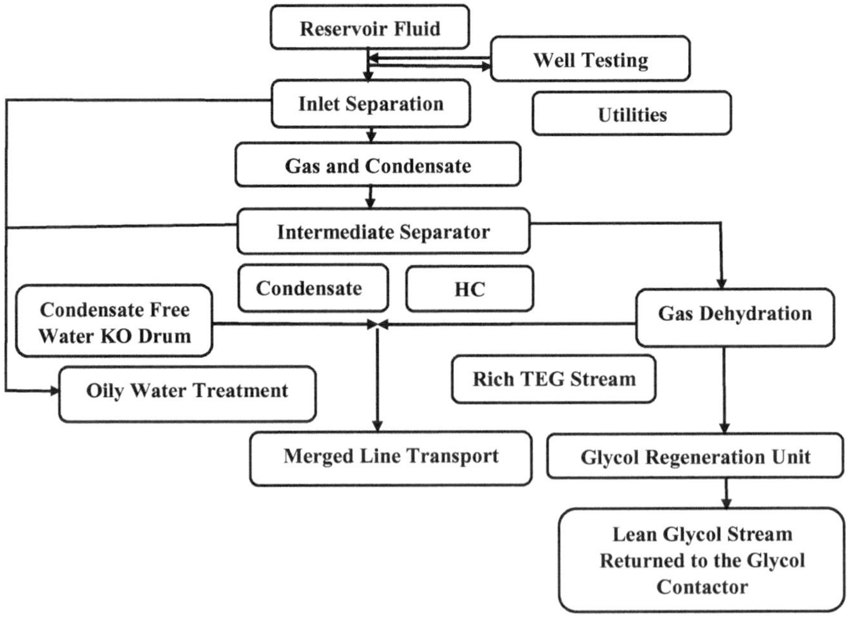

FIG. 10.1 The offshore central processing center.

hydrate formation within the subsea pipeline. The primary function of a CPP facility is to produce crude oil, gas, associated produced water, and sediment. A more complex CPP facility may include a gas sweetening system that utilizes a chemical solution to remove corrosive gases such as H_2S and CO_2. The value chain of oil and gas contains different stages that aim to explore, develop, produce, process, transport, store, refine, and market the produced crude. These stages of the value chain are generally classified into:

Upstream: this involves the initiation of the process; it concentrates on developing the wells, determining their depth, and specifying the best operating conditions to achieve the best income out of the hydrocarbon recovery. It includes exploration, drilling, and production.

Midstream: it is concerned mainly with the transportation and storage of hydrocarbons and petroleum products via tankers and pipelines.

Downstream: the final step, it means everything involved in turning crude oil and gas into thousands of valuable finished products. It includes refining, petrochemicals, liquefied natural gas (LNG) and gas to liquid (GTL) products, distribution, and marketing.

The offshore process facilities are relatively simplistic and of proven technology. The high reservoir pressure eliminates the need for any compressor. The main phase separator is the inlet separator and intermediate three-phase separator, where gas, condensate, and water are separated and routed to the gas dehydration unit, condensate free water knockout drum, and water treatment, respectively. The rich tri-ethylene glycol (TEG) stream leaves the lower section of the gas dehydration column in the glycol regeneration unit for reuse as a lean glycol stream. This approach is environmentally friendly and economically sound, as it eliminates direct routing to wastewater treatment systems or discharge to the sea and has lower operating costs.

10.3 Environmental standards and regulations compliance requirements

It is common that investors in the oil and gas industry, in general, tend to establish a comprehensive and effective management system to protect their commercial interests, maintain the marine ecosystem, and sustain healthy air, soil, and groundwater environment. Community engagement and protection of human health and safety are considered essential to ensure compliance with International Fund Corporation (IFC), World Bank (WB) (World Bank Group, 2007), and State Environmental Regulations (Meezan, 2002, 2005) in addition to other applicable standards. The exploration and production of oil and gas are associated with some environmental and societal concerns to different extents, depending on the environmental sensitivity and nearby communities. For example, the State of Qatar Law No. 4 of 1983 on the Exploitation and Protection of Living Aquatic Resources in Qatar (Meezan, 1983) states that permission is a must to perform the following acts:

a. Drilling or dredging of any debris or plant material from fisheries or onshore water in such quantities or by such methods that may harm living aquatic resources.

b. Removing or utilizing any seaweed and aquatic plants that are useful for aquatic organisms.

Several provisions are provided in the State Environmental Protection. The law includes protecting the environment and preventing all types of pollution and its adverse effects and achieving the ultimate goal of ensuring harmony between economic growth, social development, and environmental protection (Planning, 2008). The owner of a project is required to maintain a register of the effects of any pollution caused by the activities undertaken in relation to the project and to immediately rectify any breach, mitigate, and compensate as applicable for any environmental impact caused by project activities against noncompliance with the applicable regulations limit.

Table 10.1 lists key applicable international, regional, and national regulations and legislation that address issues related to marine pollution, air pollution, hazardous waste management, and safety of life, as well as combat pollution from oil and other harmful substances. The listed applicable international, regional, and national decrees and laws form the legal basis for monitoring and compliance.

TABLE 10.1 List of applicable regulations and standards.

International conventions, regulations, and standards	• International Convention for the Prevention of Pollution from Ships and its Annexes (MARPOL 73/78) (MARPOL attempts to reduce marine pollution by controlling or prohibiting discharges from ships) • International Convention for the Safety of Life at Sea (SOLAS, 1988). Entered into force on 15 April 2020 • International Convention on Oil Pollution Preparedness, Response and Co-operation 1990 (OPRC 1990) • Framework Convention on Climate Change (FCC) (Ascension Island, April 1996) • Convention for the Protection of the Ozone Layer, which established protocols on the use of ozone depleting substances (ODSs) (ratified 1996) • The Protocol on Preparedness, Response, and Cooperation to Pollution Incidents by Hazardous and Noxious Substances (HNS Protocol, 2000) • Kyoto Protocol, 2005 • World Health Organization (WHO) • PARIS Climate Change Agreement, 2015 • United Nations Convention on Biological Diversity Rio de Janeiro (1993)

TABLE 10.1 List of applicable regulations and standards.—Cont'd

	• The Convention on Civil Liability for Oil Pollution Damage 1992 (CLC 1992) • International Convention for the Control and Management of Ships' Ballast Water and Sediments, 2004 (the BWM Convention) • Basel Convention on the Control of Transboundary Movement of Hazardous Waste and Their Disposal (1989) ratified 9 August 1995 • International Convention Relating to Intervention on the High Seas in Cases of Oil Pollution Casualties and Its Annexes—Brussel 1969 • USEPA Environmental Standards—Oil and Gas Extraction Sector (NAICS 211)—(https://www.epa.gov/regulatory-information-sector/oil-and-gas-extraction-sector-naics-211) • World Bank EHS Guidance, April 30, 2007 • EU Guidelines on the Assessment of Indirect and Cumulative Impacts
Regional protocols and conventions	• ROPME Protocol on the Control of Marine Transboundary Movement of Hazardous Waste and Other Wastes (1998) • Convention on the Protection and Conservation of Wildlife and Natural Habitats in GCC Countries • Protocol Concerning Marine Pollution Resulting from Exploration and Exploitation of the Continental Shelf—Kuwait City 1989 and its guidelines • Protocol Concerning Transboundary Movement of Hazardous Wastes and on the Control of Marine Waste (1998) • International Convention about Civil Responsibility for Oil Pollution (1992) was ratified in 1993 • Kuwait Regional Convention for Cooperation on the Protection of the Marine Environment from Pollution • Protocol concerning Regional Co-operation in Combating Pollution by Oil and Other Harmful Substances in Cases of Emergency (1978) • Protocol for the Protection of the Marine Environment Against Pollution From Land Based Sources (1990) • Regional Conference of Plenipotentiaries on the Protection and Development of the Marine Environment and the Coastal Waters of Bahrain, I.R. Iran, Iraq, Kuwait, Oman, Qatar, Saudi Arabia and the United Arab Emirates
Regional coordination	• Marine Emergency Mutual Aid Centre (MEMAC) • Protocol on the control of Marine Transboundary Movement and Disposal of Hazardous Wastes and Other Wastes, Tehran, March 17, 1998

10.4 Offshore drilling and seabed dredging

Onshore oil and gas rigs are replaced by floater rigs such as drillships, semisubmersibles, submersibles, and jack-ups. The design of the floaters varies depending on the water depth and hence some of them operate in shallow and deep water. The types that are suitable for shallow water can be set on the seabed to attain stability before well drilling begins, whereas deepwater types obtain their stability by using different means, such as those using thrusters linked to satellite navigational aids to remain on station over the drill site or by being partially submerged (in the case of semisubs). During the drilling operation, the drilling fluid is circulated down the well via the drill string and through the drill bit and then returns to the surface through the annulus carrying the drilled cuttings. A shale shaker is used on the surface to remove the cuttings; the drilling fluid might be degassed, desilted, or desanded before being returned to the tanks or mud pits for recirculation. The drilling fluid exerts pressure on the formation and creates a cake layer to prevent the uncontrolled flow of the underground fluids. A blowout preventer (BOP) is installed on the top of the well below the drill floor to shut in the well if an uncontrollable flow has emerged during the drilling.

The drilling and seabed dredging activities will have impacts on the marine environment and seawater quality. The main impact of increasing suspended solids and their deposition on the marine environment constitutes a noncompliance to regulations limits of suspended solids within a defined distance from the area of activity. Other drilling and dredging impacts are related to the potential of clogging fish gills; smothering marine plant and animal species, including eggs and larva; the abundance of fish communities in the area may decline; marine mammals will be impacted and abandon the area (Chen et al., 2022). Changes in sediment quality may affect the quality of habitat for marine biological species; organic or inorganic contamination in sediments may be ingested by a benthic organism or become biologically available if resuspended in the water column. The most effective mitigation recommendations are using a silt screen around the activity areas and maintaining monitoring and control for total suspended solid (TSS) level, and selecting alternative locations and rerouting the subsea pipeline to avoid sensitive marine areas.

10.5 Environmental threats from offshore oil and gas operations

Oil and gas industry operations occur all over the world, in a diverse range of habitats and ecosystems. Both offshore and onshore operations place pressure on the local terrestrial and marine environment and related inhabitants. Fire, explosion, gas releases, and wastewater discharged in manned and unmanned platforms are an added risk to oil and gas facilities. Existing necessary protection measures can be applied, such as firewater deluge systems, emergency shutdown protocols, emergency response strategies, and evacuation of manned platforms procedures (abandonment of manned platform) to ensure appropriate protection. However, there are still gaps due to lack of appropriate education and awareness of operations people for various operational aspects of the facilities and understanding of the roles and responsibilities of established mutual aid systems and support.

While there are national, regional, and international regulations for best practices, many of the risks oil and gas industries face are site specific. These kinds of risks cannot be eliminated using a generalized framework and hence they require precautionary measures and in-depth research to address these concerns. In-house operational practices and risk management systems must be developed by oil and gas companies to decrease harmful environmental incidents and influences. The industry can attain socially useful outcomes, while avoiding potential disasters and more stringent legislation, by including environmental concerns into all aspects of daily operations. Implementing environmental proactivity is vital because unsustainable operating practices represent serious threats to the environment at both global and local levels. Oil and gas exploration often might produce harmful air emissions, cause biodiversity loss, and destroy habitats (Corporation, 2014). Oil spills and incidents could contaminate freshwater, groundwater, soil, and marine environments. The occurrence and severity of these accidents vary, depending on the stage of operation.

The estimated cost of environmental (terrestrial, marine, soil, and air quality) and public health damage given in UN reports was about $2.2 trillion per year. Moreover, it is expected that the global cost might exceed $28 trillion by 2050 (Jowit, 2010). The estimated cost of damage to the environment by the business sector revealed, among others, that the oil and gas contribution totaled $175 billion (GHGs, acid rain, smog, water abstraction, and others). In comparison, utilities account for $420 billion of damage (water abstraction, GHGs, acid rain and smog, dust and particles, and others). This cost is divided equally between local air pollution and the anthropogenic greenhouse gas emissions (Nidumolu et al., 2009). High CO_2 emissions have been proven to cause alterations in precipitation patterns as well as increase sea levels and atmospheric temperatures (Solomon et al., 2009). The incidents that arise from these multinational oil operations threaten more than just environmental quality, causing dramatic changes in the condition of the exploited habitat. Table 10.2 provides the number of oil pollution incidents and reasons for incidents in ROPME Sea Area (RSA) for different ships, as given by MEMAC. The RSA map shown in Fig. 10.2 highlights the oil spill response and salvage coordination units for dealing with oil spills and emergency response using mutual aid center protocol.

These territories are important to the native residents and traditional lifestyles and cultures are influenced. Exploring oil and gas resources might change the patterns of utilizing the lands in hunting, logging, fishing, and agriculture. The integration of environmental issues with societal safety and

TABLE 10.2 Reasons for oil pollution in RSA (1965–2020), Marine Emergency Mutual Aid Center (MEMAC), 1982, Manama, Bahrain.

Incident type	Tankers	Barges	Other V/L	Others	Total
Grounding	4	—	3	—	7
Collision	11	2	5	8	26
Explosion/fire	7	—	2	2	11
Structural failure	1	—	—	1	2
Mechanical failure	3	—	—	1	4
Sinking	3	1	9	2	15
Military action	43	—	4	20	67
International discharge	13	—	2	3	18
Leakage	18	—	2	70	90
Human error	—	—	—	1	1
Unspecified	9	4	6	28	47
Unknown/not responded	14	—	4	71	89
Total					*347*

health could create major concerns for the whole of society. To reduce undesirable impacts of these operations on the environment and society, petroleum companies should develop their objectives and practices to include all of the costs linked with environmental threats.

10.6 Integrated health safety environment and risk management system

For purposes of national environmental protection in Qatar, for example, the key binding environmental regulations are overseen by the Ministry of Environment and Climate Change and include:

- The Environment Protection Law issued via the Decree-Law No. 30 of 2002
- Environmental Protection Standard, 2005 (the executive by-laws for Law 30 of 2002)

The Environmental Societal Health Impact Assessment (ESHIA) and Risk Management study of projects usually identify the potential aspects and assess

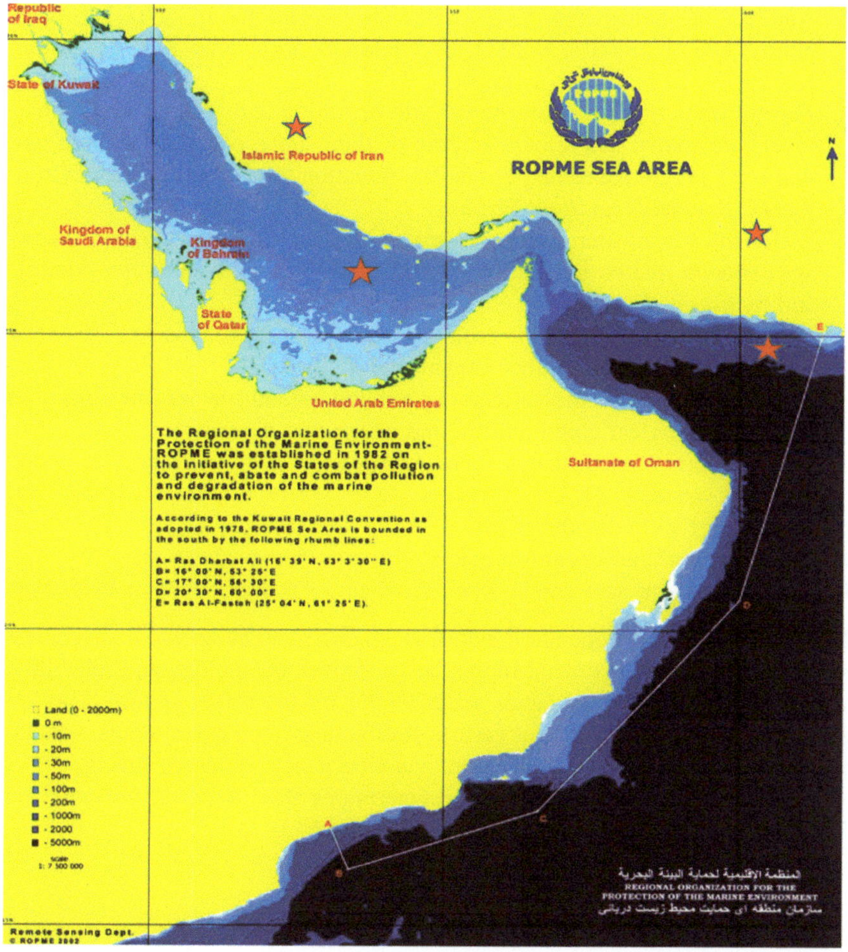

FIG. 10.2 ROPME sea area map (★) proposed oil spill response and salvage coordination units.

those aspects that have a significant impact associated with the various stages of the project, including planning, site selection, front end engineering design (FEED), construction, commissioning and start-up, normal operations and maintenance, abnormal operations, malfunction conditions, emergency scenarios, and decommissioning. The ESHIA study usually covers offshore facilities, utilities, power supply, flaring system, wastewater, cooling water, oil/gas and water separator, dehydration unit, sensitive environmental receptors, operation support facilities including warehouse, and field office, etc.

The ESHIA Study report addresses project overview, policy, legal and administrative framework, ESHIA methodology, project description and alternatives, aspects and impacts identification and quantification, mitigation and monitoring, and conclusions. Detailed environmental assessments include:

a. Ambient air quality impact assessment (assessment during the construction, normal and abnormal operation, air emissions, applicable VOC emissions, and accidental release assessment)
b. Flaring (radiation level on the off-site operation and process units, noise level, flammable gas release in case of flare flame-out, risk of toxic gas, and stack height based on permissible radiation level as per State Environmental Standards and/or API 521)
c. Water and wastewater discharge (process water, ballast water, treated sewage effluent, treated industrial effluent, produced water, seawater cooling, and hydrotested water)
d. Marine environment (marine ecology—marine biology, coral reef, seagrass, oyster bed, algae, and seabed organisms)
e. Biodiversity (affects the variety of plant and animal species, the variety of habitats, biological communities, and ecosystem diversity)
f. Waste generation and management (hazardous waste, nonhazardous waste, and inert wastes)
g. Noise level and odor threshold
h. Socioeconomic and health impacts
i. Quantitative risk assessment when the detailed design becomes available

In offshore oil and gas operation industries, to ensure an integrated management system is in place it is crucial to emphasize proven technology and reliability, with a common reliability factor set at 99.9% for the safety system. The fire and gas system is set to control various firefighting automatic systems, ventilation, fire dampers, electrical isolation, and emergency shutdown functions.

Offshore Exploration Drilling and Operations HSE Aspects, Impacts, and Required Mitigations are presented in Table 10.5.

10.7 Risk management approach

Risk management is fundamental to avoid threatening project safety and development, ensure achieving industry investment goals and business continuity, prevent work-related injury and illness, and help to achieve the cultural and operational discipline requirements. The risk management approach is represented by a detailed risk management roadmap, presented in Fig. 10.3. The roadmap comprises five steps: risk identification; risk assessment and ranking; identification of risk mitigation options; selection of the most feasible option based on SWOT analysis for the shortlisted options; and, finally, determination of the mitigation measures for the most feasible selected option. A comprehensive analysis of each of the five steps is provided to achieve successful, sustainable, and operational excellence and ultimately maintain asset integrity and ensure business continuity. Overall, the oil and gas industry's ability to successfully implement, monitor, and enforce regulations and policies will significantly improve the achievement of their risk management practices.

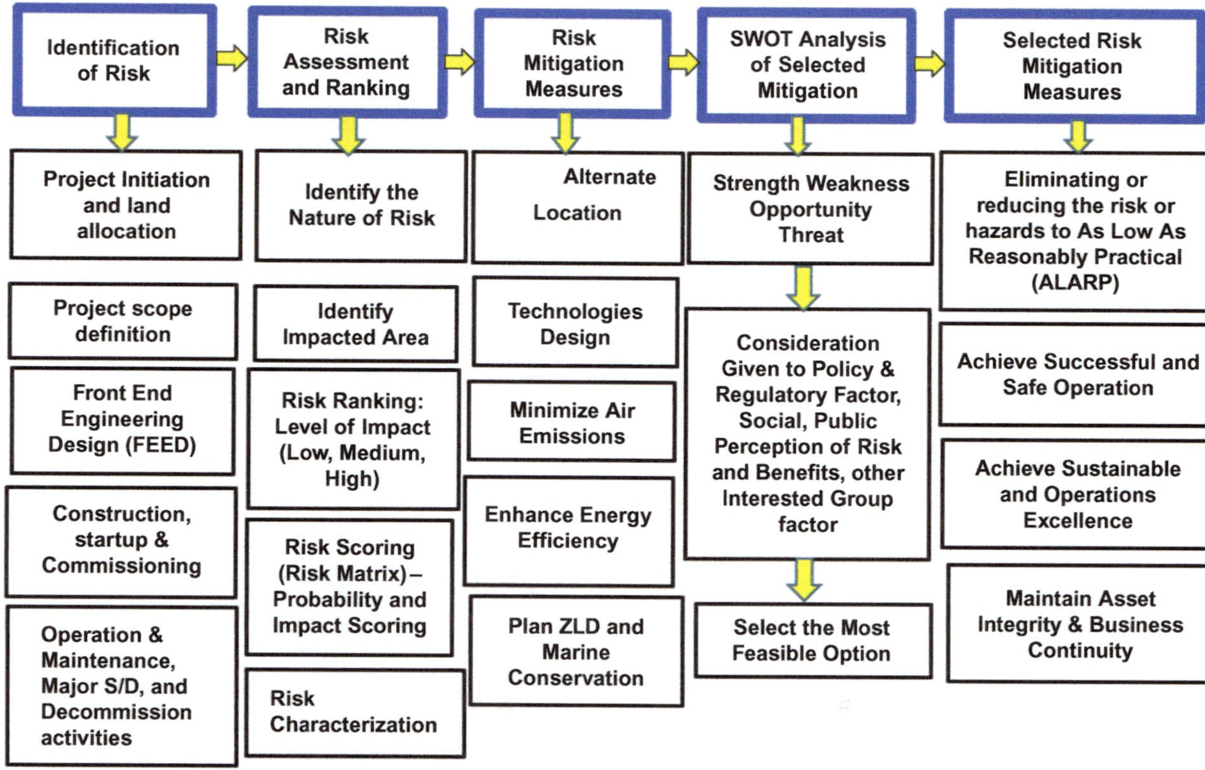

FIG. 10.3 Risk management roadmap flow diagram.

10.8 Benefits of integrated HSE and risk management system

Based on ESHIA and risk assessment studies, reliability and safety factors, and fire and gas control systems, the key advantages of comprehensive HSE and risk management systems can be summarized as follows:

a. Contribute to reduction of the serious and irreversible environmental impacts during the different stages of the project cycle.
b. Ensure system reliability and high control of process incidents and personal injury, thereby increasing operation time and minimizing associated cost.
c. Conduct risk assessment, identifying and implementing various control measures.
d. Provide roadmap to achieve regulatory compliance.
e. Biodiversity conservation and sustainable management of living natural resources.
f. Stakeholder engagement and interaction, demonstrating corporate social responsibility (CSR).
g. Maximize emissions reduction, minimizing the impact on climate change.
h. Early identification of the most feasible risk mitigation measures.
i. Optimization of energy use and efficiency.
j. Inform decision-making and condition setting.

10.9 Produced water and effluent discharge limits

Environmental regulations encourage oil and gas production companies to comply with legal effluent discharge requirements and improve the produced water treatment efficiency. Subject to the distance of the operations structures from the shoreline, the limits of the regulations may differ. The oil content removal from produced water varies based on the treatment stage, with the final treated effluent quality meeting regulations requirements. The quality and presence of specific constituents in produced water (dissolved salts, HCs, heavy metals, NORM, etc.), expected use of the effluent liquid, adherence to applicable environmental regulations, and available technologies will dictate the selection of pretreatment, treatment, and posttreatment systems suitable for achieving water quality goals. However, with technology combinations, it is possible to reduce the pollutants in produced water to or below regulation levels. The main treatment focus is commonly oil and suspended solid removal.

The PW effluent quality related to oil and grease is 42 mg/L daily maximum, and 29 mg/L monthly average (USEPA). Other monitoring and compliance requirements include toxicity, visual sheen-free oil on the surface of receiving water when discharging (shall collect and monitor marine organism tissue), and radioactivity (Radium 226 and Radium 228). The Ocean Plan of California (US) establishes a minimum criteria for the discharge of produced water into the ocean. The following lists typical wastewater effluent limitations before dilution by the ocean, where a minimum dilution of 100:1 must be achieved by an ocean outfall: Oil & Grease (<25 mg/L); Total Suspended Solids (<60 mg/L); and pH (6.0–9.0). The European Legislation related to produced water for offshore is based on the Paris Commission (PARCOM) standard target of 40 ppm for aliphatic oil. The Commission recalled that all new platforms should be equipped with separators to reduce the average oil content of a discharge to within the range of 30–50 ppm. The Commission also adopted the following recommendation from the Oil Working Group to supplement the provisional target standard: "Each national authority should set a limit on the total amount of wastewater permitted to be discharged from each platform. This limit should be fixed for each platform individually, taking into account its own treatment capacity."

For the Gulf Council countries, Iraq, and Iran, the Protocol Concerning Marine Pollution Resulting from Exploration and Exploitation of the Continental Shelf, 1989—Article IX (Regional Organization for Protection of Marine Environment (ROPME)) protocol limits the oil content of produced water to 100 mg/L daily maximum and 40 mg/L monthly average. The ROPME region uses treatment technologies such as oil/water separators and hydrocyclones to reduce oil content and comply with the protocol. The State Environmental Authority condition as part of the Consent to Operate Permit requirements states that produced water must be treated, reducing oil content to below 40 ppm, and injected into the Umm Er Radhuma formation. Table 10.3 summarizes the effluent quality standard of oil in produced water based on the region.

10.10 Ballast water discharge legal requirements

Ballast water exchange (BWE) is an effective practical method of decreasing the risk of the transfer of unwanted organisms. BWE should be considered a beneficial ballast water management practice to decrease invasion risk and species transfers. BWE is a valuable measure for all ships, especially vessels and ships before the installation of ballast water management systems. However, ships constructed in or after 2012 with a ballast water capacity of 5000 m^3 or more shall conduct ballast water management that at least meets the ballast water performance standard. The Ballast Water Management (BWM) Convention includes Regulation B-4, specifying that ballast water exchange must be undertaken within open ocean areas and two performance standards for the discharge of ballast water, D-1 and D-2.

10.11 Requirements highlights

Regulation B-4: Ballast water exchange shall be conducted at least 200 nm from shoreline and 200 m depth, or if not possible 50 nm and 200 m depth,

TABLE 10.3 Oil content of produced water effluent and discharge method

Regulations/ legislation	Oil content of produced water effluent (mg/L)	Discharge/ disposal method	Notes
ROPME	100 daily maximum, 40 monthly average	Minimum of 12 km from the shoreline (Offshore)	ROPME region uses treatment technologies such as oil separator and hydrocyclone to reduce oil content and comply with ROPME limits
MoECC (State of Qatar)	40 ppm	Injection into Um Er Radhuma Formation	Oil–water separator and hydrocyclone treatment system
USEPA	42 daily maximum	Discharge to sea (up to 4 nautical miles from shoreline)	
	29 monthly average		
	72 daily maximum	Offshore discharge	
	48 monthly average	Injection/ disposal well	
	<20 (free oil)		
	<5 (suspended solid)		
European[a]	30–50 ppm (average oil content)	Discharge to sea	Using oil–water separation system

[a]*Produced water for offshore is based on the PARIS COMMISSION (PARCOM) Standards.*

or if not possible in areas designated by the Port State. A ship shall not be required to deviate from its intended voyage and the master may decide not to carry out an exchange.

Regulation D-1: Ballast Water Exchange Standard shall meet the 95% volumetric exchange, or three times pumping through the volume of each BW tank.

10.12 Regulation D-2: Standard (all ships must meet D-2 standards by 2024)

The D-2 standard covers approved ballast water treatment systems and specifies that vessels can only discharge ballast water that meets the criteria listed in Table 10.4, where Qatar is included as an example.

The area is the largest recipient of the ship's ballast water. Annually, more than 50,000 vessels, as reported in 2016, visit this area and discharge a large amount of ballast water. With the increased amount of oil and gas development, the number of ships has likely escalated to a much higher number.

TABLE 10.4 The IMO ballast water and state of Qatar standards

Parameters	Law 30 of 2002—BW standards	Ballast water management convention (2004) (IMO BWM convention's D-2 performance standard)
Ammonia, as N	3 mg/L	—
Biochemical oxygen demand	50 mg/L	—
pH	6–9 pH	—
Chemical oxygen demand	250 mg/L	—
Visible oil and grease	Nil	—
Suspended solids	35 mg/L	—
Total oil (hexane extractable)	15 mg/L	—
Total organic carbon	100 mg/L	—
Oil and grease	15 mg/L	—
Viable organisms (Plankton)	—	<10 viable organisms per m^3 ≥50 μm and
		<10 viable organisms per mL 10–50 μm size class
Toxicogenic *Vibrio cholerae*	—	<1 colony forming unit (cfu) per 100 mL or <1 g (wet weight) zooplankton samples
Escherichia coli	—	<250 cfu per 100 mL
Intestinal enterococci	—	<100 cfu per 100 mL

The amount of ballast water is approximated by 0.5 trillion tons of ballast water discharged to sea (MEMAC Director, Personal communication). The RSA is a semienclosed water body with relatively high salinity, extensive water and air temperature fluctuation, and short cool winters and intensely hot summers. It is also characterized by a low exchange of water with open waters and high turbidity. Therefore there is also a need to manage and control the spread of harmful aquatic species in ship's ballast water by implementing a set of measures such as ballast water exchange outside the RSA (findaport.com, 2021).

The D-1 standard concerns ballast water exchange undertaken within open ocean area.

The D-2 standard covers approved ballast water treatment systems and monitoring requirements.

However, the additional requirements, in line with the ROPME protocol, state that any water discharged into RSA should be free from substances that:

a. Result in impact on the adjacent Gulf water quality and injure or are toxic or produce adverse physiological responses in humans, marine animals, or plants;
b. Produce objectionable color, odor, or turbidity;
c. Float as debris, scum, oil, or any matter to form a nuisance;
d. Settle or form objectionable deposits.

The Ballast Water Management Convention addresses one critical risk related to the introduction of invasive marine species. It is therefore important to take the necessary action to prevent or mitigate the introduction of invasive marine species (IMS) and also to harmonize the national or regional policy addressing such issues in the RSA, as it is considered the largest recipient of ships' ballast water.

10.13 Biodiversity: Legal requirement, mitigation, and compensation mechanism

Biodiversity is the diversity of all life in a given area, which includes the different types of animals and plants, the variety of habitats, biological communities, and ecological processes, and the variety of genes contained in all individual organisms. At present, many consumers and producers underestimate the importance of losing biodiversity in emerging schemes. Biodiversity contributes to quality of life, holds cultural significance, and yields aesthetic pleasure. Thus if biodiversity is to be preserved, economic activities should be replaced or modified and behavior changed that gives rise to biodiversity loss and degradation. One of the significant methods for biodiversity conservation is to manage and identify the balance between necessary compensation and mitigation measures taken to implement and create a biodiversity offset plan for industrial and commercial development. There is also a need for oil and gas industries to put in place a clear policy and guidelines for biodiversity conservation and to describe mechanisms to overcome practices leading to biodiversity deterioration. This section defines the biodiversity principles, means of mitigation, environmental legal requirements, and compensation mechanisms and impacts.

Biodiversity status can be used as an indicator of environmental conservation integrity, effective planning, and pollution control. Natural coral reefs shelter a variety of marine species and therefore maintain marine environment integrity and sustain growth. Biological diversity has a value for agriculture, social, economic, scientific, ecological, cultural, and aesthetic purposes. To maintain the integrity of biodiversity, it is important to identify environmental aspects of the project that have a significant impact during planning, construction, and operations. Equally important is to identify and evaluate species, habitats, and ecosystems in the project location and its surroundings and protect them from domestic and industrial activities. It is crucial to address issues that have significant impacts by avoiding or mitigating the impacts through redesign or relocation, and to compensate and offset any residual impacts that cannot be mitigated.

10.14 HSE key performance indicators (KPIs)

Performance evaluation of Health, Safety, and Environment (HSE) is the measure used for ensuring compliance with regulation/standard obligations and identifying areas of noncompliance, as well as highlighting industry achievements in HSE management. Indicators are individual or composite statistics that reflect important features of a system's performance towards environmental protection, safety, and risk of operations used to determine compliance status with the legal requirements, industry targets, and policy.

The following key performance indicators are used to track safety and environmental performance: fatalities; total hazardous waste, total discharges to water, total SOx and NOx emissions, total hydrocarbon emissions, greenhouse gas emissions, the volume of oil spilled that is not recovered, the number of spills affecting the environment, safety of workers and marine ecosystem, recordable illnesses, recordable injuries, and days away from work. In practice, asset integrity management on an offshore facility closely resembles process safety, focusing on the safety-critical elements (SCEs), which form crucial barriers of a system to prevent accidents and escalation of accidents once they occur. The focus is placed on the hard barriers of a system, such as operational parameters, corresponding design, and piping and instrumentation (Vinnem, 1998). The selected indicators provide benchmark measurements against specific goals to forecast the existence of a problem.

To summarize, the main objective of KPIs is to assess the relative position of the system against applicable regulations and standards, in addition to the industry objectives of achieving the business plan and its continuity. The KPIs are also used to determine the degree of change from where the system stands to where it will need to be in the future, based on using the established target and associated date of improvement. The HSE indicators are the vital precursor to the management of emissions, discharges, and waste generated.

The trends and analysis of the indicators will help to set priorities and can be used as a basis for outlining the policy and objectives. Therefore the KPIs are likely used to evaluate a possible course of action and select the most feasible option that mitigates the noncompliance issue and meets legal and regulatory requirements. Fig. 10.4 outlines the applicable HSE Key Performance Indicators and evaluation process using the data analysis generated from the operations and services along with compliance confirmation to related regulations and policy. In case the KPIs do not meet the target set or are in noncompliance with legal requirements, the steps provided in Fig. 10.4 shall be followed to achieve the target and legal compliance. The KPIs play an increasingly important role in the decision-making process (Table 10.5).

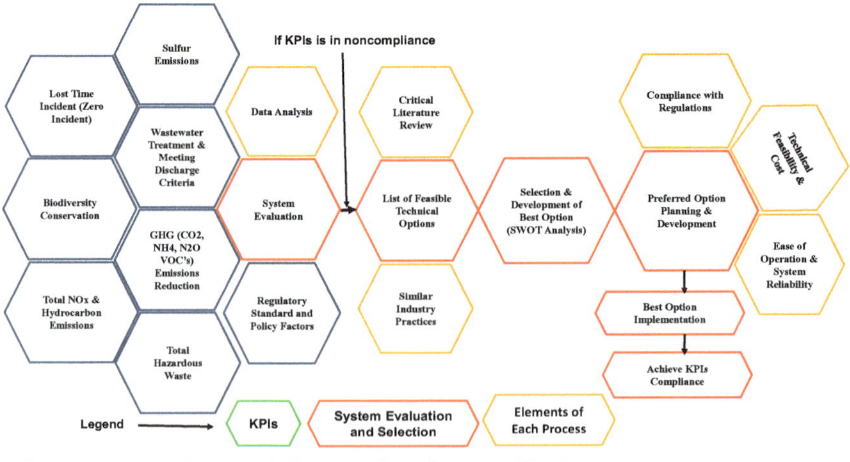

FIG. 10.4 Key performance indicators—Compliance verification.

TABLE 10.5 Offshore exploration drilling and operations HSE aspects, impacts, and required mitigations.

No.	HSE aspects and impacts	Applicable regulations and standards	Regulations and standards compliance requirements	Potential HSE negative or positive impact	Impact mitigation, compensation, and offsetting measures
1	Suspended Solids (Offshore drilling and dredging)	Qatar's Law No. 4 of 1983 on the Exploitation and Protection of Living Aquatic Resources State Environmental Standards, 2003 World Health Organization (WHO)	Permission from the competent department before performing the following acts: a. Removing or utilizing any seaweed and aquatic plants that are useful for aquatic organisms is unlawful. b. Drilling or dredging any debris or plant material from fisheries or onshore water in such quantities or by such methods that may harm living aquatic resources. c. Drilling, dredging, and disposal operations must be safely carried out without violating environmental requirements.	a. Increased suspended solids have the potential to clog fish gills b. Potential for smothering of marine plant and animal species, including eggs and larva c. The abundance of fish communities in the area may decline. Marine mammals will be impacted and abandon the area. d. Changes in sediment quality may affect the quality of habitat for marine biological species. e. Organic or inorganic contaminations in sediments may be ingested by a benthic organism or become biologically available, if resuspended in the water column.	a. Use a silt screen around the activity areas and maintain monitoring of total suspended solids level. b. Relocate or use horizontal drilling if necessary to avoid damaging the marine ecological system. c. Rerouting the subsea pipeline to avoid marine sensitive areas d. Use a numerical model to simulate sediment plumes and enable reliable impact assessments and environmentally safe planning of dredging operations (Aarninkhof and Luijendijk, 2009, 2010).

Continued

TABLE 10.5 Offshore exploration drilling and operations HSE aspects, impacts, and required mitigations.—Cont'd

No.	HSE aspects and impacts	Applicable regulations and standards	Regulations and standards compliance requirements	Potential HSE negative or positive impact	Impact mitigation, compensation, and offsetting measures
2	Flaring	Qatar Environment Protection Law was issued via the Decree Law No. 30 of 2002 and its Bylaw No. 4 of 2005.	Legal requirements for Flaring from MoECC: a. Flare to be designed to achieve 98% destruction of H_2S & VOCs. b. Smoke densities > No. 2 on the Ringleman scale, greater than 5 min shall be reported. c. Flaring of acid gases greater than 60 min shall be reported. d. Target a total maximum of 0.3% of the annual production of sweet gas, including initial and subsequent startups.	Radiation level exceeds the permissible level, air emissions and air quality deterioration impacts on off-site operation and process units, high noise level, flammable gas release in case of flare flame out, and risk of toxic gas.	Offshore, the flare boom or tower and/or high pressure, cold vent stacks should be located downwind of the Safe Area.
		Environmental Protection Standard, 2005 (executive bylaws for Law 30 of 2002).			Ensure that the sterile area that is the distance downwind of the stack is designed to a defined radiation or noise limit. The calculations were made at a defined elevation in accordance with the API-521 limits.
		Ministry of Environment and Climate Change (MoECC) Consent to Operate Permit Conditions.			Adapt the artificial intelligence (AI) platform and provide holistic flare management solution, shift from automated reports to predictive maintenance for decision-making.
		API-521			
3	Hydrogen Sulfide (H_2S) and Sulfur Dioxide (SO_2)	The Environment Protection Law issued via the Decree-Law No. 30 of 2002.	The H_2S must be reduced to low levels,	Hydrogen sulfide (H_2S) is a common contaminant and byproduct in the oil & gas industry.	Common treatment/mitigation options include: incineration (high toxic SOx emissions); acid gas injection; seawater stripping of H_2S; sulfur recovery (converts toxic gaseous H_2S into solid elemental sulfur that could be useful as a feedstock in other industries).
		The executive bylaws (2005) for Law 30 of 2002.	H_2S less than 3.5 ppm (volume) in order to meet the specifications of hydrocarbon products.	H_2S is poisonous in addition to being corrosive when moisture is present.	Use of H_2S Scavenger.
		OSHA	H_2S must be disposed of safely.		
		Best Industry Practices.			

4	Oil Spill	Executive Bylaw # 4 Standards for Environment Protection (2005) – Section 3, Part 21 (Contingency Plan to Deal with Environmental Disaster)	Oil spill emergency response plan. Records for oil discharges are to be maintained at the offshore oil platforms/ terminals. A record should include the following: ● Name and location of the terminal/platform ● The license issued for its operation	Noncompliance with environmental standards and legal regulations requirements	Improve response time & containment preparation
		State of Qatar Environmental Legislation (Law No. 19 of 2004 – Appendix 1, 2, & 3)		The offshore oil spill raises health and safety concerns for workers, the environment, and wildlife	
		The Environment Protection Law No. 30 of 2002, Section 3, Part 1. Pollution from Ships (Oil Pollution)	The owner/operator of the terminal/platform ● The activity is carried out at the terminal/platform. ● List the oil discharge treatment, control, and monitoring systems. ● Quantity and composition of liquids and substances which it is authorized to discharge annually and the rate of the discharge. ● The actual quantity of substances and liquids discharged.	Impact seawater quality and contamination of coastal habitats including beaches,	Revisit response effectiveness (conduct exercises)
		Qatar National Institution		marine environment, and ecological system.	Improve monitoring and control system
		(Article 33) IMO & IPIECA (International Petroleum Industry Environmental Conservation Association)		Impact on image and reputation.	Brainstorm the business continuity plan and ensure its comprehensiveness and effectiveness.
				May result in major cost implications (impact on tourism during response and cleanup operation)	

Continued

TABLE 10.5 Offshore exploration drilling and operations HSE aspects, impacts, and required mitigations.—Cont'd

No.	HSE aspects and impacts	Applicable regulations and standards	Regulations and standards compliance requirements	Potential HSE negative or positive impact	Impact mitigation, compensation, and offsetting measures
5	Energy Use	Qatar National Vision 2030.	Minimize the use of direct energy.	An increase in GHG emissions negatively contributes to climate change.	Cogeneration power desalting plants replacing the multistage flash (MSF) desalting system with the energy-efficient seawater reverse osmosis system
		United Nations Sustainable Development Goals.	Ensure minimum emission of GHGs.	Energy efficiency achieves saving for the customer.	Invest in heat recovery units in existing and new facilities, increasing energy efficiency and reducing fuel consumption.
		Saudi Arabia's National Energy Efficiency Program (NEEP), for example, studies the possibility of implementation of energy efficiency measures in Saudi Arabia and has set targets for reducing the country's energy intensity by 2030	Diversification of energy mix, promotes economic development.	Energy efficiency is an important utility system resource that also reduces greenhouse gas emissions, achieving savings for customers	Develop Energy Intensity Policy (Gigajoules/ton of product).
					Select percentage of energy intensity reduction target of the current figure.
					Diversify energy mix using renewable energy (solar, electrical, and hydrogen energy).
					Target to switch to the renewable energy source and to be set at (20%–25%) from the current energy source.
					Encourage residential and commercial energy efficiency.
					Conduct a study to find opportunities to reduce direct and indirect energy consumption at the Company's facilities.

6	Drilling Fluid Discharge	The Environment Protection Law was issued via the Decree-Law No. 30 of 2002 and the Bylaw No. 4 of 2005.	Industrial minerals used in drilling shall not contain more than 1 mg/kg (dry weight) of mercury and 3 mg/kg (dry weight) of cadmium as determined by TCLP analysis.	It is likely that drill muds and cuttings wastes will increase the pollution problems in an aquatic environment, with smothering effects on corals and benthos ecosystems and heavy metal pollution, thereby causing stress for fish and other aquatic organisms. Therefore disposal of cuttings and drilling wastes into the offshore environment should be totally discouraged and avoided (Adewole et al., 2010).	Control bringing the rock cuttings to the surface. Remove cuttings by vibrating screen, desand the mud, and desilt or degas before returning to the mud pits or tanks for recirculation.
		Regional Organization for the Protection of the Marine Environment (ROPME) of 1978.			Biosorption of drilling waste materials to absorb heavy metals and diesel components to reduce the effect on marine organisms may prove beneficial.
		World Health Organization standards.	Other compounds used in drilling would meet the specifications laid down by ROPME from time to time.	However, drilling mud exerts pressures that help to keep underground (oil, gas, and water) pressures under control.	Product substitution, solids control equipment, land-based treatment and disposal, and onsite subsurface injection (USEPA) (Adewole et al., 2010).
		USEPA, 2000			Oil and gas offshore companies shall evaluate the project area for the presence of corals, seagrass, oysters, etc. Also, it is important to adhere to the regulatory authorities' conditions in maintaining appropriate separation distance from any drilling discharges
7	Energy Efficiency	Qatar National Vision 2030.		Efficiency in energy use will have a positive impact on the level of GHG emissions and the cost of production, minimize the impact on the environment, and eventually minimize contribution to climate change.	a. Implement heat recovery units in existing and new facilities, increasing energy efficiency and reducing fuel consumption. b. Efforts directed at reducing flaring (such as the implementation of a flare management system and JBOG project implementation), and GHG reduction (through flaring reduction, GHG Life Cycle Assessment, among others) are directed at mitigating this impact on a regional and global level.
		United Nations Sustainable Development Goals			
		Best industry practices.			

Continued

TABLE 10.5 Offshore exploration drilling and operations HSE aspects, impacts, and required mitigations.—Cont'd

No.	HSE aspects and impacts	Applicable regulations and standards	Regulations and standards compliance requirements	Potential HSE negative or positive impact	Impact mitigation, compensation, and offsetting measures
8	Process Safety	Occupational Health and Safety Administration (OSHA) Process Safety Management (PSM)	The OSHA PSM focuses on the protection of workers within the facility, while the EPA RMP protects the public and the environment.	Process safety is an important component of overall safety management.	Improves the management of critical safety procedures and integrity programs (such as inspection of equipment and piping, flanges, gaskets, corrosion, dead-leg, and vibration surveys).
		Environmental Protection Agency (EPA) Risk Management Plan (RMP)			Revalidate safety studies to ensure effective controls are in place to help mitigate operational risk.
		Control of Major Accident Hazards (COMAH) Regulations - UK		Inappropriate process safety may result in major incidents or hazards (fires, explosions) that could arise from the release of hazardous materials and cause equipment damage.	Conduct Quantitative Risk Assessment Study
		Industries vary in their process safety management system application between using procedures, enforcement of regulations, and emergency plans.			Develop an integrated Operations Risk Process and Integrated Safety Procedure.
				Human or process safety errors hurt the health and safety of the workforce, the company's assets, and the local communities, and cause property damage, loss of production, and environmental damage.	Ensure process safety by enforcing compliance with the stringent guidelines that secure hazardous material through a safe containment process.
					Implement surveillance over all processes and operations units as well as regular inspections and testing in accordance with strict performance expectations.

9	Normally Occurring Radioactive Materials (NORM)	The State of Qatar executive regulations of the decree-law no. 31 of 2002 concerning radiation protection.	Adhere to the regulations compliance in terms of safe handling, storage, monitoring, and reporting.	Exposure to NORM creates risks to health. In addition, the spread of NORM causes a risk of contamination to assets and the environment.	Following the principle of ALARA (As Low As Reasonably Achievable), risk reduction procedures shall be developed and implemented, at a minimum for operation sources both offshore and onshore.
		Basel Convention on the Control of Transboundary Movement of Hazardous Materials and Their Disposal			Limit the exposure from the radiation sources to levels that are as low as reasonably achievable and always within the established dose limits.
		Normally Occurring Radioactive Materials (NORM), Ministry of Municipality and Environment Decision No. 45 of 2013 on Waste Management Producers of NORM.	Adherence to the regulations promotes compliance with the Decree Number 31 of 2002 on Radiation Protection and Executive Regulations in the state of Qatar		Store NORM in confined and shaded areas as per Ministry of Environment requirements.
					The storage facility shall be equipped with a continuous radiation monitoring device to monitor the radiation level inside the storage area.
10	Produced Water	Environmental Protection Standards (2003)	Oil and grease 40 mg/L (monthly average) and 100 mg/L (maximum)	Produced water provides a substantial challenge from economic and environmental perspectives.	Different means of mitigation of produced water impacts, subject to its constituents, intended use, regulations requirements, and available technologies.
		Protocol Concerning Marine Pollution Resulting From Exploration and Exploitation of the Continental Shelf. Adopted in Kuwait on 29 March 1989 (Article IX).			Reinjection of produced water into the reservoir for EOR or deep aquifer.

Continued

TABLE 10.5 Offshore exploration drilling and operations HSE aspects, impacts, and required mitigations.—Cont'd

No.	HSE aspects and impacts	Applicable regulations and standards	Regulations and standards compliance requirements	Potential HSE negative or positive impact	Impact mitigation, compensation, and offsetting measures
11	Ballast Water and Transportation of Marine Invasive Species	Regional Organization for the Protection of the Marine Environment (ROPME)	Problem Statement:	Ballast water quality may vary depending on uptake location (e.g., from estuaries, shallow water, deep water, high or low salinity waters, rich in aquatic life, clean or contaminated).	Ballast water is an essential requirement to stabilize ships during voyages
		Ballast Water Management Convention (2004)	Ballast water arrives at the RLC port from various sources around the world		
		Marine Emergency Mutual Aid Centre (MEMAC)	Invasive species transferral through ballast water is recognized by IMO as one of the four greatest risks threatening the world's oceans today.	An estimated variety of 700–10,000 marine organisms are transported worldwide from one area to another through ballast water transportation (Qatargas, 2022).	
		The Environment Protection Law issued via the Decree-Law No. 30 of 2002	Restrictive legislation is being pursued by IMO to curtail the spread of invasive species.		
		Environmental Protection Standard, 2005 (executive bylaws for Law 30 of 2002)	Qatar is signatory to various national and international treaties and conventions for the protection of the marine environment, i.e., MARPOL 73/78 and various ROPME conventions		
			An increasing number of vessels are visiting the RLC port to take on the product	When discharged into new environments, these species may become invasive, severely disrupting the native ecology with major impacts on the health of the Gulf in general and economy and human health in particular.	

12	Biodiversity	World Heritage Convention, Convention on Biological Diversity, Framework Convention on Climate Change, and International Oil Pollution Damage Convention	Qatar National Biodiversity Strategy and Action Plan (2015–2025).	With the rapid loss of biodiversity, human-induced environmental destruction has eliminated habitats, killed living organisms, reduced genetic diversity, and caused the rate of species extinction to increase dramatically (26).	Use alternative drilling and piping routings such as directional drilling that allows deviation in preferred directions through strata, even horizontally, over considerable distances and depths.
		MARPOL Requirements			
		UN Convention on The Law of Sea (UNCLOS)	The MoECC & Qatar Energy and affiliates are committed to preserving and promoting biodiversity in all areas where they work or manage. This includes the conservation of high-value areas and habitats, and reclamation of marine sites and areas disturbed.	*Pseudo-nitzschia* has been associated with the increasing number of harmful algal blooms occurring in the UAE. Environment Agency Abu Dhabi (EAD) believes this species was introduced via ballast water discharged by passing ships.	Habitat enhancement, restoration or recreation of the same habitat nearby or elsewhere.
		Agenda 21 - United Nations Conference on Environment & Development Rio de Janeiro, Brazil, June 1992.			
		State of Qatar Environmental Legislation (Law No. 19 of 2004 – Appendix 1, 2, & 3)			
		The Environment Protection Law was issued via the Decree-Law No. 30 of 2002.			
		Qatar National Institution (Article 33)	Agenda 21: Section II. Conservation and management of Resources for Development (Chapter 15. Conservation of biological diversity)	Large blooms of *Noctiluca miliaris* and *scintillans* have appeared off the coast of Oman. A *Noctiluca* red tide caused a massive fish kill in Oman in 1988, which triggered the beginning of a phytoplankton monitoring program in Oman	Provide opportunities for sustainable management of biodiversity.

Continued

TABLE 10.5 Offshore exploration drilling and operations HSE aspects, impacts, and required mitigations.—Cont'd

No.	HSE aspects and impacts	Applicable regulations and standards	Regulations and standards compliance requirements	Potential HSE negative or positive impact	Impact mitigation, compensation, and offsetting measures
		Qatar National Vision 2030 – 4th pillar "Environmental Management" considering the four principal subjects of Biodiversity, Climate Change, Water & Wastewater, and Waste Management.		*Gyrodinium instiatum* was present in large numbers during a massive fish kill in Kuwait in 2001	Enhance awareness and appreciation of the value and benefits of conservation and sustainable use of biodiversity.
		ROPME (Regional Organization for the Protection of Marine Environment).		*Gymnodinium catenatum* has been associated with an increasing number of harmful algal blooms occurring in the UAE. This species was the dominant strain during many of the HABs in the UAE. Environment Agency Abu Dhabi (EAD) believes this species was introduced via ballast water discharged by passing ships.	Agenda 21 - Promote the rehabilitation and restoration of damaged ecosystems and the recovery of threatened and endangered species.

13	Chemical Use and Discharge	Environmental Protection Standards (2003)	• Storage/use of the following chemicals should be avoided: mercury, cadmium, cyanide, chlorophenols, polychlorinated biphenyl, terphenyl, and other persistent organo-halogens, organo-tins, and other organo-metallic compounds. • Preparation of a Chemicals Use Plan is required. Operators must notify the competent state authority if an additional chemical is to be used and has the potential to escape to the marine environment.	Cause potential hazards of improperly handled chemicals. Chemicals pose a risk to the product, marine environment, and personnel.	A process chemical inventory shall be maintained for each storage and use area.
		Protocol Guideline on the use and storage of chemicals in offshore operations (Protocol Concerning Marine Pollution Resulting from Exploration and Exploitation of the Continental Shelf, ROPME, Feb 17, 1990).			
		Kuwait Regional Convention for Cooperation on the Protection of the Marine Environment from Pollution, 1978.			Material Safety Data Sheet shall be made available for reference on site.
		The Safe Disposal of Hazardous Wastes, The Special Needs of Developing Countries, Volume 1 of 3, A joint study sponsored by the World Bank, the World Health Organisation, and the United Nations Environment Programme (UNEP), 1989, Roger Balstone et al.			

Continued

TABLE 10.5 Offshore exploration drilling and operations HSE aspects, impacts, and required mitigations.—Cont'd

No.	HSE aspects and impacts	Applicable regulations and standards	Regulations and standards compliance requirements	Potential HSE negative or positive impact	Impact mitigation, compensation, and offsetting measures
14	Mud and Drill Cutting Impact on Benthic Biota/ seabed Environment	Protocol Guideline on the Voluntary Regulation of the Disposal of Drill Cuttings on the Seabed, ROPME, Feb. 17, 1990	Use appropriate nontoxic drilling fluid and mud type.	Impact on marine sensitive areas.	Rerouting the subsea pipeline(s)
					Use drilling waste management hierarchy of waste minimization and recycling
					Treat fluid produced and dispose of it using different treatments and disposal methods, such as thermal incineration, thermal desorption, bioremediation, solvent extraction, etc.
		Various regulations are imposed by various regulatory bodies according to the discharge limit for various chemicals, grease, and oil and discharge limits, locations, and boundaries.	Meeting discharge limits of the country related regulations	Discharge drill cuttings near to sea surface to facilitate dispersion and minimize deposition thickness on the seabed.	Maintain monitoring tor TSS level.
					Use a silt screen around the activity area and ensure its effectiveness is maintained
					Conduct a marine ecological survey around the drilling well during the project period.
15	Noise Impact on Marine Biota and Platform Personnel During Drilling	(National Geographic, 2020)		Noise pollution is considered to be any unwanted or disturbing sound that affects the health and well-being of humans and other organisms.	Prepare a marine mammal protection plan – which includes a provision for protection from noise/vibration.
				The most common health problem it causes is noise-induced hearing loss (NIHL).	Industry to conduct marine mammal watches prior to initiation of drilling.
				Noise pollution also impacts the health and well-being of wildlife.	

13	Pipeline Corrosion	Corrosion standard and control in pipeline industry (ASTM International, 1973)	Provide corrosion prevention and control procedure.	Uncontrolled corrosion can cause rupture or leak of a pipeline resulting in the release of hazardous substances.	Use effective pigging and chemical inhibition programs. Pigging of pipeline segments to remove liquids, solids, and debris.
		National Association for Corrosion Engineers (NACE), 2022 National Association for Corrosion Engineers (NACE) Standards and specifications related to oil and natural gas infrastructures (iii)-Examples	Establish a corrosion management system.	Corrosion could have consequences for the environment, humanity, and production loss. Corrosion hazards can cause pipeline failure and put at risk the safety and well-being of both employees and the public.	Control process parameters such as pH, temperature, and the pressure of the pipelines.
		a. Corrosion Control of Submerged Areas of Permanently Installed Steel Offshore Structures Associated with Petroleum Production- SP0176–2007;	The integrity of oil & gas pipelines is of utmost importance.	Corrosion can lead to severe damage to process or unit operation.	Apply corrosion inhibitor to provide a protective barrier on the inside surface of the pipe.
		b. Internal Corrosion Direct Assessment Methodology for Pipelines Carrying Normally Dry Natural Gas (DG-ICDA)-SP0206–2016;	Corrodents in the O&G industry that need control include CO_2, H_2S, bacteria, oxygen, and acids.	Corrosion of subsea pipelines may have an effect on other subsea pipeline systems and can result in significant environmental damage and financial consequences.	Biocides are the most effective chemicals used to terminate microbiologically influenced corrosion.
		c. Wet Gas Internal Corrosion Direct Assessment Methodology for Pipelines-SP0110–2018;	The main corrosion types that need to be addressed are sweet corrosion (CO_2), sour Corrosion (H_2S), and microbiologically induced corrosion		Oxygen corrosion in pipelines can be mitigated by the use of gas blanketing, vacuum deaeration, and oxygen scavengers.
		d. Internal Corrosion Direct Assessment Methodology for Liquid Petroleum Pipelines-SP0208–2008;			Adequate risk analysis during design, construction, and inspection stages.
		e. Petroleum and natural gas industries - Materials for use in H_2S-containing environments in oil and gas production-ANSI/NACE MR0175/ISO 15156-2015 (Jayasinghe, 2021)			Implementation of appropriate risk-based management methodology

Continued

TABLE 10.5 Offshore exploration drilling and operations HSE aspects, impacts, and required mitigations.—Cont'd

No.	HSE aspects and impacts	Applicable regulations and standards	Regulations and standards compliance requirements	Potential HSE negative or positive impact	Impact mitigation, compensation, and offsetting measures
17	Oil, Condensate, LNG, GTL, Elemental Sulfur Transportation Fleet	IMO Regulations No. 13 & 14, of MARPOL Annex VI,	All ships to meet the new requirements of MARPOL - Annex VI, Reg.14 for using Low Sulfur Fuel Oil (0.5% m/m) on board.	NOx, SOx and VOCs emissions to the atmosphere cause deterioration of air quality and health impact.	Ensure compliance with the new MARPOL – Annex VI, Regulation 14 requirements.
		Qatar Energy Ports Regulations (Ras Laffan Industrial City & Mesaeed Industrial City)		Negatively contribute to climate change.	Implement advanced solutions that ensure efficient hydrocarbon products and by-product transportation across the world in a manner that generates minimal environmental impact.
		Related MoECC Consent to Operate Permits requirements.			Implement the use of LNG as a bunker fuel. The LNG fuel has major potential for diminishing the environmental impact of shipping, as vessels typically used to burn fuel oil, which is less friendly to the environment.

18	Waste Management	The Environment Protection Law was issued via the Decree-Law No. 30 of 2002.	The waste management mandate shall focus on the development and implementation of cost-effective technology and alternative process for reducing waste, optimizing resource use and improving production efficiency, recovering, reusing, and recycling wastes.	Increased operating costs, impact on public health and environmental damage, increased risk of worker health and safety elevating the risk of liability, damage to corporate image, and lowered global competitiveness.	Policy statement and set target
		Bylaw No. 4 of 2005 Environment Protection Law c) Bylaw.	Addressing the by-product recovery and pollution prevention technologies shall be aimed at reducing, recovering, reusing, and recycling wastes		Review and analyze current waste flows including residential, commercial, and industrial wastes
		Annexes 2005 of Environment Protection Law.			Provide alternatives for waste management based on environmental impact assessment findings, cost implications, and socioeconomic factors
		USEPA Integrated Waste Management Plan.			Adopt the developed breakthrough related technologies as a strategic option to handle an integrated waste management system that is environmentally sound, responsible, and sustainable with a minimal detrimental effect on human health and life and on the ecosystem.
		The Basel Convention, Secretariat of the Basel Convention, Geneva Executive Centre, Switzerland, 1992.			
		US EPA Leachate tests to determine the toxicity of a waste product, US Environmental Protection Agency in 40CFR Part 261.24 as Method 1311 and subpart D, section 261.35.			Plan and implement a Comprehensive Environmental Control, Monitoring, and Audit Program.

Continued

TABLE 10.5 Offshore exploration drilling and operations HSE aspects, impacts, and required mitigations.—Cont'd

No.	HSE aspects and impacts	Applicable regulations and standards	Regulations and standards compliance requirements	Potential HSE negative or positive impact	Impact mitigation, compensation, and offsetting measures
19	Sewage	The Environment Protection Law was issued via the Decree-Law No. 30 of 2002.	Ships and offshore marine platforms are prohibited from discharging untreated sewage into the internal waters, the territorial sea, or the exclusive economic zone.	Impact on seawater quality and biological diversity, impact on the fishing zone, recreation areas, natural reserve, tourism, and economic impact.	Comply with environmental legal requirements and MoECC Consent to Operate Permit conditions.
		Environmental Protection Standard, 2005 (The executive bylaws for law 30 of 2002).	Dispose of sewage subject to the following standards and procedures:		
			a. The ship or offshore platform shall be provided with a sanitary treatment unit.		
			b. Treated sewage effluent is to be discharged at a distance not less than (4) nautical miles off the coast.		
			c. Untreated sewage effluent can only be discharged beyond 12 nautical miles from the shore.		
			The ship or offshore platform shall discharge sewage effluent accumulated in the tanks gradually and at		

			moderate rates. The ship is to be navigating at a speed of not less than 4 knots at the time of discharge.		
			The discharge should not result in floating objects nor change the color of the water.		
			All treated sewage must comply with discharge standards in Annex (3/2) and Annex (4) of the Final Executive By-Law of 2005 (Standard for Environmental Protection).		
20	Reduction in Biological Diversity	The United Nations Convention on Biological Diversity, 1993.	Preservation of coastal ecosystems and sustainability of marine resources	Negative effect on biota health,	Comply with the national strategy plan ensuring the protection of biodiversity and the ecosystem.
		Nagoya Protocol, October 2010 (Supplementary agreement to the convention on biological diversity).			
		World Heritage Convention, Convention on Biological Diversity.	Preservation of key terrestrial biodiversity through effective management of protected areas	Loss of rare or endangered species, and	Having biodiversity conservation incorporated into national planning processes for sustainable development.
		Framework Convention on Climate Change, and	Protection of genetic resources and ensuring their equitable sharing	Loss of critical productive habitats	Increased awareness and education enhance local capacity for biodiversity conservation.
		International Oil Pollution Damage Convention.			
		ROPME (Regional Organization for the			

Continued

TABLE 10.5 Offshore exploration drilling and operations HSE aspects, impacts, and required mitigations.—Cont'd

No.	HSE aspects and impacts	Applicable regulations and standards	Regulations and standards compliance requirements	Potential HSE negative or positive impact	Impact mitigation, compensation, and offsetting measures
		Protection of Marine Environment).			
		IMO/MARPOL Requirements			
		State of Qatar Environmental Legislation (Law No. 19 of 2004 – Appendix 1, 2 & 3)			
		The Environment Protection Law was issued via the Decree-Law No. 30 of 2002.			
		Qatar National Institution, Article 33			
		Qatar National Biodiversity Strategy and Action Plan: 2015–2025			
		Qatar National Vision 2030–4th pillar: Environmental Management Considering the four principal subjects of Biodiversity, Climate Change, Water & Wastewater, and Waste Management.			

21	Ship Collisions with Platform	Country-specific policy, procedures, management plan, and follow appropriate guidelines for the design of offshore structures against ship collisions and conducting risk assessment study requirements.	Effective traffic monitoring.	Major hazards in connection with the production of hydrocarbons from offshore platforms.	Use computer simulation techniques to ensure proper prediction of the impact forces, impact resistance of concrete as well as steel platform base structure.
					Adequate preventative and protection measures.
			Apply proper design, fendering, and operational limits.	Collision is one of the significant risks contributing potentially to an accident of very serious consequences.	Test lifeboats and conduct evacuation drills.
			Using manned platform radar system.	Endangering human life safety, structure, and marine environment.	Ensure rescue by standby vessel or navy support.

10.15 Conclusions

The environmental impacts of offshore oil and gas exploration, drilling, production, and operation are addressed and the importance of controlling and managing emissions and discharges, and disposal strategies for avoiding, eliminating, or minimizing impacts and consequences are emphasized. In particular, the impact of drilling and dredging activities on the marine environment is highlighted as one of the main threats to the marine ecosystem. Moreover, the key related regulations and legislation and safety and risk management approaches are detailed. The chapter emphasizes that risk management is fundamental to avoid threats to project safety and development and ensure achieving industry investment goals and business continuity. The risk management approach represented by a detailed risk management analysis is the roadmap to achieving regulatory compliance. Analysis of each of the risk assessment steps is provided to achieve successful, sustainable, operational excellence and ultimately maintain asset integrity and ensure business continuity. The key advantages of the HSE and risk management systems are summarized.

Produced water and ballast water effluent discharge requirements are discussed, and ballast water exchange and ballast water management convention requirements highlighted. The key performance indicators are presented and the steps required to conduct further system evaluation are explained, as the KPIs are likely used to evaluate possible courses of action and to select the most feasible option that mitigates the noncompliance issues and meets legal regulatory requirements. HSE aspects, impacts and required mitigations, compensation, and offsetting measures are thoroughly presented, addressing concerns from exploration, drilling, and dredging to subsea pipeline and ship collision with a platform, for example.

Abbreviations

BWE	ballast water exchange
BWM	ballast water management convention
CPP	central processing platform
FCC	framework convention on climate change
GHG	greenhouse gas
HAB's	terrestrial and aquatic habitats
IFC	International Fund Corporation
IMO	International Maritime Organization
KO	knock out drum
MARPOL	International Convention for the Prevention of Pollution from Ships
MEMAC	Marine Emergency Mutual Aid Center
MoECC	Ministry of Environment and Climate Change (Qatar)
NOx	nitrogen oxides

OPRC	International Convention on Oil Pollution Preparedness, Response and Co-Operation
ROPME	Regional Organization for Protection of Marine Environment
RSA	ROPME sea area
SOLAS	International Convention for the Safety of Life at Sea
SOx's	sulfur oxides
TCLP	toxicity characteristic leaching procedure
TSS	total suspended solid
VOC's	volatile organic compounds
ZLD	zero liquid discharge

References

Aarninkhof, S.G.J., Luijendijk, A.P., 2009. Safe Disposal of Dredged Material in Sensitive Environment Based on Innovative Plume Prediction, CEDA Dredging Days. Central Dredging Association Conference - Netherland.

Aarninkhof, S., Luijendijk, A., 2010. Safe disposal of dredged material in an environmentaly sensitive environment. Port Technol. Int. 47, 39–45.

Adewole, G.M., Adewale, T.M., Ufuoma, E., 2010. Environmental aspect of oil and water-based drilling muds and cuttings from Dibi and Ewan off-shore wells in the Niger Delta, Nigeria. African J. Environ. 4, 284–292. https://doi.org/10.4314/ajest.v4i5.56361.

Chen, Y.-S., Kang, C.-K., Liu, T.-K., 2022. Ballast water management strategy to reduce the impact of introductions by utilizing an empirical risk model. Water 14, 981. https://doi.org/10.3390/w14060981.

Corporation, C, 2014. 2014 Corporate Responsibility Report Highlights.

Corrosion standard and control in pipeline industry – ASTM International, 1973. Peabody, AW. Ebasco Services Incorporated, New York, NY. https://www.astm.org/stp37899s.html.

drilling formulas.com, 2017. What are Central Processing Platforms (CPP)?.

findaport.com, 2021. General Information for Qatar.

Jayasinghe, D., 2021. Internal Corrosion of Pipelines Carrying Crude Oil [WWW Document]. URL. https://www.corrosionpedia.com/internal-corrosion-of-pipelines-carrying-crude-oil/2/7282. (accessed 5.28.22).

Jowit, J., 2010. World's Top Firms Cause $2.2tn of Environmental Damage, Report Estimates [WWW Document]. Guard. https://www.theguardian.com/environment/2010/feb/18/worlds-top-firms-environmental-damage. (accessed 5.25.22).

Marine Emergency Mutual Aid Center (MEMAC), 1982. Manama, Bahrain.

Meezan, A., 1983. Law No. 4 of 1983 on the Exploitation and Conservation of Living Aquatic Resources in Qatar [WWW Document]. https://almeezan.qa/LawView.aspx?opt&LawID=3981&language=en. (accessed 5.27.22).

Meezan, A., 2002. Al Meezan—Qatary Legal Portal | Legislations | Law No. 30 of 2002 Promulgating the Law of the Environment Protection [WWW Document]. URL. https://www.almeezan.qa/LawPage.aspx?id=4114&language=en. accessed 5.27.22.

Meezan, A., 2005. Al Meezan - Qatary Legal Portal | Legislations | Resolution No. 4 of 2005 by the Chairperson of the Supreme Council of the Environment and Natural Reserves (SCENR) issuing executive bylaw for law No. 30 of 2002 on environment protection | Related Files [WWW Document]. URL. https://www.almeezan.qa/LawOtherAttachments.aspx?id=2108&language=en. (accessed 5.27.22).

National Association for Corrosion Engineers (NACE), 2022. Standards and Specifications Related to Oil & Natural Gas Infrastructure.

National Geographic, 2020. Noise Pollution | National Geographic Society [WWW Document]. Natl. Geogr. Mag. https://education.nationalgeographic.org/resource/noise-pollution. (accessed 5.28.22).

Nidumolu, R., Prahalad, C.K., Rangaswami, M.R., 2009. Why sustainability is now the key driver of innovation. Harv. Bus. Rev. 57–64.

Planning, G.S.F.D, 2008. July 2008 General Secretariat For Development Planning.

Pricing environmental damage: US$ 28 trillion by 2050. (November 17, 2014) https://conscioustourism.files.wordpress.com/2014/11/pricing-environmental-damage-us-28-trillion-by-2050-principles-for-responsible-investment.pdf, 2022. Qatargas - Sustainability [WWW Document]. URL https://www.qatargas.com/english/sustainability (accessed 5.28.22).

Solomon, S., Plattner, G.K., Knutti, R., Friedlingstein, P., 2009. Irreversible climate change due to carbon dioxide emissions. Proc. Natl. Acad. Sci. U. S. A. 106, 1704–1709. https://doi.org/10.1073/pnas.0812721106.

USEPA, 2000. Development Document for Final Effluent limitations. Guidelines and Standards for Synthetic-Based Drilling Fluid and Other Non-Aqueous Drilling Fluids in the Oil and Gas Extraction Point Source Category. 518.

Vinnem, J.E., 1998. In: Lysersen, S., Hansen, G.K., Sandtorv, H. (Eds.), Use of Performance Indicators for Monitoring HSE Operating Achievement., Safety and Reliability: Proceedings of the European Conference on Safety and Reliability (IESRE' 98). CRC Press, Trondheim, Norway.

Wikipedia, 2019. Upstream (Petroleum Industry) - Wikipedia [WWW Document]. https://en.wikipedia.org/wiki/Upstream_(petroleum_industry. (accessed 5.17.22).

World Bank Group, 2007. Environmental, Health, and Safety (EHS) Guidelines.

Index

Note: Page numbers followed by "*f*" indicate figures and "*t*" indicate tables.

D

E

O

P

Q

R

S

T

U

V

W

Y

Kolibri 360 GmbH
Ferdinand-Jühlke-Straße 7
99095 Erfurt, Deutschland
produktsicherheit@kolibri360.de